SAFECOMP '93

W0106590

INTERNATIONAL PROGRAM COMMITTEE

J. Górski (*Chairman*)	PL		
R. E. Bloomfield	UK	J. Lindeberg	N
S. Bologna	I	S. L.-Hansen	DK
W. Cellary	PL	M. J. P. van der Meulen	NL
G. Cleland	UK	G. Rabe	D
B. Cronhjort	S	F. Redmill	UK
P. Daniel	UK	M. Rodd	UK
B. K. Daniels	UK	B. Runge	DK
G. Dahl	N	G. H. Schildt	A
W. Ehrenberger	D	E. Schoitsch	A
H. Frey	CH	I. C. Smith	UK
K. Kanoun	F	A. Toola	SF
F. Koornneef	NL	W. M. Turski	PL
H. Krawczyk	PL	U. Voges	D
R. Lauber	D	W. Zamojski	PL

National Organizing Committee

J. Górski (*Chairman*),
T. Doroszewski, L. Gimzicka, K. Ratajczak, E. Rosłanowski

SAFECOMP '93

The 12th International Conference on Computer Safety, Reliability and Security

Poznań-Kiekrz, Poland
27-29 October 1993

Edited by
Janusz Górski

Sponsor

*European Workshop on Industrial Computer Systems
Technical Committee 7 (EWICS TC7)*

Co-Sponsored by

IFIP Technical Committee 5 WG 5.4.

*Franco-Polish School of New Information and
Communication Technologies*

Polish Council for Scientific Research

European Research Office, USARDSG-UK

OCG, the Austrian Computer Society

Organized by

*Franco-Polish School of New Information and
Communication Technologies*

Springer-Verlag
London   Berlin   Heidelberg   New York
Paris   Tokyo   Hong Kong
Barcelona   Budapest

Janusz Górski
Franco-Polish School of New Information and
Communication Technology
60-854 Poznań
Poland

ISBN-13: 978-3-540-19838-3   e-ISBN-13: 978-1-4471-2061-2
DOI: 10.1007/978-1-4471-2061-2

British Library Cataloguing in Publication Data
A catalogue record for this book is available from the British Library

Library of Congress Cataloging-in-publication Data
A catalogue record for this book is available from the Library of Congress

Apart from any fair dealing for the purposes of research or private study, or
criticism or review, as permitted under the Copyright, Designs and Patents Act
1988, this publication may only be reproduced, stored or transmitted, in any form
or by any means, with the prior permission in writing of the publishers, or in the
case of reprographic reproduction in accordance with the terms of licences issued
by the Copyright Licensing Agency. Enquiries concerning reproduction outside
those terms should be sent to the publishers.

© Springer-Verlag London Limited 1993

The use of registered names, trademarks etc. in this publication does not imply,
even in the absence of a specific statement, that such names are exempt from the
relevant laws and regulations and therefore free for general use.

The publisher makes no representation, express or implied, with regard to the
accuracy of the information contained in this book and cannot accept any legal
responsibility or liability for any errors or omissions that may be made.

Typesetting: camera-ready by author

Preface

The safe operation of computer systems continues to be a key issue in many applications where people, environment, investment, or goodwill can be at risk. Such applications include medical, railways, power generation and distribution, road transportation, aerospace, process industries, mining, military and many others.

This book represents the proceedings of the 12th International Conference on Computer Safety, Reliability and Security, held in Poznań, Poland, 27-29 October 1993. The conference reviews the state of the art, experiences and new trends in the areas of computer safety, reliability and security. It forms a platform for technology transfer between academia, industry and research institutions. In an expanding world-wide market for safe, secure and reliable computer systems SAFECOMP'93 provides an opportunity for technical developers, users, and legislators to exchange and review the experience, to consider the best technologies now available and to identify the skills and technologies required for the future. The papers were carefully selected by the International Program Committee of the Conference. The authors of the papers come from 16 different countries. The subjects covered include formal methods and models, safety assessment and analysis, verification and validation, testing, reliability issues and dependable software technology, computer languages for safety related systems, reactive systems technology, security and safety related applications. As to its wide international coverage, unique way of combining participants from academia, research and industry and topical coverage, SAFECOMP is outstanding among the other related events in the field.

The reader will get insight into the basic status of computer safety, reliability and security (through invited presentations) and will receive a representative sample of recent results and problems in those fields presented by experts from both industrial and academic institutions.

The response to the Call for Papers produced many more good papers than could be included in the programme. I must thank all the authors who submitted their work, the presenters of the papers, the International Program Committee and National Organising Committee, the Sponsor and Co-sponsors for their efforts and support. Through their strong motivation and hard work the Conference and this book have been enabled.

Janusz Górski   Poznań, Poland
                                                          August 1993

Contents

List of Contributors

S. Al-Karaawy
Faculty of Electronics
Technical University of
Gdańsk
Poland

S. Anderson
Department of Computer
Science
University of Edinburgh
Edinburgh EH9 3JZ, UK

T. Anderson
Department of Computing
Science
University of Newcastle
Newcastle upon Tyne
NE1 7RU, UK

J.-P. Banâtre
Inria-Rennes/Irisa
Campus de Beaulieu
35042 Rennes cedex, France

R. Baumann
Siemens Integra
Verkehrstechnik AG
8304 Wallisellen, Switzerland

C. Bernardeschi
Department of Information
Engineering
University of Pisa
Via Diotisalvi 2
56100 Pisa, Italy

A. Bondavalli
CNUCE-CNR
Via S. Maria 36
56126 Pisa, Italy

J. Brummer
Institute for Safety
Technology (IST)
Gesellschaft für Anlagen- und
Reaktorsicherheit (GRS)mbH
Garching, Germany

G. Bruns
Department of Computer
Science
University of Edinburgh
Edinburgh EH9 3JZ, UK

W. Cellary
Franco-Polish School of New
Information and
Communication Technology
60-854 Poznań, Poland

M.F. Chudleigh
Cambridge Consultants
Limited
Science Park, Milton Road
Cambridge CB4 4DW, UK

J.N. Clare
Cambridge Consultants
Limited
Science Park, Milton Road
Cambridge CB4 4DW, UK

B. Cole
Software Metrics Laboratory
Glasgow Caledonian
University
Glasgow, Scotland, UK

P. Collinson
Department of Chemical
Pathology
Mayday University Hospital
Croydon
Surrey CR7 7YE, UK

M. Colnarič
Faculty of Technical Sciences
University of Maribor
Maribor, Slovenia

S.S. Dhanjal
Lloyd's Register of Shipping
29 Wellesley Road
Croydon
Surrey CR0 2AJ, UK

H. Fierz
Institut für Technische
Informatik und
Kommunikationsnetze
ETH-Zentrum
CH-8092 Zurich, Switzerland

R. Fink
West Middlesex University
Hospital
Twickenham Road
Isleworth
Middlesex TW7 6AF, UK

I. M. Galkin
Computer Center
Academy of Sciences
Minsk
25 Scorina str.
Republic of Belarus 220072

C.J. Goring
August Systems Limited
Jenner Road
Crawley
West Sussex RH10 2GA, UK

W. A. Halang
Fern Universität
Department of Electrical
Engineering
P.O. Box 940
D-58084 Hagen, Germany

J. R. Hoelscher
General Railway Signal
Corporation
Rochester, NY 14620, USA

S. Hughes
Lloyd's Register of Shipping
29 Wellesley Road
Croydon
Surrey CR0 2AJ, UK

M.R. Inggs
Department of Electrical
Engineering
University of Cape Town
Private Bag Rondebosch
South Africa, 7700

M. Kaâniche
LAAS-CNRS
7 Avenue du Colonel Roche
31077 Toulouse Cedex, France

K. Kanoun
LAAS-CNRS
7 Avenue du Colonel Roche
31077 Toulouse Cedex, France

H. Krawczyk
Faculty of Electronics
Technical University of
Gdańsk
Poland

M. Kündig-Herzog
Siemens Integra
Verkehrstechnik AG
8304 Wallisellen
Switzerland

W. Kuhn
Austrian Research Center
Seibersdorf
Department of Information
Technology
A-2444 Seibersdorf
Austria

C. E. Landwehr
Center for High Assurance
Computing Systems
Naval Research Laboratory
Washington D.C., USA

J.-C. Laprie
LAAS-CNRS
7 Avenue du Colonel Roche
31077 Toulouse Cedex, France

R. de Lemos
Department of Computing
Science
University of Newcastle
Newcastle upon Tyne
NE1 7RU, UK

Z. Liu
Department of Computer
Science
University of Warwick
Coventry CV4 7AL, UK

B. Malcolm
Malcolm Associates Ltd.
Savoy Hill
London, UK

V. Manoni
SASIB Signalamento
Ferroviario
40128 Bologna, Italy

L. Mé
Laboratoire d'Informatique
SUPÉLEC
Avenue de la Boulaie
B.P. 28
F-35511 Cesson Sévigné
Cedex
France

M.J.P. van der Meulen
Department of Industrial
Safety
Institute for Environmental
and Energy Research
The Netherlands
Organization for Applied
Scientific Research TNO
Apeldoorn, The Netherlands

H. Müller
Institut für Technische
Informatik und
Kommunikationsnetze
ETH-Zentrum
CH-8092 Zurich, Switzerland

C. C. Michael
Reliable Software
Technologies Corporation
Penthouse Suite
1001 North Highland Street
Arlington, VA 22201 USA

K. W. Miller
Department of Computer
Science
College of William & Mary
Williamsburg, VA 23187 USA

P. Molinaro
Ecole Centrale de Nantes
Université de Nantes
Laboratoire d'Automatique de
Nantes
URA 823
Nantes, France

G.A. Mutone
AEG Transportation Systems,
Inc.
Pittsburgh, PA, USA

S. Netos
Institut für Technische
Informatik und
Kommunikationsnetze
ETH-Zentrum
CH-8092 Zurich, Switzerland

J. Nordahl
Department of Computer
Science
Technical University of
Denmark
DK-2800 Lyngby, Denmark

S. Oppert
Department of Clinical
Biochemistry
West Middlesex University
Hospital
Isleworth
Middlesex, UK

R. Posch
Institute for Applied
Information Processing
Graz University of
Technology
Klosterwiesgasse 32/I A-8010
Graz, Austria

J. Rainer
Austrian Federal Test &
Research Centre Arsenal
(BVFA)
Department System
Reliability & Traffic
Electronics
A-1030 Vienna
Faradaygasse 3, Austria

F. Redmill
Redmill Consultancy
and Co-ordinator of the
Safety-Critical Systems Club
22 Onslow Gardens
London N10 3JU, UK

H.E. Rhody
RIT Research Corporation
Rochester Institute of
Technology
Rochester, NY 14623, USA

M. Rothfelder
Institute for Software,
Electronics,
Railroad Technology (ISEB) of
TÜV Rheinland
P.O. Box 91 09 37
Cologne
D-5000 Koeln 91, Germany

O. H. Roux
Ecole Centrale de Nantes
Université de Nantes
Laboratoire d'Automatique de
Nantes
URA 823
Nantes, France

A. Saeed
Department of Computing
Science
University of Newcastle
Newcastle upon Tyne NE1
7RU, UK

J. Scheepstra
Rijksuniversiteit Groningen
Department of Computing
Science
P.O. Box 800
NL-9700 AV Groningen
The Netherlands

E. Schoitsch
Austrian Research Center
Seibersdorf
Department of Information
Technology
A-2444 Seibersdorf, Austria

M.B. Schrönen
Department of Electrical
Engineering
University of Cape Town
Private Bag Rondebosch
South Africa, 7700

G. Sen
Reactor Control Division
Bhabha Atomic Research
Centre
Bombay, India

R. E. Seviora
Department of Electrical and
Computer Engineering
University of Waterloo
Ontario,
Canada N2L 3G1

L. Simoncini
Department of Information
Engineering
University of Pisa
Via Diotisalvi, 2
56100 Pisa, Italy

E. V. Sørensen
Department of Computer
Science
Technical University of
Denmark
DK-2800 Lyngby, Denmark

T. Stålhane
SINTEF DELAB
O.S. Bragstads Plass
NTHN-7034 Trondheim
Norway

B. Stamm
Siemens Integra
Verkehrstechnik AG
K517
Industriestr. 24
8304 Wallisellen, Switzerland

J. M. Voas
Reliable Software
Technologies Corporation
Penthouse Suite
1001 North Highland Street
Arlington, VA 22201 USA

N.J. Ward
TA Consultancy Services Ltd
The Barbican
East Street
Farnham
Surrey GU9 7TB, UK

M. Witte
Fern Universität
Department of Electrical
Engineering
P.O. Box 940
D-58084 Hagen, Germany

INVITED PAPER

Safety - status and perspectives

Tom Anderson
Department of Computing Science
The University of Newcastle upon Tyne, NE1 7RU, UK

Abstract

Safety can be all things to all men - that is, different people in different situations will, quite legitimately, interpret the term "safety" in different ways. This paper expresses a personal perspective on safety as an engineering concern.

1 Introduction

Delegates will be aware that this is the 12th occasion of presenting SAFECOMP, a conference which, under the auspices of EWICS Technical Committee 7, has laid stress on the importance of *safety* in the context of computing systems since the very first SAFECOMP in 1979. Consequently, the event has an enviable lineage with respect to a topic that is recognised to be of rapidly increasing significance, commensurate with the growth in automatic control of critical applications. It seems inevitable that these trends will continue and accelerate, given current projections for the semiconductor and telecommunication industries. Over the past 15 years, work on both research and system development has enhanced our understanding of the issues and techniques relating to safety in computing systems. However, much remains to be done, in further advancing the discipline and in more widely promulgating the current state of the art. In this brief perspective I have taken the opportunity to make some elementary observations on the tenets of safe computing systems; if any of these are considered provocative or unsound I welcome correction.

2 Definition

Because "safe" and "safety" are words in everyday use, they have dictionary definitions and popular interpretations. These interpretations can differ widely: for the general public, for politicians, for professionals (lawyers, engineers, regulators etc), across industrial sectors, and over time (especially after a major accident). A scientist or engineer recognises the range of interpretations, but must nevertheless adopt a specific working definition - and thus accepts the consequence that because others may select an alternative definition, conflicts may need to be resolved if confusion is to be avoided.

The usual starting point for a definition of safety is that a system is safe if it will not kill anyone. However, numerous points then need clarification, such as "what about multiple deaths?", "what about injuries, severe and minor?", "what about environmental damage, with implications for human well-being?", "what about vast financial losses, with implications for the well-being of some?". An (inadequate) escape route is to assert that a system is safe if it will not harm anyone. But does this mean *never* harm anyone, under *any* possible circumstances? Only when these, and other, questions are answered would we have a semblance of a definition. (There is, of course, no single "correct" definition, so these questions will not be answered here!) One way forward is to define a system to be safe if it will not cause an accident, thereby postponing (albeit briefly) the definition of what constitutes an accident. Even given an agreed definition of a safe system, it is then vital to examine how degrees of unsafeness should be characterised, which leads on to the notion of risk to capture the likelihood and magnitude of losses incurred through use of the system.

From an engineer's viewpoint, ensuring that these issues are addressed and resolved is much more important than the details of their resolution in a particular case.

3 Misconceptions

Despite, or perhaps because of, the widespread use of safety concepts, a number of misconceptions are frequently encountered in the wider computing community - SAFECOMP delegates will, I trust, concur with my critique of the following aberrant assertions.

(a) *Safety is paramount.* If this were true, then in almost all cases, the proper course of action would be not to implement the system, or at least not to operate it. Safety is an attribute of a system which frequently conflicts with other desirable attributes. The design engineer has the difficult task of striving to achieve the optimum compromise between safety and the other required characteristics for the system, all within budgetary and other resource constraints.

(b) *Safety is an absolute.* The notion of absolute safety can be formulated and discussed if necessary, but the real engineering issues concern *levels* of safety and tradeoffs between safety and other system properties. Consider the following questions: How safe should the system be designed to be? How unsafe could the system be and still be considered adequately safe? How safe is the implemented system? How safe has the system been during operation? By comparison the question "Is the system absolutely safe?" seems pointless.

(c) *Safety can't be quantified* (less extreme versions: safety ought not to be quantified; avoid quantification in safety analyses). On the contrary, it is essential that safety be quantified - to the extent that this is feasible, and fully acknowledging the limitations and imprecision of measurement techniques. Quantified analysis of safety should be viewed as the normal engineering goal, and consequently the inability to quantify safety should be recognised as a deficiency - in which case subjective rankings or objective comparisons may be employed as a weaker alternative.

(d) *Safety must be guaranteed.* Since safety does not equate to death or taxes such a guarantee must be regarded as a forlorn hope, other than in the sense of a warranty establishing corporate liability.

(e) *Safety is unique.* Safety is a highly significant system attribute because of the importance we rightly attach to the lives of others. Nevertheless, it has very much in common with other system attributes such as reliability and security, and safety engineering can and does benefit greatly from the techniques developed for other aspects of dependability in systems - and vice-versa of course. [A personal aside. At SAFECOMP'83 in Cambridge I asserted (as a panellist) that the concepts of safety and reliability were essentially identical, differing only in the criterion which specified success. Although I still believe this to be true, I have learned a little in the last ten years, and do not expect to reiterate this academic and potentially misleading observation in Poznan at SAFECOMP'93.]

4 Axioms

In contrast to the above, the following truths are held to be self-evident.

(a) *Safety is a system attribute.* This is sometimes taken to imply that safety is solely a property of the overall application system (e.g. nuclear power plant) operating in the real-world environment; a very narrow interpretation then misleads by inferring that subsystems do not have this property (contradicted by axioms *b* and *c* below). A more generic use of the term *system* is much to be preferred, encompassing subsystems, units, modules, components etc., in which case axiom *a* is almost tautological.

(b) *Computing systems can kill.* See Leveson and Turner [2].

(c) *Software can kill.* See Leveson and Turner [2]. Obviously, the software directs the computing system which in turn acts via the controlled equipment - analogously, most murderers make use of a weapon.

(d) *Perfection is unattainable.* Samuel Butler advised "Strive for imperfection - there's some change of getting it". Dijkstra warned "Testing can show the presence, but never the absence of faults". Lebesgue cautions "Logic makes us reject certain arguments, but it cannot make us believe any argument". Juvenal asked "But who is to guard the guards themselves?". Brookes summed it all up - "There is inherently no silver bullet".

(e) *There's safety in numbers.* Although this is a well known English phrase it is perhaps a little too ambiguous to be axiomatic. A literal interpretation is unusual and the benefits of quantification have already been suggested; here I wish to take the standard usage, which suggests that members within a group are less exposed to attack than isolated individuals, and thereby make the standard argument in favour of redundancy. Any single entity can fail, and to avoid a single point of failure alternative mechanisms should be available (eg. retry, or a spare, or diversity, or fail-safe).

5 Engineering Safe Computing Systems

The tasks of safety engineering are clearly manifold: to establish the safety requirements for the system and its subsystems, to formulate safety policies, specifications and strategies, to design for safety, to conduct hazard and safety analyses, to compose the safety case and gain certification for the system, to implement, install, operate and maintain the system in accordance with all of the preceding. All are of vital importance (literally), which makes prioritisation rather difficult. I would place particular emphasis on achieving safety, and feel that the specific topics of requirements, validation and fault-tolerance deserve special mention - but this may merely be a consequence of personal prejudice. In any case, the above list of topics is driven by system life-cycle stages, and we should also include: management, procedures, documentation, standards, human factors and real-time considerations.

My position in 1989 was stated as:

> "I would commend three attributes to those involved in the construction of [safe] computing systems. First, *vigilance*, in avoiding and eliminating faults; second, *diversity*, to provide protection against the consequences of faults; and third, *simplicity*, the hand-maiden of dependability" [1].

Almost five years on, the only change I wish to make is to reverse the ordering.

Lastly, I would like to refer readers to the most enjoyable text on system safety I have encountered [3], which happens to be in the domain of railway safety and the lessons to be learnt from accidents; as well as being highly instructive, the book provides this closing quotation to emphasise that even safety engineers can learn from their mistakes:

> Out of this nettle, Danger
> We pluck this flower, Safety
>
> *Henry IV (Part I)*

References

1. Anderson T (ed). Safe & Secure Computing Systems - Preface. Blackwell Scientific, Oxford, 1989

2. Leveson NG, Turner CS. An Investigation of the Therac-25 Accidents. IEEE Computer 1993; 26,7:18-41

3. Rolt LTC. Red for Danger (3rd edition). Pan Books, London, 1976

FORMAL METHODS
AND
MODELS

Chair: G. Cleland
University of Edinburgh, UK

Data Flow Control Systems: an Example of Safety Validation

Cinzia Bernardeschi, Luca Simoncini
Department of Information Engineering, University of Pisa
Pisa, Italy

Andrea Bondavalli
CNUCE-CNR
Pisa, Italy

Abstract

In this paper a methodology to develop safety-critical control systems is proposed. These systems continuously interact with the physical environment, and those admitting at least one failure causing a catastrophe are classified as safety-critical. Our methodology takes into account both the control system (*controller*) and the physical environment (*plant*). After the requirements analysis, the system is developed following data flow model, i.e., described as a static data flow network of nodes executing concurrently and communicating asynchronously. The plant is used as the test case for the validation of the controller and their composition is analysed to show whether hazards are reached. To this purpose we apply a transformation from data flow networks to LOTOS specifications. The transformation preserves the semantics of the original network and data flow network properties can be derived and proved on the LOTOS specification using available support tools. A train set example for the contact-free moving of trains on a circular track divided into sections is shown as an application of the methodology.

1 Introduction

Control systems are computing systems which continuously interact with the physical environment, e.g. traffic control or industrial process control systems. Many control systems are safety-critical, i.e. systems for which at least one failure exists that can cause a catastrophe. Therefore, in addition to their functional capabilities, these systems require specified levels of dependability. In the framework of safety-critical systems, one approach to improve the level of dependability is to use formal specification and verification in conjunction with other methods of software development such as testing and fault tolerance. The analysis of the critical issues of a control system plays a vital role in the development of safety-critical systems. Critical issues address what the system should not do and allow to concentrate on the elimination and control of the hazards. The study of the critical issues of the system, allows us to derive the constraints necessary to guarantee a safe behaviour of the system (safety constraints) and the strategies to realise it (safety strategies) [1]. The validation phase is as important as requirements analysis. Validation is the activity that aims to check that the actual behaviour of the developed system is as expected.

Data flow is a paradigm for concurrent computations. A data flow network is composed by a set of nodes (or processes) all executing concurrently and asynchronously. They communicate by exchanging messages, representing data items, over asynchronous communication channels (following a FIFO policy). The computation proceeds in a data driven manner: a node of the network is ready to execute as soon as the required data tokens are available. Data flow is receiving great attention being known for its suitability for achieving a high degree of execution parallelism, thus allowing to improve performance, but has other useful characteristics as well. A data flow network is usually very close to the intuitive representation of a control system, that is the translation from the conceived system to a data flow graph is straightforward, as well as to inspect the data flow graph to determine which aspects of the system are represented [2], [3]. This makes data flow generally recognised as a convenient programming paradigm for the development of control systems. The referential transparency property admitted when nodes compute functions, by which two executions of the same node with the same input data produce equal output results, makes data flow "inherently fault tolerant": it is possible to tolerate simple failures by re-evaluating the same function on the same input data [4], [5]. If a non deterministic behaviour of nodes is allowed, still the strong isolation and information hiding enforces a good confinement useful for setting error confinement areas around modules by means of appropriate consistency checks. The property of composability which puts in direct relation the general behaviour of a system from its constituent parts [6], [7] helps verification and validation. Lastly, structural models for software reliability assessment can be applied since all data necessary to their use can be obtained by a simple instrumentation of software code [8].

In this paper a systems development methodology is proposed. After the requirements analysis, the system is developed following the computational model based on the Jonsson's formalism [7]. In the validation phase, the specification of the physical environment is assumed as the test case for the control system: the *plant* and the *controller* are composed and the resulting behaviour is analysed to be sure that hazards are never reached in the system. To this purpose, we apply a transformation from data flow networks to LOTOS (Language Of Temporal Ordering Specification) [9] specifications. The transformation maintains the data flow network properties which can be derived and proved on the LOTOS specification. Available LOTOS software support tools are then used [10]. The adequacy of the proposed methodology is shown through the design and the validation of a simple control system: a train set example for the contact-free moving of trains on a circular track divided into sections [1], [11]. The rest of this paper is as follows. Section 2 is devoted to the definition of our methodology, including a description of the data flow formalism adopted, the transformation and its properties. Section 3 develops the example of the train set to show how the methodology can be applied. Lastly, Section 4 contains our conclusion.

2 System Development Methodology

The proposed development methodology takes into account the parallel interaction between a *plant* and a *controller* which must eliminate unsatisfactory behaviours of the plant. The interface between the *plant* and the *controller* contains sensors and actuators. Sensors detect events in the plant and send signals to the controller. Upon reception of the signals the controller can take actions by issuing appropriate control commands through actuators. The analysis of the critical issues addressing what the system should not do, allow to define the hazards for the system into consideration and their elimination and control. The analysis is performed in two phases: the first phase to identify the real world properties relevant to the critical behaviour of the system and the second phase to specify the system behaviour required at the interface with the environment, i.e. the sensors and actuators. Thus the constraints necessary to guarantee a safe behaviour of the system (safety constraints) and the strategies to realise it (safety strategies) may be derived.

Then the system realising the safety strategy is developed following a data flow computational model. Since we shall use the specification of the physical environment as the test case for the control system in the validation, we shall model also the plant. As previously mentioned we adopt the formalism for the specification of data flow network proposed in [7] in which the semantics of the networks is based on *traces*. Here we give some definitions and a brief explanation on this model. Given a data flow network N, let V be the set of data items exchanged over the channels. We denote by V* the set of finite sequences on V and by <> the empty sequence.

Definition: A data flow **node** P is a tuple $<I_P, O_P, S_P, s^0_P, R_P, FAIR_P>$ where:

I_P is the set of input channels;

O_P is the set of output channels with $(I_P \cap O_P) = \emptyset$;

S_P is the set of states; s^0_P is the initial state, $s^0_P \in S_P$;

R_P is the set of firings. A firing F is a tuple $F=<s, \chi_{in}, s', \chi_{out}>$ where $s, s' \in S_P$, χ_{in} is a mapping from I_P to V* and χ_{out} is a mapping from O_P to V*.

$FAIR_P \subseteq \mathcal{P} (R_P)$ is a finite collection of fairness sets. If $FAIR_P=R_P$, then the node executes firings until no more data are present on the input channels. ♦

For the sake of this paper, the meaning of a firing $<s, \chi_{in}, s', \chi_{out}>$ can be assumed as follows: when the node is in state s and for each input channel $inp \in I_P$ the sequence $\chi_{in}(inp)$ is a prefix of the content of the channel (i.e. the firing is *executable*), then these sequences may be consumed, while the node changes its state to s' and the sequence $\chi_{out}(out)$ is produced on each output channel $out \in O_P$. Note that the empty sequence <> is a prefix of each sequence of data.

A data flow **network** N consists of a set P_N of data flow nodes such that in P_N each channel occurs at most once as an input channel and at most once as an output channel. The network is obtained connecting input channels to output channels with the same name and a network transition can be generated by the firing of a node or by a communication event, where a **communication event** can be either an input event or an output event. Communication events occur when a data item is inserted (removed) into (from) an input (output) channel of the network. C_N denotes the set of all the channels of the network. A **computation** of the network is a sequence of transitions of the network. Informally a computation of the network is a complete run of the network in which all nodes perform firings according to their definition and all channels behaves like unbounded FIFO channels. The semantics of the network is the set of its **traces**; a trace represents the interleaving of the communication events during a computation.

The use of information about the presence/absence of data items and the data driven asynchronous execution of data flow nodes in data flow networks, make reasoning about these networks and their semantics very difficult. To perform the semantic analysis of data flow networks, we apply a transformation from data flow networks to process algebras specifications using the LOTOS formal specification language [9]. LOTOS represents recent work on the combination of CCS (with some extension) [12] to describe the behaviour of the system and an algebraic formalism for the definition of data types. Software support tools have been developed allowing the simulation, the compilation and the proof of properties of a LOTOS specification [10].

The transformation is obtained by mapping each node and each channel of the network into a process in the process algebras and then all the processes are composed in parallel with synchronisation on the proper set of actions to realise the global behaviour of the network [13]. The names of gates in the specification are directly derived from the names of the channels. For each channel "a"$\in C_N$, "a#" is the gate corresponding to get a data from the channel "a" while "a" is the gate corresponding to put a data on the same channel "a". Let CP be the process which simulates the behaviour of a channel "a" of N (CP behaves like a FIFO buffer) and nodeP be the process that realises the behaviour of the data flow node P, the specification of the network is:

<u>specification</u> net$_N$[Egates$_N$] : noexit

<data type definition>

<u>behaviour</u>

<u>hide</u> I[Cgates$_N$-Egates$_N$]I <u>in</u>

(CP[a, a#] III ...<$\forall c \in C_N$>... III CP[b, b#])

I[Cgates$_N$-Egates$_N$]I (nodeP[Ip#, Op] III ...<$\forall Q \in P_N$>... III nodeQ[I$_R$#, O$_R$])

<u>endspec</u> (* net$_N$ *)

where Cgates$_N$ are the gates corresponding to get (put) from (onto) the whole sets of channels of N, Egates$_N$ are the gates corresponding to get (put) from (onto) the input (output) external channels of N. Furthermore, the notation Ip# (Op) is used to denote the set of "a#" ("a") gates for the input (output) channels of the node P. The set of processes associated to channels execute disjoint actions, so they are put in parallel with an empty set of synchronisation gates (III operator). The same applies to the set of processes associated to the nodes. These two sets of processes synchronise on the set of all the actions defined for the two behaviour expressions. The network specification has the same behaviour of the original data flow network and the formal verification methods of the process algebras can be applied to prove properties of the original network. Interested readers may find more details on the transformation itself and a prove that the transformation preserves the data flow network properties, i.e., the LOTOS specification has the same behaviour of the network from which it has been derived, in [13]. The previous transformation is defined for a class of data flow networks in which the firings of the nodes do not require sophisticated synchronisation mechanisms between the processes associated to the channels and the processes which simulates the behaviour of the nodes. The transformation for general networks is described in [14] .

To summarise, our methodology is based on:
- modelling the physical environment as a part of the overall system (*plant*) ;
- executing the requirements analysis for both the mission and the critical issues of the system;
- specifying safety constraints and a safety strategy for the system to eliminate hazards;
- developing the control system in the data flow computational model;
- applying the transformation to the data flow specification of the system (both the control system and the plant) obtaining a LOTOS specification which maintains all the relevant properties (and doing some expression transformation if necessary for their automatic analysis);
- verifying the correct behaviour of the system composed by the *plant* and the *controller* through an automatic analysis of the LOTOS resulting expression using the available tools.

3 The Train Set Example

The train set example consists of a simple control system for the contact-free moving of trains on a circular track [1], [11]. Suppose to have one directional moving of two trains on a circular track divided in six sections, with the constraint that trains are less than one section in length. Hazardous states are the states in which a train may be involved in a collision. In our system, a state is hazardous if the front of one train is in the same or adjacent section as the front of another train. They are avoided in a system if the following condition (safety condition) always holds: the heads of the trains differ at least by 2 sections. The concept of reserved section is introduced and our safety strategy is based on: 1) a section can be reserved by only one train; 2) for any train the section of the front of the train and the section behind must be reserved; 3) a train must always reserve a section before entering it. We use $\ominus$ and $\oplus$ to represent the operation of subtraction modulo 6 and the operation of addition modulo 6, respectively.

We divide the system under development into the physical *plant* and the *controller* which communicate by sending control signals and then we apply the data flow model based on the Jonsson's formalism [7]. The *plant* is composed by six sections {Sect$_0$, ..., Sect$_5$} shown in Figure 1 (a). In each section a sensor detects a train entering in the section and an actuator has the task to stop a train before leaving the section when necessary. We model the flow of the train by messages sent by one section to the next (channel sn_i). On receipt of this message the section sends a signal to the controller to notify the passage of a train (channel es_i). Then before the train is allowed to move on the section, it waits for a message from the controller with the meaning that the train is allowed to leave the section (channel go_i). On receipt of this message, the section sends a message to the next section in the circular track to simulate the movement of the train (channel $sn_{i\oplus1}$).

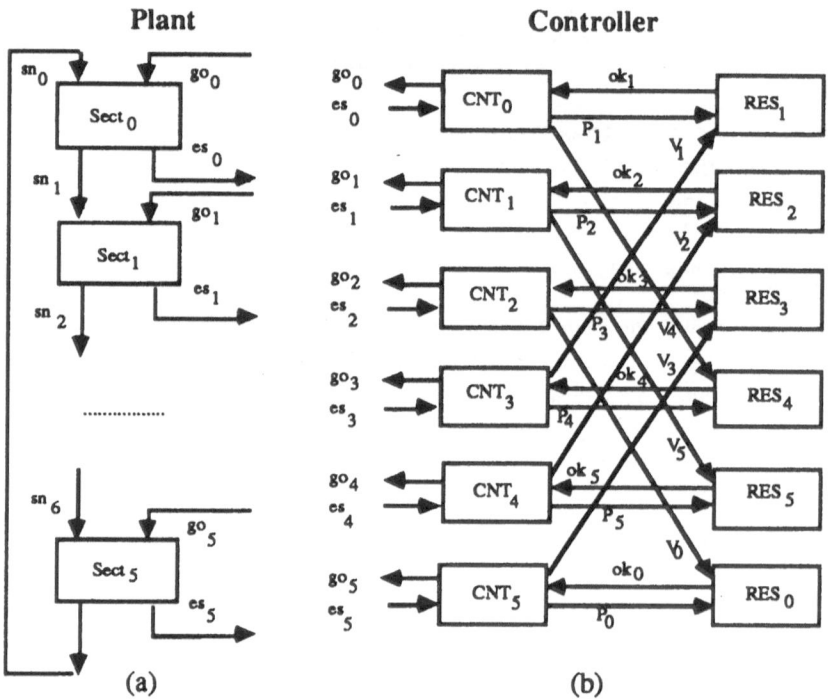

Figure 1: The Data Flow Network of the Train Set System.

The *controller* interacts with the *plant* and is composed by the data flow nodes reported in Figure 1 (b): six CNT_i nodes and six RES_i nodes. Each CNT_i realises the communication with the section $Sect_i$ of the *plant* while each RES_i implements the correct reservation mechanism of the corresponding section $Sect_i$. The CNT_i node, after having received a signal from section $Sect_i$ that a train has arrived (channel es_i), sends a signal to $RES_{i\ominus2}$ to mark section $Sect_{i\ominus2}$ as free (channel $V_{i\ominus2}$), and then it tries to book the section ($i\oplus1$) for the train sending a signal to $RES_{i\oplus1}$ (channel $P_{i\oplus1}$). CNT_i waits for a positive answer from $RES_{i\oplus1}$ (the next section has been reserved) (channel $ok_{i\oplus1}$); and then it sends a signal to $Sect_i$ for allowing the train to leave section $Sect_i$ (channel go_i). Each RES_i node controls the status of the corresponding section which can be reserved for one train or free. It receives signals from the $CNT_{i\ominus1}$ (channel P_i) and reserves the section by sending an acknowledgement (channel ok_i). After the section has been reserved it accepts only a signal through the channel V_i to free the section before accepting (and making) any further reservation.

The resulting data flow network N, composed by the *controller* and the *plant*, is shown in Figure 1. Let A and B be natural numbers representing the identifiers for the trains running over the track, we suppose an initial state with train A in section $Sect_1$ and train B in section $Sect_5$; it follows that the sections 1 and 0 (for train A) and the sections 5 and 4 (for train B) must be reserved. The initialisation is used to define the initial state of the data flow nodes. For those communications which are signals we associate the dummy value 1 in defining of the firings of the nodes (another way is to allow any data value). The definition of the data flow nodes is:

Section node $Sect_i$

$I_{Sect_i} = \{sn_i, go_i\}$ $\qquad$ $O_{Sect_i} = \{es_i, sn_{(i \oplus 1)}\}$ $\qquad$ $S_{Sect_i} = \{s, s_a, s_b\}$

$R_{Sect_i} = \{F1, F2, F3, F4\}$ $\qquad$ $FAIR_{Sect_i} = R_{Sect_i}$

$F1 = <s, [sn_i \to A], s_a, [es_i \to 1]>$ $\qquad$ $F3 = <s_a, [go_i \to 1], s, [sn_{(i \oplus 1)} \to A]>$

$F2 = <s, [sn_i \to B], s_b, [es_i \to 1]>$ $\qquad$ $F4 = <s_b, [go_i \to 1], s, [sn_{(i \oplus 1)} \to B]>$

$s^0_{Sect_1} = s_a$, $s^0_{Sect_5} = s_b$ and $s^0_{Sect_i} = s$ for $i = \{0,2,3,4\}$.

Controller node CNT_i

$I_{CNT_i} = \{es_i, ok_{(i \oplus 1)}\}$ $\qquad$ $O_{CNT_i} = \{go_i, P_{(i \oplus 1)}, V_{(i \ominus 2)}\}$ $\qquad$ $S_{CNT_i} = \{s, s'\}$

$R_{CNT_i} = \{F5, F6\}$ $\qquad$ $FAIR_{CNT_i} = R_{CNT_i}$

$F5 = <s, [es_i \to 1], s', [V_{(i \ominus 2)} \to 1, P_{(i \oplus 1)} \to 1]>$ $\quad$ $F6 = <s', [ok_{(i \oplus 1)} \to 1], s, [go_i \to 1]>$

$s^0_{CNT_i} = s'$ for $i = \{1,5\}$ and $s^0_{CNT_i} = s$ for $i = \{0,2,3,4\}$.

Controller node RES_i

$I_{RES_i} = \{P_i, V_i\}$ $\qquad$ $O_{RES_i} = \{ok_i\}$ $\qquad$ $S_{RES_i} = \{s, s'\}$

$R_{RES_i} = \{F7, F8\}$ $\qquad$ $FAIR_{RES_i} = R_{RES_i}$

$F7 = <s, [P_i \to 1], s', [ok_i \to 1]>$ $\qquad$ $F8 = <s', [V_i \to 1], s, []>$

$s^0_{RES_i} = s'$ for $i = \{0, 1, 4, 5\}$ and $s^0_{RES_i} = s$ for $i = \{2,3\}$.

To apply the transformation we specify the maximum size of the channels which may be assumed equal to two, while the signal communications can be transformed in pure synchronisation action in LOTOS. We give here the LOTOS process definition for the single data flow nodes obtained applying the transformation described in Section 2. The process definitions for the Sect, CNT and RES nodes and that for the CP which simulates a FIFO buffer of length two are:

process nodeSect[sn, es, go, nextsn](actstate: state) : noexit :=

$\qquad$ ([actstate=s] -> (sn?X:nat [X=A]; i; es!1; nodeSect[sn, es, go, nextsn](s_a)

$\qquad\qquad\qquad\qquad$ [] sn?X:nat [X=B]; i; es!1; nodeSect[sn, es, go, nextsn](s_b))

$\qquad$ [] [actstate=s_a] -> go?X:nat; i; nextsn!A; nodeSect[sn, es, go, nextsn](s)

$\qquad$ [] [actstate=s_b] -> go?X:nat; i; nextsn!B; nodeSect[sn, es, go, nextsn](s))

endproc (* nodeSect *)

```
process nodeCNT[es, P, ok, V, go](actstate:state) : noexit :=
    ([actstate=s] -> es?X:nat; i; V!1; P!1; nodeCNT[es, P, ok, V, go](s')
    [] [actstate=s'] -> ok?X:nat; i; go!1; nodeCNT[es, P, ok, V, go](s))
endproc (* nodeCNT *)
process nodeRES[P,ok,V](actstate:state) : noexit :=
    ([actstate=s] -> P?X:nat; i; ok!1; nodeRES[P,ok,V](s')
    [] [actstate=s'] -> V?X:nat; i; nodeRES[P,ok,V](s))
endproc (* nodeRES *)
process CP[inp, out]: noexit :=
    hide mid in oneslot[inp, mid] |[mid]| oneslot[mid, out]
where
    process oneslot[a, b] : noexit := a?X:nat ; b!X; oneslot[a, b]
    endproc (* oneslot *)
endproc (* CP *)
```

Since LOTOS specifications belonging to the subset of LOTOS without data (basic LOTOS) can be completely analysed by the verification tools, while for specifications with data values we can only simulate and/or compile and run them, we will restrict ourselves to basic LOTOS whenever possible without loosing properties. The LOTOS behaviour analyser AUTO [15], allows us to build the automaton of a basic LOTOS specification to prove strong and weak bisimulation between specifications. Although it fails when running on large specifications, simple ones like ours can be successfully run and the LOGIC CHECKER tool [16] can be used to prove action-based logic formulas ACTL, over the specification. To this purpose we make some manipulations of the specification obtained directly by the data flow to LOTOS transformation, trying to synchronise processes and to hide actions as soon as possible. This allows AUTO to reduce the number of the states during the generation of the automaton of the specification. The LOTOS "Regrouping Parallel Processes" correctness preserving transformation can be applied automatically by the LOTOS structure editor to regroup processes differently. The transformation preserves the strong bisimulation equivalence. All the previous tools are included in the LOTOS integrated tool environment Lite [10] developed inside the LOTOSPHERE ESPRIT project. Since all the nodeSect, nodeCNT and nodeRES processes execute all the actions in state s and then the actions in the state s' (nodeSect executes actions either in s_a or s_b) before repeating, we assume s as the initial state and rewrite the processes as:

```
process nodeSect[sn, es, go, nextsn]: noexit :=
    (sn?X:nat [X=A]; i; es!1; go?X:nat; i; nextsn!A; nodeSect[sn, es, go, nextsn]
    [] sn?X:nat [X=B]; i; es!1; go?X:nat; i; nextsn!B; nodeSect[sn, es, go, nextsn])
endproc (* nodeSect *)
process nodeCNT[es, P, ok, V, go]: noexit :=
    es?X:nat; i; V!1; P!1; ok?X:nat; i; go!1; nodeCNT[es, P, ok, V, go]
endproc (* nodeCNT *)
```

process nodeRES[P,ok,V]: noexit :=
    P?X:nat; i; ok!1; V?X:nat; i; nodeRES[P,ok,V]
endproc (* nodeRES *)

To keep into account the initial position of trains, the corresponding processes must contain a prefix behaviour expression representing the action to be performed at system start. This lead to the definition of the following processes: nodeSectA and nodeSectB for the sections where train A and train B are at the beginning, respectively; nodeICNT for the controllers that have to reserve the next section for allowing the trains to move (1 and 5 in our case) and nodeIRES for the sections that are reserved at the beginning (0, 1, 4 and 5 in our case). We have:

process nodeSectA[sn, es, go, es, nextsn]: noexit :=
    go?X:nat; i; nextsn!A; nodeSect[sn, es, go, nextsn]
endproc (* nodeSectA *)
process nodeSectB[sn, es, go, es, nextsn]: noexit :=
    go?X:nat; i; nextsn!B; nodeSect[sn, es, go, nextsn]
endproc (* nodeSectB *)
process nodeICNT[es, P, ok, V, go]: noexit :=
    P!1; ok?X:nat; i; go!1; nodeCNT[es, P, ok, V, go]
endproc (* nodeICNT *)
process nodeIRES[P,ok,V]: noexit :=
    V?X:nat; i; nodeRES[P,ok,V]
endproc (* nodeIRES *)

We can now map our specification into basic LOTOS. Lite, provides many mappings from a full LOTOS specification onto a basic LOTOS one. They differ for the data value information that are removed. We can apply the simplest transformation named "trans_np0" where all data are dropped, keeping simply the original gate identifiers as basic LOTOS actions. The transformation can be directly invoked by the behaviour analysis menu entry. In order to apply this mapping without loosing information, we modify the specification defining one gate for train A and another one for train B when they run over the track (i.e. substituting each action sn_i with two actions asn_i and bsn_i). The new process nodeSect is simply a non deterministic choice between the actions corresponding to the passage of the two trains. This is the only communication channel where data are important, in all the others the value of the data are not significant and can be dropped. The basic LOTOS specification of the section is:

process nodeSect [asn, bsn, es, go, nextasn, nextbsn] :noexit :=
    (asn; i; es; go; i; nextasn; nodeSect [asn, bsn, es, go, nextasn, nextbsn]
    [] bsn; i; es; go; i; nextbsn; nodeSect [asn, bsn, es, go, nextasn, nextbsn])
endproc (* nodeSect *)

The behaviour expression of the whole specification of the system is reported in the Appendix; where the observable actions are the actions corresponding to the movement of the trains over the track (gates $asn_i\#$ and $bsn_i\#$). Note that there are not external channels of the network and the set of processes associated to the nodes must synchronise with the set of channel processes on the whole set of gates. The LOTOS behavioural analyser AUTO can be run over the specification allowing to easily prove our safety strategy. The automaton (considering the weak bisimulation equivalence) has 18 states and 24 transitions and it is deadlock free. We proved automatically, by using the LOGIC CHECKER over the automaton, the following logic formulas to be true for train A:

1) train A can enter any section: $A[true\{true\}U\{asn_i\#\}true]$;

2) train A can only move from section i to section $i\oplus 1$:

$AG([asn_i\#]A[true\{cond\}U\{\sim asn_{(i\oplus 1)}\#\}true])$;

where $cond=((\sim asn_0\#)\&(\sim asn_1\#)\&(\sim asn_2\#)\&(\sim asn_3\#)\&(\sim asn_4\#)\&(\sim asn_5\#))$;

3) for each path such that train A enters section i, train B cannot enter section $(i\ominus 1)$ until train A enters section $(i\oplus 1)$:

$AG([asn_i\#]A[true\{\sim bsn_{(i\ominus 1)}\#\}U\{bsn_{(i\ominus 1)}\#\}A[true\{\sim bsn_{(i\ominus 1)}\#\}U\{asn_{(i\oplus 1)}\#\}true]])$.

The same formulas can be proved to be true for the train B.

From these we have that when train A is in section i, train B is never in section $i\ominus 1$, i, $i\oplus 1$. This holds also for train B, thus satisfying the safety condition.

4 Conclusions

In this paper we have presented a methodology which can be used for the design of safety-critical systems and for the validation of the design. Quite apart the modelling of the physical environment as a part of the overall system which can be used as test case for the control system, the use of the data flow computational model for the description of the system specification allows the designer to use notations which are very natural and which can be made even more user friendly by the use of development tools like a graphical editor [4]. The transformation into process algebras specification allows the use of the analysis tools available in LOTOS, making the entire process from specification to verification and validation fully automated.

The proposed approach has been applied to a simple control system where advantage could be taken by the use of the basic LOTOS tools like the behavioural analyser AUTO for the generation of the automaton and the LOGIC CHECKER. The extension of the proposed approach to the validation of control systems LOTOS specifications with data value involves the use of the simulator tool [10] and the compiler available in the full LOTOS environment, which allows to derive the possible traces of execution of the original data flow network. This extension is anyway limited by the fact that tracing the behaviour of a general network may be very lengthy and unfeasible in case of infinite input sequences. Nevertheless for control systems where the possible input sequences are constrained either on data value or on periodicity, the proposed approach can be used for problems of larger size than that presented in this paper.

References

1. Saeed A, de Lemos R, Anderson T. The role of formal methods in the requirements analysis of safety-critical systems: a train set example. Proc. of FTCS-21, Montreal, Canada, 1991, pp. 478-485

2. Kavi K, Buckles B, Bhat U. Isomorphism between Petri nets and data flow graphs. IEEE TSE 1987; SE-13: 1127-1134

3. Bondavalli A, Strigini L, Simoncini L. Data-flow like languages for real-time systems: issues of computational models and notation. Proc. of SRDS-11, 11th Symposium on Reliable Distributed Systems, Houston, Texas, USA, 1992, pp. 214-221

4. Bondavalli A, Simoncini L. Functional paradigm for designing dependable large-scale parallel computing systems. Proc. of ISADS 93 International Symposium on Autonomous Decentralized Systems, Kawasaki, Japan, 1993, pp. 108-114

5. Jagannathan R, Ashcroft E A. Fault tolerance in parallel implementations of functional languages. Proc. of FTCS-21, Montreal, Canada, 1991, pp. 256-263

6. Kahn G. The semantics of a simple language for parallel programming. Proc. of IFIP 74, 1974, pp. 471-475

7. Jonsson B. A fully abstract trace model for data flow networks. Journal of ACM 1989; 36: 155-165

8. Mellor P. Modular structured software reliability modelling. Private communication, 1992

9. Bolognesi T, Brinskma E. Introduction to the ISO specification language LOTOS. In: The Formal Description Technique LOTOS. Elsevier Science Publishers B.V. (North-Holland), 1989, pp. 23-73

10. van Eijk P. The Lotosphere integrated tool environment LITE. Proc. of IFIP TC6/WG6.1 4th International Conference on Formal Description Techniques for Distributed Systems and Communication Protocols - FORTE 91, Sydney, Australia, 1991, pp. 473-476

11. Genrich H J. Predicate/transition nets. In: LNCS 254. Springer Verlag, 1986, pp. 207-247

12. Milner R. Communication and concurrency. Prentice Hall, Englewood Cliffs, NJ, 1989

13. Bernardeschi C. An approach to the analysis of data flow networks by LOTOS. Proc. of Congresso annuale AICA'93 (to appear), Lecce, Italy, 1993

14. Bernardeschi C, Bondavalli A, Simoncini L. From data flow networks to process algebras. Proc. of PARLE 93, Munchen, Germany, 1993.

15. Madeleine E, Vergamini D. AUTO: a verification tool for distributed systems using reduction of finite automata networks. Proc. of IFIP TC6 2nd International Conference on Formal Description Tecniques for Distributed Systems and Communication Protocols - FORTE 89, Vancouver, B.C., Canada, 1989, pp. 61-66

16. De Nicola R, Fantechi A, Gnesi S, Ristori G. An action-based framework for verifying logical and behavioural properties of concurrent systems. Computer Networks and ISDN Systems 1993; 25: 761-778

Appendix

specification SYSTEM [asn0#, asn1#, asn2#, asn3#, asn4#, asn5#,

bsn0#, bsn1#, bsn2#, bsn3#, bsn4#, bsn5#] : noexit

behaviour

hide

asn0, asn1, asn2, asn3, asn4, asn5, bsn0, bsn1, bsn2, bsn3, bsn4, bsn5, es0, es1, es2, es3, es4, es5, es0#, es1#, es2#, es3#, es4#, es5#, go0, go1, go2, go3, go4, go5, go0#, go1#, go2#, go3#, go4#, go5#, P0, P1, P2, P3, P4, P5, P0#, P1#, P2#, P3#, P4#, P5#, ok0, ok1, ok2, ok3, ok4, ok5, ok0#, ok1#, ok2#, ok3#, ok4#, ok5#,V0, V1, V2, V3, V4, V5, V0#, V1#, V2#, V3#, V4#, V5#

in

(nodeSect[asn0#,bsn0#,es0,go0#,asn1,bsn1] ||| nodeCNT[es0#,P0,ok0#,V0,go0] |||
nodeSectA[asn1#,bsn1#,es1,go1#,asn2,bsn2] ||| nodeICNT[es1#,P1,ok1#,V1,go1] |||
nodeSect[asn2#,bsn2#,es2,go2#,asn3,bsn3] ||| nodeCNT[es2#,P2,ok2#,V2,go2] |||
nodeSect[asn3#,bsn3#,es3,go3#,asn4,bsn4] ||| nodeCNT[es3#,P3,ok3#,V3,go3] |||
nodeSect[asn4#,bsn4#,es4,go4#,asn5,bsn5] ||| nodeCNT[es4#,P4,ok4#,V4,go4] |||
nodeSectB[asn5#,bsn5#,es4,go0#,asn1,bsn1] ||| nodeICNT[es5#,P5,ok5#,V5,go5] |||
nodeIRES[P0#,ok0,V0#] ||| nodeIRES[P1#,ok1,V1#] ||| nodeRES[P2#,ok2,V2#] |||
nodeRES[P3#,ok3,V3#] ||| nodeIRES[P4#,ok4,V4#] ||| nodeIRES[P5#,ok5,V5#])
|| (* full synchronisation *)
(CP[asn0,asn0#] ||| CP[asn1,asn1#] ||| CP[asn2,asn2#] ||| CP[asn3,asn3#] |||
CP[asn4,asn4#] ||| CP[asn5,asn5#] ||| CP[bsn0,bsn0#] ||| CP[bsn1,bsn1#] |||
CP[bsn2,bsn2#] ||| CP[bsn3,bsn3#] ||| CP[bsn4,bsn4#] ||| CP[bsn5,bsn5#] |||
CP[es0,es0#] ||| CP[es1,es1#] ||| CP[es2,es2#] ||| CP[es3,es3#] ||| CP[es4,es4#] |||
CP[es5,es5#] ||| CP[go0,go0#] ||| CP[go1,go1#] ||| CP[go2,go2#] ||| CP[go3,go3#]
||| CP[go4,go4#] ||| CP[go5,go5#] ||| CP[P0,P0#] ||| CP[P1,P1#] ||| CP[P2,P2#] |||
CP[P3,P3#] ||| CP[P4,P4#] ||| CP[P5,P5#] ||| CP[ok0,ok0#] ||| CP[ok1,ok1#] |||
CP[ok2,ok2#] ||| CP[ok3,ok3#] ||| CP[ok4,ok4#] ||| CP[ok5,ok5#] ||| CP[V0,V0#] |||
CP[V1,V1#] ||| CP[V2,V2#] ||| CP[V3,V3#] ||| CP[V4,V4#] ||| CP[V5,V5#])

where

<process definitions>
endspec (* SYSTEM *)

Validating Safety Models with Fault Trees

Glenn Bruns and Stuart Anderson

Department of Computer Science
University of Edinburgh
Edinburgh EH9 3JZ, UK

Abstract. In verifying a safety-critical system, one usually begins by building a model of the basic system and of its safety mechanisms. If the basic system model does not reflect reality, the verification results are misleading. We show how a model of a system can be compared with the system's fault trees to help validate the failure behaviour of the model. To do this, the meaning of fault trees are formalised in temporal logic and a consistency relation between models and fault trees is defined. An important practical feature of the technique is that it allows models and fault trees to be compared even if some events in the fault tree are not found in the system model.

1 Introduction

Safety-critical systems often have mechanisms designed to prevent, detect, or tolerate system system faults. To ensure that these mechanisms work as intended, a model of the system can be built from two parts: a model of the basic system and a model of the safety mechanisms (see Figure 1). Important properties of the system are then verified of the model. For example, if a component failure occurs, then it is detected.

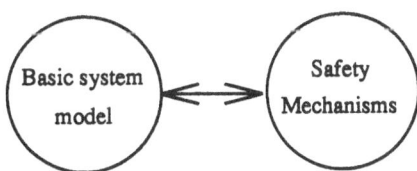

Fig. 1. A Model of a Safety-Critical System

For the verification results to be valid, the basic part of the model should reflect the true connection between component failures and system faults in the system. We are aware of a study of a rail interlocking system in which the preliminary system model allowed only one train per track section, thus making collisions impossible. Less obvious problems may be harder to discover, such as when a particular combination of failures leads to a system fault in the real system but not the system model.

We propose a validation technique in which a system model is compared to its fault trees. If a system model and its fault trees are not consistent in a sense that we will define, then the system model may not be valid. Fault trees are well suited for this purpose because they are specifically intended to capture the relationship between component failure and system faults.

The two main sections of the paper cover the precise meaning of fault trees and our proposed relationship between fault trees and system models. First, however, we present an example.

2 Example

To make discussion of the problem more concrete, we present a simple boiler system example (see Figure 2).

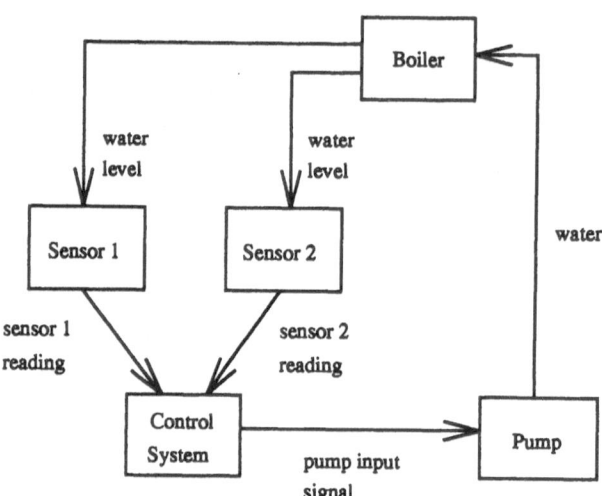

Fig. 2. A Simple Boiler System

Steam is produced by water contained in the boiler vessel. The water level in the vessel is read by two sensors, which pass their readings to a control system. If the readings are below a certain value, the pump is turned on, delivering water to the vessel. If the level readings are above a certain value, the pump is turned off.

One safety-critical fault of the system is a boiler level that is too high. A fault tree for this fault is given in Figure 2.

A fault tree represents how events in a system can lead to a particular system fault. The event symbols used here are either *basic* events (which are drawn as circles and represent component failures) or *intermediate* events (which are drawn as rectangles and represent events which occur because of lower-level

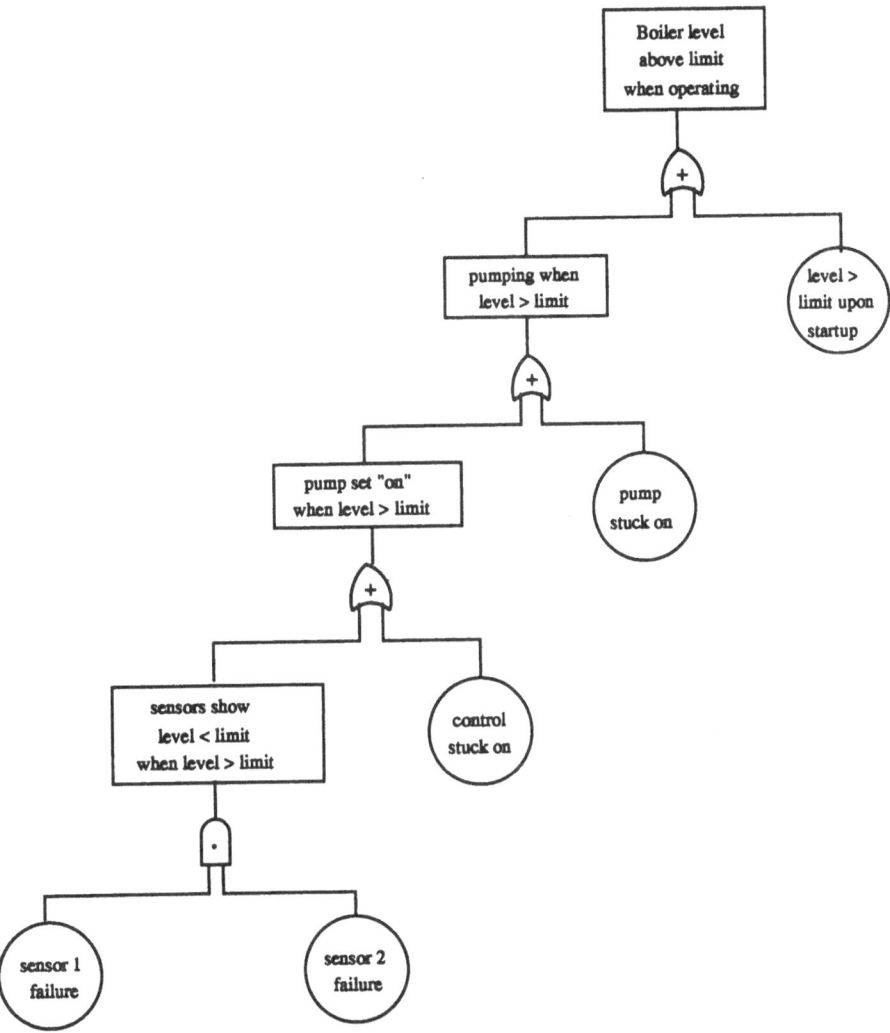

Fig. 3. A Fault Tree for the Boiler System

events). The system fault is shown as the event at the root of the tree. Event symbols are connected in the tree by gate symbols, which are either *and*-gates or *or*-gates.

The full fault tree notation has many more event and gate symbols, but if we do not consider the probabilistic meaning of fault trees then the symbols we have described are enough.

3 Fault Tree Semantics

If we are to compare fault trees and system models, we need to understand precisely what a fault tree means. Unfortunately, even the most definitive sources (e.g., the Fault Tree Handbook [5]) are vague on some critical points.

One issue is the *nature of events*. Are they to be regarded as conditions having duration or as instantaneous occurrences? The example event "contacts fail to open" from the Fault Tree Handbook suggests the former, but the example "timer reset" suggests the latter.

The second issue is the *gate condition*: does "and" mean that both input events happen at once, or only that one happens and then the other?

A third issue is the *nature of causality*. A gate models a *sufficient* cause if the output must occur if the gate condition is satisfied by the inputs. A gate models a *necessary* cause if the gate condition must be satisfied by the inputs if the output occurs. According to the Fault Tree Handbook, fault trees model sufficient and necessary causes. However, Figure IX-10 of the Handbook shows an event labelled "wire faults in K3 relay & comp. circuitry" as a cause of "K3 relay contacts fail to close", but one can imagine circumstances in which wire faults occur in such a way that the relay contacts do not fail to close. Therefore the cause as stated is not a sufficient one.

Causes of an event are also supposed to be *immediate*. This term seems related to the notion of flow, and may not be relevant in systems that cannot be captured easily with flow models. All examples in the Fault Tree Handbook are illustrated with flow diagrams. Immediacy also suggests time. For our purposes, a gate models an immediate cause if no time passes between a cause and its effect.

We now present a formal semantics for fault trees. Events are treated as conditions having duration, and the gate condition is taken to be that both inputs to an and-gate must occur at once. Three different formalisations of gates are given, corresponding to different stances on the issue of gate causality.

Formally, fault trees are interpreted as formulas of temporal logic. We use the modal mu-calculus (see Appendix A), but nearly all temporal logics are expressive enough for our purposes. Similarly, the kinds of structures that temporal logics are interpreted over are very general. We assume only that a system model can be represented as a transition system or as a set of sequences of states.

Events are formalised as atomic propositions, which are interpreted as sets of states. For example, the event "sensor failure" could be modelled as the atomic proposition SF, which is interpreted as all states in which the sensor has failed. This formalisation of events fits with most of the examples of the Fault Tree Handbook, and is consistent with the meaning of the term "event" in probability theory. Since fault tree are subject to probabilistic analysis, a consistent view of events is desireable.

Next we will formalise the meaning of gates. We will let $+(in_1, in_2, out)$ stand for an or-gate with inputs in_1 and in_2 and output out. Similarly, $\bullet(in_1, in_2, out)$ stands for an and-gate. The semantics of a gate g, denoted $[\![g]\!]$, gives the logical relationship between the input and output events of g.

3.1 A Propositional Semantics for Gates

Formalising gates with propositional logic is a simple approach that is reasonably close to the informal description of gates in the Fault Tree Handbook. In terms of the issues just discussed, this interpretation requires and-gate inputs to occur at the same time for the gate condition to be satisfied, and takes causality to be necessary, sufficient, and immediate. The subscript p on the semantic function stands for "propositional".

$$[+(in_1, in_2, out)]_p \overset{\text{def}}{=} out \Leftrightarrow in_1 \lor in_2$$

$$[\bullet(in_1, in_2, out)]_p \overset{\text{def}}{=} out \Leftrightarrow in_1 \land in_2$$

Informally, the first statement says that the output of an and-gate is true whenever both inputs are true. Remembering that events are treated as sets of states, the statement alternatively says that the set of states denoted by out is the intersection of the sets denoted by in_1 and in_2. The concept of causality here is truly immediate: whenever both causes are present the effect is also present.

3.2 Two Temporal Semantics for Gates

The greatest weakness of the propositional interpretation of fault trees is the assumption that no time can pass between cause and effect. This assumption violates a common intuition about causality. Since the examples in the Fault Tree Handbook mostly concern examples in which flow is virtually instantaneous (as in an electric circuit), the problem rarely arises there. In cases where flow is not instantaneous, events are modelled so that causes can be made immediate, albeit somewhat unnaturally. For example, in the pressure tank analysis of Chapter VII continuous pump operation can lead to a pump failure. This cause is modelled as the event "tank ruptures due to internal over-pressure caused by continuous pump operation for t > 60 sec". Since the idea of a cause leading to an event is natural, it is worthwhile to try to view fault trees in this way.

Our first temporal semantics requires that and-gate inputs occur at the same time to satisfy the gate condition, and takes causality to be only sufficient, not necessary or immediate. This means that once the gate condition is satisfied, the gate output must eventually occur. The temporal logic operator **even** is used to express the temporal condition of eventuality. Thus **even**(ϕ) means that the property expressed by formula ϕ will hold in the future.

The temporal relation between input and output events for gates can be defined as

$$[+(in_1, in_2, out)]_{t1} \overset{\text{def}}{=} (in_1 \lor in_2) \Rightarrow \textbf{even}(out)$$

$$[\bullet(in_1, in_2, out)]_{t1} \overset{\text{def}}{=} (in_1 \land in_2) \Rightarrow \textbf{even}(out)$$

The first definition says that it is always the case that if input events in_1 and in_2 occur together, then eventually output event out will occur.

Our second temporal semantics treats causality as only necessary. The temporal operator $\mathbf{prev}(\phi)$ means that the property expressed by formula ϕ held in the past.

$$[+(in_1, in_2, out)]_{t2} \stackrel{\text{def}}{=} out \Rightarrow \mathbf{prev}(in_1 \vee in_2)$$

$$[\bullet(in_1, in_2, out)]_{t2} \stackrel{\text{def}}{=} out \Rightarrow \mathbf{prev}(in_1 \wedge in_2)$$

However, these definitions allows the gate output out to occur many times for a single occurrence of $in_1 \wedge in_2$. A better interpretation might require that if out happens, then $in_1 \wedge in_2$ must have happened at least as recently as the previous occurrence of out.

There are other possible interpretations based on other choices about the basic semantic issues. For example, combining the two temporal semantics we have presented would give one modelling sufficient and necessary causality.

Fault tree gates have been interpreted temporally before (see [1]), but the use of temporal logic here allows much simpler semantics. This simplicity makes comparison between alternative interpretations easier.

3.3 Putting Gates Together

We now present the semantics of a fault tree t based on the set of gates contained in the tree (written as $gates(t)$). We use the temporal operator $\mathbf{always}(\phi)$, which means that the property expressed by ϕ holds in every state.

$$[t] \stackrel{\text{def}}{=} \mathbf{always}(\bigwedge_{g \in gates(t)} [g])$$

In English, this definition says that it is always the case that every gate condition is satisfied. Note that the meaning of a fault tree is given in terms of the meaning of its gates.

The propositional semantics has some great advantages over the temporal ones. Because a gate output is defined in the propositional case to be logically equivalent to the disjunction or conjunction of its inputs, the fault tree can be manipulated according to the laws of propositional logic. This property allows internal events of a fault tree to be removed by simplification, giving a relation between only the primary failures and the system fault (as is found in minimal cut set interpretations of fault trees [5]).

A further advantage of the propositional interpretation of fault trees is that the meaning is given as an *invariant* property – a property that can be checked by looking at states in isolation. Invariant properties are an easy class of temporal logic formulas to prove.

The main advantage of the temporal semantics is their ability to model richer notions of causality. Unfortunately, it is no longer possible to eliminate internal events by simplification, and thus minimal cut sets cannot generally be obtained. Furthermore, this formalisation of fault trees uses the temporal property of eventuality, and is therefore a *liveness* property. This class of temporal logic formulas are generally more difficult to prove than invariant formulas.

The best interpretation of a fault tree probably depends on the system being studied. In some cases one might want to choose different interpretations for different kinds of gates. For example, or-gates could be interpreted propositionally and and-gates interpreted temporally. Alternatively, a wider variety of gate types could be defined, and their use mixed in a single fault tree.

The material in the next section can be applied independently of choice of semantics for fault trees.

4 Relating Fault Trees to System Models

The last section showed that a fault tree expresses a property of failure events in a system. We might therefore expect a model of the system to have the properties expressed by its fault trees. We will attempt to make this relationship precise.

Let $\mathcal{F}$ stand for the set of system faults for which fault trees have been developed, and let $ft(F)$ be the fault trees for fault F. Given a system model $\mathcal{M}$ with an initial state s_0, we write $s_0 \models_{\mathcal{M}} \phi$ if the system model has the property expressed by formula ϕ. The condition expressing that a model $\mathcal{M}$ of a system is consistent with the set of fault trees for the system is

$$s_0 \models_{\mathcal{M}} \bigwedge_{F \in \mathcal{F}} [ft(F)]$$

This condition is too strong, however, because usually a system model will capture only certain aspects of a system. One way to weaken the relation above is to require a system model to satisfy the property expressed by a fault tree only if the system fault of the tree is found in the system model. Letting $faults(\mathcal{M})$ stand for the system faults in a model $\mathcal{M}$, the new consistency condition is

$$s_0 \models_{\mathcal{M}} \bigwedge_{F \in \mathcal{F} \cap faults(\mathcal{M})} [ft(F)]$$

This relation is still quite strong, however. If a system model only captures certain failures, then it probably would not satisfy this condition. It would be useful to know the weakest relation that should definately be expected to hold between a model of a system and the fault trees of a system. Our approach is to assume that we know nothing about events not given in a system model. As an example, suppose that we have a single or-gate, $+(B, C, A)$, which by the propositional interpretation gives the relation $A \Leftrightarrow B \vee C$ between events A, B, and C. Also suppose that we know nothing about event B. Then we will still expect that $C \Rightarrow A$. Logically this amounts to the projection of the relation $A \Leftrightarrow B \vee C$ onto the atomic propositions A and C. The projected relation is arrived at by taking the disjunction of the cases where B is true and B is false. In other words, the disjunction of $A \Leftrightarrow true \vee C$ and $A \Leftrightarrow false \vee C$ is equivalent to the formula $C \Rightarrow A$. In the general case, where more than one event might be missing, we need to consider all combinations of possibilities for the missing events.

To formalise this idea, let $\phi[\phi'/Q]$ be the formula ϕ with every occurrence of atomic proposition Q within ϕ replaced by ϕ'. For example, $A \vee B[false/B]$ gives $A \vee false$. For multiple substitutions, let $\phi[\phi_1/Q_1, \ldots, \phi_n/Q_n]$ be the formula ϕ with occurrences of $Q_1, \ldots, Q_n$ in ϕ simultaneously replaced by $\phi_1, \ldots, \phi_n$ (no atomic proposition is allowed to occur twice in the substitution list). For example, $A \vee B \Leftrightarrow C[true/A, false/C]$ gives $true \vee B \Leftrightarrow false$. We will write $S_1 \times S_2$ for the cross product of sets S_1 and S_2, i.e., $S_1 \times S_2 \stackrel{\text{def}}{=} \{(x, y) \mid x \in S_1 \text{ and } y \in S_2\}$.

Let $Bool$ be the set $\{true, false\}$ of boolean constants, and let $Bool^n$ be the n-fold product of $Bool$. The interpretation of a fault tree in the absence of a set of events $\mathcal{E} = \{a_1, \ldots, a_n\}$ is defined to be:

$$[t - \mathcal{E}] \stackrel{\text{def}}{=} \bigvee_{\{b_1, \ldots, b_n\} \in Bool^n} [t][b_1/a_1, \ldots, b_n/a_n]$$

Let $events(\mathcal{M})$ be the set of events in the system model $\mathcal{M}$, and let $events(t)$ be the set of events in fault tree t. Then $events(t) \backslash events(\mathcal{M})$ is the set of events found in the fault tree t but not the model $\mathcal{M}$. The condition expressing that a model $\mathcal{M}$ of a system is consistent with the set of fault trees for the system is now

$$s_0 \models_{\mathcal{M}} \bigwedge_{F \in \mathcal{F} \cap faults(\mathcal{M})} [ft(F) - (events(ft(F)) \backslash events(\mathcal{M}))]$$

5 Conclusions

This paper contains three contributions to the study of safety-critical systems. First, it presents the idea that fault trees can be used to check the validity of safety-critical system models. Second, it contains three formal semantics for fault trees. These semantics are an improvement on earlier work by expressing the meaning of fault trees with temporal logic, by expressing events as sets of states, and by identifying four elements of the meaning of gates: gate condition, sufficiency, necessity, and immediacy. Finally, the paper defines a consistency condition between a model of a system and the system's fault trees that works even for models that contain only some of the failure events in the their fault trees.

Tool support for checking the consistency condition exists in the form of *model checkers*, which automatically show whether a finite-state model satisfies a temporal logic formula [3]. Proof tools (such as [2]) are available in case the model is not finite-state.

The work described here should be regarded as a first step towards a complete understanding of fault trees and their relation to system models. As mentioned in the section on the semantics of gates, the formalisation here of necessary causes may be too simplistic. The consistency condition given might need to be strengthened to ensure that an event representing a component failure can always occur provided it has not already occurred. Our consistency condition handle the case in which a single failure event in a system model represents several failure events in a fault tree.

A A Temporal Logic

We use an extended form of the modal mu-calculus [4, 6] as a temporal logic to express behavioural properties. The syntax of the extended mu-calculus is as follows, where L ranges over sets of actions, Q ranges over atomic sentences, and Z ranges over variables:

$$\phi ::= Q \mid \neg\phi \mid \phi_1 \wedge \phi_2 \mid [L]\phi \mid Z \mid \nu Z.\phi$$

The operator νZ binds free occurrences of Z in ϕ, with the syntactic restriction that free occurrences of Z in a formula ϕ lie within an even number of negations.

Let S be a set of states and Act a set of actions. A formula ϕ is interpreted as the set $\|\phi\|_{\mathcal{V}}^{\mathcal{T}}$ of states, defined relative to a a fixed transition system $\mathcal{T} = (S, \{\overset{a}{\rightarrow} \mid a \in Act\})$ and a valuation $\mathcal{V}$, which maps variables to sets of states. The notation $\mathcal{V}[S'/Z]$ stands for the valuation $\mathcal{V}'$ which agrees with $\mathcal{V}$ except that $\mathcal{V}'(Z) = S'$. Since the transition system is fixed we usually drop the state set and write simply $\|\phi\|_{\mathcal{V}}$. The definition of $\|\phi\|_{\mathcal{V}}$ is as follows:

$$\|Q\|_{\mathcal{V}} = \mathcal{V}(Q)$$
$$\|\neg\phi\|_{\mathcal{V}} = S - \|\phi\|_{\mathcal{V}}$$
$$\|\phi_1 \wedge \phi_2\|_{\mathcal{V}} = \|\phi_1\|_{\mathcal{V}} \cap \|\phi_2\|_{\mathcal{V}}$$
$$\|[L]\phi\|_{\mathcal{V}} = \{s \in S \mid \text{if } s \overset{a}{\rightarrow} s' \text{ and } a \in L \text{ then } s' \in \|\phi\|_{\mathcal{V}}\}$$
$$\|Z\|_{\mathcal{V}} = \mathcal{V}(Z)$$
$$\|\nu Z.\phi\|_{\mathcal{V}} = \bigcup\{S' \subseteq S \mid S' \subseteq \|\phi\|_{\mathcal{V}[S'/Z]}\}$$

A state s satisfies a formula relative to a model $\mathcal{M} = (\mathcal{T}, \mathcal{V})$, written $s \models_{\mathcal{M}} \phi$, if $s \in \|\phi\|_{\mathcal{V}}^{\mathcal{T}}$.

Informally, $[L]\phi$ holds of a state $\dot{s}$ if ϕ holds for all states s' that can be reached from s through an action a in L. A fixed point formula can be understood by keeping in mind that $\nu Z.\phi$ can be replaced by its "unfolding": the formula ϕ with Z replaced by $\nu Z.\phi$ itself. Thus, $\nu Z.\psi \wedge [\{a\}]Z = \psi \wedge [\{a\}](\nu Z.\psi \wedge [\{a\}]Z) = \psi \wedge [\{a\}](\psi \wedge [\{a\}](\nu Z.\psi \wedge [\{a\}]Z)) = \ldots$ holds of any process for which ψ holds along any execution path of a actions.

The operators $\vee$, $\langle a\rangle$, and μZ are defined as duals to existing operators (where $\phi[\psi/Z]$ is the property obtained by substituting ψ for free occurrences of Z in ϕ):

$$\phi_1 \vee \phi_2 \overset{\text{def}}{=} \neg(\neg\phi_1 \wedge \neg\phi_2)$$
$$\langle L\rangle\phi \overset{\text{def}}{=} \neg[L]\neg\phi$$
$$\mu Z.\phi \overset{\text{def}}{=} \neg\nu Z.\neg\phi[\neg Z/Z]$$

These additional basic abbreviations are also convenient:

$$[a_1, \ldots, a_n]\phi \stackrel{\text{def}}{=} [\{a_1, \ldots, a_n\}]\phi$$
$$[-]\phi \stackrel{\text{def}}{=} [Act]\phi$$
$$true \stackrel{\text{def}}{=} \nu Z.Z$$
$$false \stackrel{\text{def}}{=} \neg true$$

Common operators of temporal logic can also be defined as abbreviations:

$$\textbf{always}(\phi) \stackrel{\text{def}}{=} \nu Z.\phi \wedge [-]Z$$
$$\textbf{even}(\phi) \stackrel{\text{def}}{=} \mu Z.\phi \vee (\langle - \rangle true \wedge [-]Z)$$

To define a *previously* operator, a reverse modal operator $\overline{[L]}$ must be added to the logic.

$$\|\overline{[L]}\phi\|_\nu = \{s \in S \mid \text{if } s' \stackrel{a}{\rightarrow} s \text{ and } a \in L \text{ then } s' \in \|\phi\|_\nu\}$$

The previously operator is just the reverse version of **even**:

$$\textbf{prev}(\phi) \stackrel{\text{def}}{=} \mu Z.\phi \vee (\overline{\langle - \rangle}true \wedge \overline{[-]}Z)$$

References

1. R.E. Bloomfield, J.H. Cheng, and J. Gorski. Towards a common safety description model. In J.F. Lindeberg, editor, *SAFECOMP '91*, 1991.
2. J.C. Bradfield. A proof assistand for symbolic model checking. In *Proceedings of CAV '92*, 1992.
3. Rance Cleaveland, Joachim Parrow, and Bernhard Steffen. The concurrency workbench: A semantics based tool for the verification of concurrent systems. Technical Report ECS-LFCS-89-83, Laboratory for Foundations of Computer Science, University of Edinburgh, 1989.
4. D. Kozen. Results on the propositional mu-calculus. *Theoretical Computer Science*, 27:333–354, 1983.
5. N.H. Roberts, W.E. Vesely, D.F. Haasl, and F.F. Goldberg. *Fault Tree Handbook*. U.S. Nuclear Regulatory Commission, 1981.
6. C. Stirling. Temporal logics for CCS. In J.W. de Bakker, W.-P. de Roever, and G. Rozenberg, editors, *Linear Time, Branching Time and Partial Order in Logics and Models*. Springer Verlag, 1989. Lecture Notes in Computer Science, 354.

Composition and Refinement of Probabilistic Real-Time Systems *

Zhiming Liu[1] , Jens Nordahl[2] and Erling Vagn Sørensen[2]

[1] Department of Computer Science, University of Warwick, England
[2] Department of Computer Science, Technical University of Denmark

Abstract. In this paper we illustrate, by way of examples, composition, analysis and refinement of systems modelled by means of probabilistic automata behaving as Markov chains over discrete time. For a formalised and general treatment of these ideas the reader is referred to [6].

1 Introduction

Dependability analysis of failure-prone real-time systems or performance analysis of real-time systems interacting with stochastic environments is frequently based on the use of *Probabilistic Automata*, PA's, with Markov properties as computation models.

A recent formalism using this approach is *Probabilistic Duration Calculus* PDC [5], an extension of *Duration Calculus*, DC, which in turn was developed for specification and verification of embedded real-time systems [7, 1].

For a given discrete-time PA representing a design PDC makes it possible to calculate and reason about the probability that the PA satisfies a DC-formula (expressing a requirement or design decision) during the first t time units.

In this paper we consider *parallel composition* of component PA's into larger component PA's or into a system PA. Each component PA may depend on states in the other components, (the PA is then said to be open), but the system PA is independent of external states (the PA is then said to be closed). Closedness is a condition for analysis by means of PDC.

We also consider *probabilistic refinement* with respect to a DC formula. This means that with two consecutive system designs, if for all $t \geq 0$ the second design satisfies the requirement with higher probability than the first design, then the second design is said to refine the first design with respect to the requirement. Simple examples of probabilistic refinement are included.

Compositionality has also been treated in probabilistic extensions of CSP- and CCS-like process algebras [4, 2], but none of these approaches cover the

---

* This research was supported by the Danish Technical Research Council under project **Co-design**. The research of Zhiming Liu was also supported in part by research grant GR/H39499 from the Science and Engineering Research Council of UK.

dependencies between components referred to above. A notion of probabilistic refinement different from ours is described in [3]. In this work probabilistic specifications prescribe permissible intervals for the target probabilities, and refinement refers to the narrowing of these intervals in subsequent specifications.

2 Probabilistic Automata Over Discrete Time

The behaviour of probabilistic systems having a finite number of states is conveniently modelled by means of finite probabilistic automata over discrete time. These automata are defined by the set of states, the set of initial state probabilities and the set of transition probabilities per time unit.

As a running example we consider a Gas Burner consisting of a Burner (an abstraction of the gas-valve, the ignition device and the control box) and a Detector (an abstraction of the mechanism for detection of unburnt gas). We assume that the gas is turned on at $t = 0$ and remains on.

2.1 States

The Burner and Detector components are characterised by disjoint sets of primitive Boolean states. For simplicity these sets, denoted A_B and A_D respectively, are assumed to be the singleton sets

$$A_B = \{Flame\} \quad \text{and} \quad A_D = \{Act\}$$

where *Flame* asserts that the flame exists and *Act* asserts that the Detector is able to detect unburnt gas. (If the gas was not permanently on, we would have to define A_B as $\{Gas, Flame\}$ where the additional primitive state *Gas* asserts, that gas is released.)

The set of component states S_B and S_D are defined as subsets of the set of minterms over A_B and A_D respectively.

$$S_B = \{\neg Flame, Flame\} \subseteq 2^{A_B} \quad \text{and} \quad S_D = \{\neg Act, Act\} \subseteq 2^{A_D}$$

(where, in this case, all minterms are possible states). Accordingly the Burner makes transitions between the states ¬*Flame* and *Flame* (corresponding to alternation between successful flame ignitions and unintended flame extinctions) while the Detector makes transitions between *Act* and ¬*Act* (corresponding to alternation between failure and repair).

For the composed system we have:

$$A = A_B \cup A_D = \{Flame, Act\}$$
$$S = S_B \times S_D =$$
$$\{\neg Flame \wedge Act, Flame \wedge Act, \neg Flame \wedge \neg Act, Flame \wedge \neg Act\} \subseteq 2^A$$

2.2 Dependency of External States

In a composite system a component can only change its own primitive states. However, the local transition probabilities may depend on external states, i.e. states which are local to other components in the composition.

This is exemplified by the Burner. For example, given that the Burner is in state $\neg Flame$ at time t, the probability that it will be in state $Flame$ at time $t+1$ is zero if the Detector is non-active and non-zero if it is active. In the latter case the actual value depends on the quality of the ignition mechanism in the Burner.

2.3 Open and Closed Probabilistic Automata

The previous discussion suggests that a component, which depends upon environmental states, should be modelled by a collection of sub-models, one for each environmental state. This is illustrated by the transition graphs for the Burner automaton in Figure 1a according to which:

- The Burner starts from state $\neg Flame$ with probability $p_1 = 1$.

- With the Detector in state Act the Burner behaves as follows

    - Given that it is in state $\neg Flame$, it remains in that state with probability p_{11} per time unit or it goes to state $Flame$ with probability p_{12} per time unit where $p_{11} + p_{12} = 1$. .

    - Given that it is in state $Flame$, it remains in that state with probability p_{22} per time unit or it goes to state $\neg Flame$ with probability p_{21} per time unit where $p_{22} + p_{21} = 1$.

- With the Detector in state $\neg Act$ the Burner behaves as follows

    - Given that it is in state $\neg Flame$, it remains in that state with probability $p_{11} = 1$ per time unit. This implies that $p_{12} = 0$, i.e. the Burner can never go to state $Flame$.

    - Given that it is in state $Flame$, it remains in that state with probability p_{22} per time unit or it goes to state $\neg Flame$ with probability p_{21} per time unit where p_{22} and p_{21} are the same as when the detector is in state Act.

p_{11} and p_{12} depend on the detector state because they characterise the ability of the Burner to establish $Flame$. In contrast to this p_{22} and p_{21} are entirely independent of the detector state because they characterise the stability of the flame.

A PA in which some transition probabilities depend on external states will be called an *open PA*.

For the Detector automaton we assume the transition graph shown in Figure 1b according to which:

- The Detector starts from state Act with probability q_1 or from state $\neg Act$ with probability q_2 $(= 1 - q_1)$.

- Given that it is in state Act, it remains in that state with probability q_{11} per time unit or it goes to state $\neg Act$ with probability q_{12} per time unit where $q_{11} + q_{12} = 1$.

- Given that it is in state $\neg Act$, it remains in that state with probability q_{22} per time unit or it goes to state Act with probability q_{21} per time unit where $q_{22} + q_{21} = 1$.

The transition probabilities of the Detector are independent of the Burner state. This reflects that the probabilities of failure or repair of the Detector are considered to be unaffected by flame- or ignition failures occurring in the Burner.

A PA in which no transition probabilities depend on external states will be called a *closed PA*.

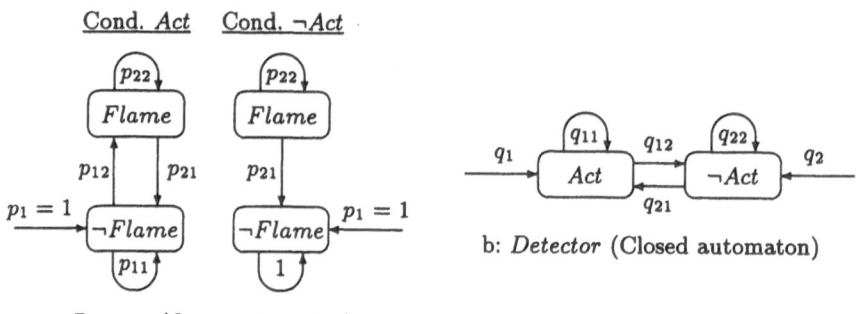

a: *Burner* (Open automaton)

b: *Detector* (Closed automaton)

Fig. 1. The probabilistic component automata: *Burner* and *Detector*

3 Parallel Composition of PA's

The PA for the Gas Burner is determined by parallel composition of the PA's for the Burner and the Detector.

$Gas\_Burner = Burner \parallel Detector$

This operation is fully formalised and generalised in [6]. The resulting PA for the Gas Burner is shown in Figure 2, (where the state numbering is arbitrary and introduced for later use). The reasoning behind this construction is illustrated informally be means of a few examples.

The probability that the Gas Burner starts in, say, state 3: $\neg Flame \wedge \neg Act$ is the product of the probability that the Burner starts in state $\neg Flame$ and the probability that the Detector starts in state $\neg Act$. According to Figure 1 this product is $p_1 * q_2 = 1 * q_2 = q_2$. Similarly we find that the initial probabilities of states 1, 2 and 4 are q_1, 0 and 0 respectively.

The probability that the Gas Burner transits from e.g. state 1: $\neg Flame \wedge Act$ to state 2: $Flame \wedge Act$ within one time unit is the product of the probability that the Burner transits from $\neg Flame$ to $Flame$, given that the Detector is active, and the probability that the Detector transits from Act to Act, both within one time unit. This product is $p_{12} * q_{11}$. On the other hand, the probability that the Gas Burner transits from state 3 to state 4 is zero because it requires a transition of the Burner from $\neg Flame$ to $Flame$ while the Detector is non-active, but according to Figure 1 this is impossible.

The other composite transition probabilities can be determined in a similar way.

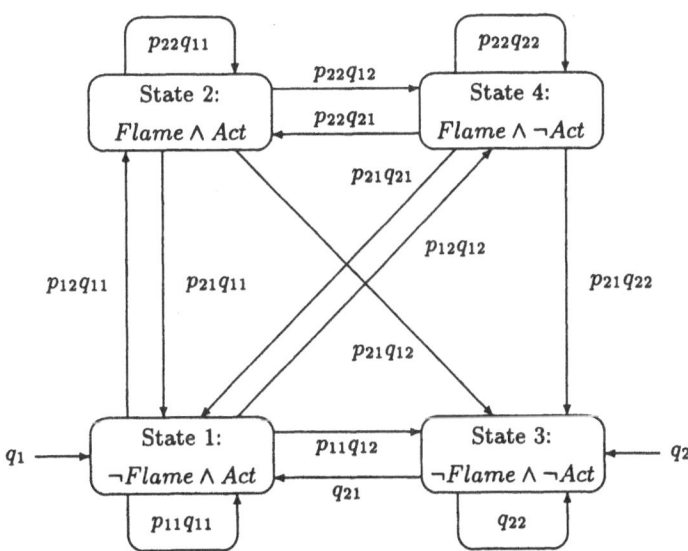

Fig. 2. The closed automaton $Gas\_Burner = Burner \parallel Detector$

We observe that even though the Burner PA is open, the composition of the Burner and the Detector PA's is a closed PA. This illustrates, that when we compose two component PA's the dependencies of one PA on primitive states in the other will be hidden in the resulting PA[3].

---

[3] This resembles the hiding of a communication between two processes in parallel composition.

In [6] the formalisation of such constructions is based on a representation of a PA as a tuple (S, C, τ_0, τ) where S is the set of states, C is a set of primitive external states which is sufficient for definition of openness, τ_0 is the initial probability function and τ is the transition probability function, which is parameterised by the elements in the condition space 2^C. A PA is closed if no transition probability depends on the conditions in 2^C. C can then be eliminated from the tuple.

The parallel operator $\|$ constructs a well formed tuple $(S_c, C_c, \tau_{0c}, \tau_c)$ from two well formed tuples $(S_a, C_a, \tau_{0a}, \tau_a)$ and $(S_b, C_b, \tau_{0b}, \tau_b)$ such that

$$S_c = S_a \times S_b \subseteq 2^{A_a \cup A_b} \quad \text{and} \quad C_c = (C_a \setminus A_b) \cup (C_b \setminus A_a)$$

where A_a and A_b are the sets of primitive states for components a and b.

4 Requirements and Satisfaction Probabilities

The Gas Burner has critical states characterised by release of gas while the flame is absent. The disjunction of these states is a state called *Leak*. Since the gas is permanently on in our example, *Leak* is identical to $\neg$*Flame* which is the disjunction of states 1 and 3 on Figure 2

One of the design decisions could be that whenever *Leak* occurs, it should be detected and eliminated within one time unit. In [7] this constraint, called **Des-1**, is expressed as the following formula in Duration Calculus:

Des-1: $\Box(\lceil Leak \rceil \implies \ell \leq 1)$

This formula reads:

$\Box$	"For any subinterval of the observation interval,"
$\lceil Leak \rceil \implies$	"if there is *Leak* in that subinterval then"
$\ell \leq 1$	"its length should not exceed one time unit".

Duration Calculus, DC, is an interval logic for the interpretation of Boolean states over time (a logic for timing diagrams). Its distinctive feature is reasoning about durations of states within any time interval without explicit reference to absolute time. It is used to specify and verify real-time requirements. The reader is referred to [7, 1]. for further details.

For a real design with failure-prone components we can not in general expect a duration formula (expressing some requirement) to hold for all times. The question is then: does it hold with sufficiently high probability over a specified observation interval $[0, t]$. This question is answered as follows.

Let G denote the closed PA modelling the design and D denote the formula. Then we must compute the probability that G satisfies D in the time interval $[0, t]$. This probability is called the *satisfaction probability* of D by G and is denoted $\mu_G(D)[t]$.

Satisfaction probabilities are computed or reasoned about by means of Probabilistic Duration Calculus, PDC [5], a recent extension of Duration Calculus[4].

The DC formula for **Des-1** belongs to a class for which the satisfaction probability can be expressed explicitly in PDC by means of the initial probability vector p and the the transition probability matrix P of G, [5]. These matrices are well known from the theory of Markov chains. For the PA of the composite Gas Burner, Figure 2, they are given by:

$$p = (q_1, 0, q_2, 0) \text{ and } P = \begin{pmatrix} p_{11}q_{11} & p_{12}q_{11} & p_{11}q_{12} & p_{12}q_{12} \\ p_{21}q_{11} & p_{22}q_{11} & p_{21}q_{12} & p_{22}q_{12} \\ q_{21} & 0 & q_{22} & 0 \\ p_{21}q_{21} & p_{22}q_{21} & p_{21}q_{22} & p_{22}q_{22} \end{pmatrix}$$

with row- and column ordering according to the chosen state numbering. It is easy to see, that for p as well as for each row of P the sum of elements is 1 (this is a well-formedness condition for probability matrices).

With D denoting a DC formula of the class referred to above, the explicit expression for $\mu_G(D)[t+1]$ is a scalar product of the form [5]:

$$\mu_G(D)[t+1] = p' \cdot (P')^t \cdot 1_c$$

where p' and P' are obtained from p and P, respectively, by replacement of certain entries (depending on D) by zeros, $(P')^t$ denotes the t'th power of P' and 1_c denotes a column vector in which all elements are 1.

For G representing the composite Gas Burner and with D given as the DC formula for **Des-1** above, $p' = p$, (i.e. no entries in p needs to be zeroed) and P' is obtained from from P by changing the entries in P with (row,column) numbers (1,1), (1,3), (3,1) and (3,3) to zero. Accordingly:

$$\mu_G(\Box(\lceil Leak \rceil \Longrightarrow l \leq 1))[t+1] = p' \cdot (P')^t \cdot 1_c =$$

$$(q_1, 0, q_2, 0) \cdot \begin{pmatrix} 0 & p_{12}q_{11} & 0 & p_{12}q_{12} \\ p_{21}q_{11} & p_{22}q_{11} & p_{21}q_{12} & p_{22}q_{12} \\ 0 & 0 & 0 & 0 \\ p_{21}q_{21} & p_{22}q_{21} & p_{21}q_{22} & p_{22}q_{22} \end{pmatrix}^t \cdot \begin{pmatrix} 1 \\ 1 \\ 1 \\ 1 \end{pmatrix}$$

Informally the rules for obtaining the primed matrices from the unprimed ones (i.e. for the zeroing of entries) are as follows:

If D is violated by behaviours which have state i as the initial state (the state in the first time unit), then the i'th entry of p should be zeroed in the

---

[4] The semantic model of PDC is the finite probability space $\langle V^t, \mu \rangle$ induced by G, where V^t is the set of behaviours (state sequences of G) of length t and μ is the probability measure which assigns a probability to each behaviour. This probability is the product of the initial probability and the transition probabilities involved in the behaviour. $\mu_G(D)[t]$ is then defined as the sum of the behaviour probabilities for the subset of behaviours which satisfy D over the first t time units.

matrix expression for $\mu(D)[t+1]$. If D is violated by a transition from state i to state j, then the entry at location (i,j) in P should be zeroed.

The DC formula for **Des-1** places no restriction on the choice of initial state, and accordingly no entry of p needs to be zeroed. However, the formula is violated for all transitions such that there is *Leak* before as well as after the transition. This is because such transitions imply existence of *Leak* states lasting for at least two time units, whereas the formula only tolerate *Leak* states lasting for at most one time unit. As previously observed, *Leak* $(= \neg Flame)$ holds for the composite states 1 and 3 on Figure 2. This implies that the offensive transitions are those associated with entries (1,1), (1,3), (3,1) and (3,3) in P.

5 Probabilistic Refinement

Let G_1 and G_2 be the closed PA's representing two designs and let D represent a common requirement for these designs. Then G_2 is said to *refine* G_1 with respect to D if, and only if:

$$\forall t_0 \geq 0 \bullet \mu_{G_2}(D)[t_0] \geq \mu_{G_1}(D)[t_0]$$

We shall now examine this for various Gas Burner designs and for **Des-1**.

First we notice that if $t_0 = 0$, then **Des-1** will be trivially satisfied for any design G, i.e. $\mu_G(\textbf{Des-1})[0] = 1$. The reason for this is that in the formula for **Des-1** the left side of the implication, i.e. $\lceil Leak \rceil$, is false for a point interval (a leak state must last for at least one time unit).

For $t_0 > 0$ we make the substitution $t_0 = t + 1$, $t \geq 0$ and compute $\mu_G(\textbf{Des-1})[t+1]$ from the matrix expression $p' \cdot (P')^t \cdot 1_c$.

As previuosly explained, with **Des-1** as the D formula, $p' = p$. This, in turn, implies that the matrix expression will evaluate to 1 for $t = 0$. The reason for this is that the sum of entries in p is 1 and $(P')^0$ is the identity matrix. This result reflects that the initial *Leak* state (caused by the necessary gas release before the first ignition) lasts for (at least) one time unit and does not violate **Des-1** during the first time unit.

We will compare four designs of the Gas Burner

- Design 1 is a poor design. It has a Burner but no Detector (or the Detector is permanently non-active). The PA: G_1 for this design is shown in Fig 3a. The probability matrices are

$$p = (1,0) \quad p' = p \quad P = \begin{pmatrix} 1 & 0 \\ p_{21} & p_{22} \end{pmatrix} \quad P' = \begin{pmatrix} 0 & 0 \\ p_{21} & p_{22} \end{pmatrix}$$

It is easy to prove (and intuitively clear), that $\mu_{G_1}(\textbf{Des-1})[t+1] = 0$ for $t > 0$. This reflects that the Gas Burner never will be able to establish flame, i.e. to eliminate the initial *Leak* caused by gas release.

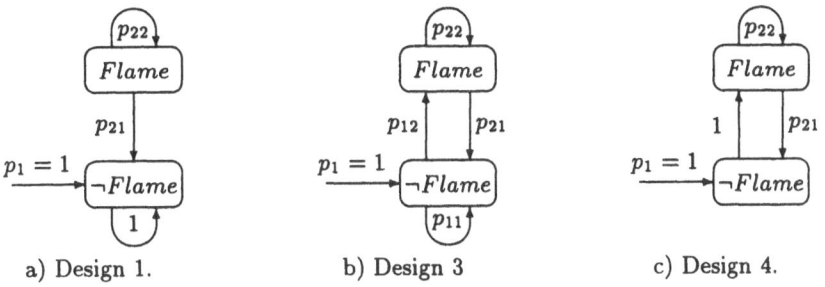

a) Design 1. b) Design 3 c) Design 4.

Fig. 3. Various designs of the Gas Burner. (Design 2 is defined by Figure 2.)

- Design 2 is the composite Gas Burner: (*Burner* ‖ *Detector*) modelled and analysed in the previous sections. Intuitively G_2 (defined by Figure 2) refines G_1 because it uses a detector which is not permanently failed. This can be validated by computation of the matrix expressions for a suitable range of t's.

- Design 3 is the composite gasburner with a permanently active Detector. *Leak* is detected immediately, but Ignition may still fail with probability p_{11} within one time unit. The PA: G_3 for this design is shown in Fig 3b. Since *Act* is always true, there are only two states $\neg Flame$ and *Flame* to consider. The probability matrices are

$$\boldsymbol{p} = (1,0) \quad \boldsymbol{p}' = \boldsymbol{p} \quad \boldsymbol{P} = \begin{pmatrix} p_{11} & p_{12} \\ p_{21} & p_{22} \end{pmatrix} \quad \boldsymbol{P}' = \begin{pmatrix} 0 & p_{12} \\ p_{21} & p_{22} \end{pmatrix}$$

Intuitively G_3 refines G_2 because it uses a permanently active detector. This can also be validated by computation of the μ's over a suitable time range.

- Design 4 is the ideal composite Gas Burner. The Detector is perfect and the ignition always succeeds within one time unit. The PA: G_4 for this design is shown in Fig 3c.

The probability matrices are

$$\boldsymbol{p} = (1,0) \quad \boldsymbol{p}' = \boldsymbol{p} \quad \boldsymbol{P} = \begin{pmatrix} 0 & 1 \\ p_{21} & p_{22} \end{pmatrix} \quad \boldsymbol{P}' = \boldsymbol{P}$$

From a theorem in [5] it follows, that with $\boldsymbol{p}' = \boldsymbol{p}$ and $\boldsymbol{P}' = \boldsymbol{P}$ the matrix expression $\boldsymbol{p}' \cdot (\boldsymbol{P}')^t \cdot \boldsymbol{1}_c$ evaluates to 1 for all values of t. Therefore $\mu_{G_4}(\textbf{Des-1})[t+1] = 1$ independent of t and accordingly G_4 refines all other Gas Burner designs with respect to **Des-1** (but of cause Design 4 is not implementable).

Verification is stronger than validation. However, proof of probabilistic refinement is a difficult area, and so far we can only offer the following theorem applicable under rather special conditions [6].

For D-formulas such that $\mu(D)_G[t+1] = \boldsymbol{p}' \cdot (\boldsymbol{P}')^t \cdot \boldsymbol{1}_c$ and for two designs G_A and G_B with the same number of states, if all elements in $\boldsymbol{p}'_B$ are greater than or equal to the corresponding elements in $\boldsymbol{p}'_A$, and the same property holds for $\boldsymbol{P}'_B$ and $\boldsymbol{P}'_A$ then G_B refines G_A with respect to D.

This theorem is applicable to Designs 1, 3 and 4 and proves that Design 4 refines Design 3 which in turn refines Design 1.

6 Conclusion

We have presented new results concerning composition, analysis and refinement of probabilistic real time systems. The technique needs further consolidation with regard to tools and theorems for validation and verification, and its practical applicability to realistic dependability problems remains to be tested.

References

1. A.P.Ravn, H.Rischel, and K.M.Hansen. Specifying and verifying requirements of real-time systems. *IEEE Trans. Software Eng.*, 19(1), 1993.

2. J.C.M. Baeten, J.A. Bergstra, and S.A. Smolka. Axiomatizing probabilistic processes: Acp with generative probabilities. In *Lecture Notes in Computer Science, volume 630*, pages 472–485. Springer Verlag, 1992.

3. B. Jonsson and K. Larsen. Specification and refinement of probabilistic processes. In *Proc. 6th Annual Symp. on Logics in Computer Science*, pages 226–277, 1991.

4. K. G. Larsen and A. Skou. Compositional verification of probabilistic processes. In *Lecture Notes in Computer Science, volume 630*, pages 456–471. Springer Verlag, 1992.

5. Z. Liu, E.V. Sørensen, A.P. Ravn, and C.C. Zhou. Towards a calculus of systems dependability. Technical report, Department of Computer Science, Technical University of Demark, 1993. To appear in *Journal of High Integrity Systems*, Oxford University Press.

6. Zhiming Liu, J. Nordahl, and E. V. Sørensen. Compositional design and refinement of probabilistic real time systems. Technical report, Institute of Computer Science, Technical University of Denmark, DK-2800 Lyngby, Denmark, 1993.

7. C.C. Zhou, C.A.R. Hoare, and A.P. Ravn. A calculus of durations. *Information Processing Letters*, 40(5):269–276, 1992.

The Application of Formal Methods for the Redevelopment of a Laboratory Information Management System

Dr. Paul Collinson
Dept. of Chemical Pathology, Mayday University Hospital
Croydon, England

Susan Oppert
Dept. of Clinical Biochemistry, West Middlesex University Hospital
Isleworth, Middlesex, England

Steven Hughes
Lloyd's Register of Shipping
Croydon, England

1.   Introduction

The overall goal of a clinical laboratory is to analyse samples of blood and other bodily fluids received from a patient, and to return the correct results to the patient's doctor within a suitable period. Automated analysers are used in most laboratories to analyse the various samples. Computer systems (LIMS[1]) are widely used in laboratories to control, support and monitor the work done in the laboratory, keeping pace with the increased analytical capability provided by these analysers. In particular, a LIMS is typically used to control and monitor (at least):

- the working of the analysers, deciding what tests need to be done for each sample by each analyser,
- the collating of requests and results, and
- the printing of the results.

The results of an analysis will directly influence the treatment of a patient - treatment that can have potentially life-threatening consequences. For example, in some types of suspected heart attacks, treatment is largely based on the results of the analysis. The patient can die if the wrong treatment is administered. In a recent case in the United Kingdom, a bank's computer system sent payments to the wrong

---

[1]Laboratory Information Management System

accounts. Consider what might happen (in the equivalent scenario) if the wrong results were sent to a patient by the LIMS.

Consequently, although a LIMS is fundamentally an information system, it must be classed as a safety critical system and developed as such. In the main, however, these safety implications have not been considered in the development of LIMS's. Furthermore, in contrast to other disciplines, little effort has been spent on the standardisation and classification of the safety aspects of using computer-based systems for medical care.

These problems will become more acute when LIMS's are linked to general hospital information systems. In the long-term, this will make the laboratory results accessible from wards in the hospital and from local GP surgeries. The aim of this is to improve patient care. However, as more and more people have (instant) access to the results, it is essential to ensure the integrity and correctness of any results that are accessible from outside the laboratory.

This paper discusses the re-development of the LIMS at the WMH,[2] undertaken as part of the MORSE[3] project and carried out jointly by WMH and Lloyd's Register. The MORSE project uses a multi-disciplinary approach to the development of safety critical systems, based on those proposed in the draft Ministry of Defence standards 00-55 [1] and 00-56 [2]. This approach combines the use of safety analysis with the use of formal development methods. This paper describes the overall approach, and concentrates on the application of RAISE [3], a particular formal development method, in the re-development of the LIMS. As space will not allow a full description of the work, the use of RAISE will be illustrated by describing the specification of certain key areas.

2. Simplified LIMS

A simplified layout of a LIMS is shown in Figure 1. This shows a complete analytical loop, with the LIMS controlling a single analyser. It indicates how:

- test requests are entered using a terminal,
- the analyser receives the samples and requests, analyses the samples and returns the appropriate results, and
- the results are printed before being dispatched to the patient.

This is only one of many possible layouts, and in practice the LIMS would be controlling several analysers. However, by using a single analyser, the description of the LIMS is simplified as all test requests will go to the one analyser. As a further simplification, the validation and archiving of results is not shown here.

---

[2]West Middlesex University Hospital
[3]Method for Object Re-use in Safety-critical Environments

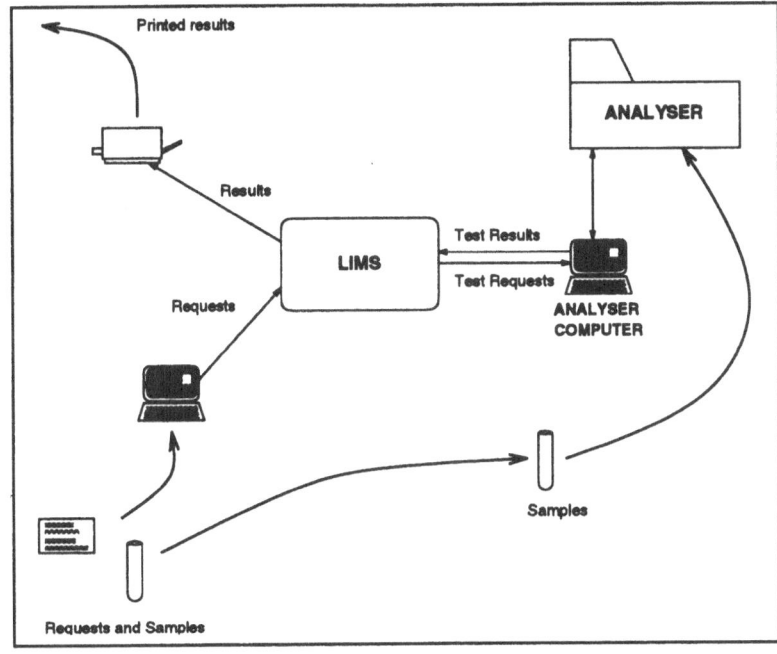

Figure 1: The layout of a simplified LIMS

3. Approach used to analyse the LIMS

The data handling in the laboratory as a whole (including the LIMS) was described and modelled before starting the safety analysis. The safety analysis produced an assessment of the safety of the data handling, and an associated list of hazards. The safety analysis of the LIMS is described in an accompanying paper and in [4]. This description of the LIMS and the list of hazards was used as the basis for the formal specification and re-development of the LIMS.

The entire LIMS was formally specified using RAISE. Safety properties that remove or at least constrain the hazards identified by the safety analysis, were also described and captured in the specification. These properties can be thought of as forming a safety case for the LIMS. As RAISE is mathematically based, these safety properties are described as constraints on the behaviour of the LIMS.

Several components of the system, identified in the specification, have been selected for further development. The selection, based on the safety assessment of the laboratory, will re-develop a complete analytical loop of the LIMS - from the input of requests to the printing of results. The development will be (rigorously) verified to ensure that the safety properties identified in the specification are maintained. It is intended that another safety analysis of the LIMS be carried out after the development to assess whether the re-development of the LIMS has improved the safety of the LIMS.

3.1. Additional safety considerations

Ensuring that the LIMS functions correctly and preserving the integrity of the data entrusted to it, is only part of what is done to ensure that the laboratory will meet it goal. The reliability of the LIMS and of the laboratory as a whole must be considered, ensuring that results will be returned to patients in time. This will include the backing up of data, the presence of standby machines, contingency plans for staff illness, etc.

The accuracy of the chemical analysis must also be considered. In the UK, this is independently assessed by a central body. In this paper, however, only the LIMS and its workings will be considered.

4. Some results of the safety analysis

In the safety analysis, several key areas were identified. Two of these will be discussed here: the identification of patients from information on the request forms, and the identification and collation of requests, samples and results.

4.1. Identification of patients

A database of all the patients who have been to the hospital is kept on a central patient administration system. Each patient in this database has a unique hospital number. If a request for a patient that has been to the hospital before, or is currently in the hospital, is received, the patient's details are retrieved from the database. If the patient's details cannot be found - if the patient has not been to the hospital, or no match can be made - a new number is assigned to the patient.

It is extremely important that this matching of details is accurate. If the wrong match is made, wrong details will be used. Consequently, the criteria for matching details are (and must be) strict. If a satisfactory match cannot be made, the patient is treated as *new*, rather allowing wrong details to be used.

4.2. Identification and collation of requests, samples and results

Samples and requests are assigned a unique label or identifier when they are received at the laboratory. This label makes it possible to distinguish samples from different patients, and is used to track requests and samples in the laboratory and for matching up results with the appropriate requests. This labelling is essential to the working of the laboratory.

5. Specification of the system

The main activities carried out in a laboratory are described below: receiving requests, analysing samples, and collation and printing results. A simple model of the LIMS is defined, describing how the LIMS supports these activities. Using this model of a LIMS, we will introduce some constraints on the system. These constraints will capture some of the safety properties of a LIMS.

This model will contain a simple information model, detailing the minimum amount of information that is needed for a LIMS to work effectively. Considerable detail is omitted by using under-specification, while still defining the essential properties of the LIMS.

In this paper, the model of the LIMS will be sketched out using RAISE, defining only the signatures of functions for the most part. The RAISE is not complete, and not all the modules have been included.

Several activities carried out in the laboratory have not been described here. In particular, the validation of the results has been omitted. During validation, the accuracy, completeness (no results missing), and internal consistency of the results is checked. The archiving of results has also been omitted. This is in no way intended to imply that these activities are not essential in the laboratory.

5.1. Receiving requests and samples

When samples and requests are received at the laboratory, they are assigned a unique label. A request will contain at least: the patient details, and a list of the tests to be carried out on the sample, as described in DATA_MODEL0.

```
scheme DATA_MODEL0 = class
  type
    PatientDetails, Sample, TestId,
    Request ==
      _(patient_details : PatientDetails, test_requests : TestId* )
end
```

Normally, requests will also contain: the name of the referring doctor, the location for of the doctor, the time and date of sampling, any relevant clinical information, the type of the specimen, and any special information or precaution relevant to specimen collection or handling.

We will also assume that each request is related to a single sample. In practice this is not be the case and a request can relate to several different samples.

5.1.1.   Identifying the patient

When a request is received, it will normally contain the patient's surname, initials, date of birth and possibly a hospital number. For this paper, a match can be made, match_found, if the patient's hospital number and surname match, or if the surname, initials and date of birth match.

    scheme DATA_MODEL1 = **extend** DATA_MODEL0 **with class**
        type HospitalNum, Date, Name, Initials
        value no_num : HospitalNum
        value patient_id : Request $\rightarrow$ HospitalNum
        value
         surname : PatientDetails $\rightarrow$ Name,
         initials : PatientDetails $\rightarrow$ Initials,
         dob : PatientDetails $\rightarrow$ Date
    end

no_num is used to denote no hospital number on the request form. The patient database is represented as a mapping from HospitalNum to PatientDetails.

    scheme LIMS1(D : DATA_MODEL1) = **class**
        variable patient_db : D.HospitalNum $-m\rightarrow$ D.PatientDetails
        value
        /* Match using hospital number. */
        match_id : D.Request $\rightarrow$ **read any Bool**
        match_id(r) $\equiv$ **let** i = D.patient_id(r) **in**
          let req_details = D.patient_details(r), db_details = patient_db(i) **in**
           D.surname(req_details) = D.surname(db_details)
         end end
         pre D.patient_id(r) $\in$ **dom** patient_db,
        /* Match using patient details.*/
        match_details : D.Request $\rightarrow$ **read any Bool**
        match_details(r) $\equiv$ ($\exists$i : D.HospitalNum $\bullet$ i $\in$ **dom** patient_db $\land$
          let req_details = D.patient_details(r), db_details = patient_db(i) **in**
           D.surname(req_details) = D.surname(db_details) $\land$
           D.initials(req_details) = D.initials(db_details) $\land$
           D.dob(db_details) = D.dob(req_details) **end**)
          pre D.patient_id(r) = D.no_num $\lor$ D.patient_id(r) $\notin$ **dom** patient_db

value
  match_found : D.Request → **read any Bool**
  match_found(r) ≡
    If D.patient_id(r) ∈ **dom** patient_db **then** match_id(r)
    else match_details(r) **end**
  axiom [*dbase_consistency1*] D.no_num ∉ **dom** patient_db
end

For the database to be consistent, no_num cannot be used to identify a patient's details.

5.1.2.   Identifying the requests and the samples

For each request and accompanying sample that is received, a new label is created and assigned to that request. The details of the request with this label are entered into the LIMS, enter_test. As the collation of results with requests depends on the label being unique, the LIMS must guarantee that different requests with the same label cannot be entered.

  theory TH_LIMS2 : **axiom**
    /* Assuming no duplicates, enter_test will not create a labelled */
    /* request which has the same label as another request. */
    In class object D : DATA_MODEL2, L : LIMS2(D)
      value /* Check that no two requests have the same label. */
        no_duplicates : **Unit** → **read any Bool**
      end |-
      ∀l : D.Label, r : D.Request • L.enter_test(l, r) ; no_duplicates() ≡
          L.enter_test(l, r) ; **true pre** no_duplicates() ∧ L.label_used(l)
  end

5.2.   Analysing the samples

After the samples are labelled, they are prepared for analysis and are passed to the analyser. For each sample received by the analyser, it must:
- identify the sample, using the label on the sample,
- get the list of tests required for the sample,
- perform the necessary tests, and
- return the results of the tests.

The result of a test must contain details of what test was done. Furthermore, each result must also be labelled with the same label as the request so that the two can be collated.

> **scheme** DATA_MODEL3 = **extend** DATA_MODEL2 **with class type**
>> AnalyserRequest == _(label : Label, test_id : TestId),
>> TestResult == _(test_id : TestId),
>> AnalyserResult == _(label : Label, test_result : TestResult)
> **end**

5.3. Collating and printing the results

After the samples are analysed, the results must be collated with the tests to ensure that: all the tests have been performed, and that only the tests wanted have been done. The results of the tests are then collated with the patient details into a report. This report is printed and is sent to the patient's doctor. This report must contain the same label as the request to check that the request has been completed.

It can be shown in this model that once a label has been assigned to a request, it is never changed. Furthermore, the same label is assigned to each test and each result associated with that request.

> **scheme** DATA_MODEL4 = **extend** DATA_MODEL3 **with class type**
>> Report == _(label : Label, patient_details : PatientDetails,
>>> test_results : TestResult*) **end**

In this model, it is assumed that reports are only printed when all the tests have been completed. In practice this is not the case as some results may be needed urgently and these will be sent to the patient's doctor as soon as they are completed. However, it can be shown for this model (proved formally if necessary) that reports are only printed after all the tests have been completed.

> **theory** TH_LIMS4_1 : **axiom**
>> **In class object** D : DATA_MODEL4, L : LIMS4(D) **end** |-
>>> $\forall$ l : D.Label • L.report_is_printed(l) $\Rightarrow$ L.tests_completed(l) **end**

A further safety property of the LIMS is that spurious reports are not generated - only reports for requests received by the laboratory will be printed.

> **theory** TH_LIMS4_2 : **axiom**
>> **In class object** D : DATA_MODEL4, L : LIMS4(D) **end** |-
>>> { D.label(x) | x : D.Report • x ∈ L.printed } ⊆
>>> { D.label(y) | y : D.LabelledRequest • y ∈ L.requests } **end**

6. Concluding comments

This paper describes how the multi-disciplinary approach advocated by the MORSE project has been applied in re-developing the LIMS at the WMH. In particular, the use of the formal method RAISE is demonstrated by defining a model of a simplified LIMS. In this model, both the information and the functionality required by the LIMS to support the work done in the laboratory are defined.

By using RAISE, the properties of the LIMS can be investigated at the specification stage of the re-development (as shown here), rather at later stages of the development. Some safety properties of the LIMS are also defined, showing how hazards can be removed (or at least reduced). Furthermore, by using RAISE, one can prove that these safety properties are maintained throughout the re-development of the LIMS - from its specification to its implementation. This ensures that hazards removed in the specification of the LIMS are removed from the implementation of the LIMS. The effectiveness of this approach will be assessed after the second safety analysis is carried out after the re-development is completed.

References

1. Ministry of Defence. The Procurement of Safety Critical Software in Defence Systems. Interim Defence Standard 00-55. April, 1991.

2. Ministry of Defence. Hazard Analysis and Safety Classification of the Computer and Programmable Electronic System Elements of Defence Equipment. Interim Defence Standard 00-56. April, 1991.

3. The RAISE Language Group. The RAISE Specification Language. The BCS Practitioner Series. Prentice Hall, London, 1992

4 Fink R, et al: Data Management in Clinical Laboratory Information Systems. In: Directions in Safety Critical Systems. Redmill F and Anderson T (eds). Springer-Verlag, 1993.

Session 2

APPLICATIONS

Chair: I. C. Smith
AEA Technology, UK

PLC-Implementation of Emergency Shut-Down Systems

Wolfgang A. Halang
FernUniversität, Department of Electrical Engineering
D-58084 Hagen, Germany

Johan Scheepstra
Rijksuniversiteit Groningen, Department of Computing Science
P. O. Box 800, NL-9700 AV Groningen, The Netherlands

Abstract

The task of safeguarding systems is to bring processes from dangerous into safe states. A special class of safeguarding systems are emergency shut-down systems (ESD), which, until now, are only implemented in inherently fail safe hard wired forms. Despite their high reliability, there is an urgent industrial need to replace them by more flexible systems. Therefore, a low complexity, fault detecting computer architecture was designed, on which a programmable logic controller for ESD applications can be based. Functional logic diagrams, the traditional graphical specification tool of ESDs, are directly supported by the architecture as appropriate user oriented programming paradigm. Thus, by design, there is no semantic gap between the programming and machine execution levels enabling the safety licensing of application software by formal methods or back translation. The concept was proven feasible by a working demonstration model.

1 Introduction

Many technical systems have the potential of disastrous effects on, for instance, the environment, equipment, employees, or the general public in case of malfunctions. An important objective of the design, construction, and commissioning of such systems is, therefore, to minimise the chances that hazards occur. One possibility to achieve this goal is the installation of a system whose only function is to supervise a process and to take appropriate action if anything in the process turns dangerous. So, to prevent hazards, many processes are guarded by these so called *safeguarding systems*. A special kind of them systems are *Emergency Shut-Down systems* (ESD), which are defined as:

> *A system that monitors a process, and only acts — i.e., guides the process to a static safe state (generally, a process shut-down) — if the safety of either human beings, the environment, or investments is at stake.*

The mentioned monitoring consists of observing whether certain physical quantities such as temperatures or pressures stay within given bounds and to supervise Boolean quantities for value changes. Typical ESD actions are opening or closing valves, operating switches etc. Structurally, ESDs are functions composed of Boolean operators and delays. The latter are required, because in start-up and shut-down sequences often some monitoring or actions need to be delayed. Originally, safeguarding systems were constructed pneumatically and later, e.g., in railway signaling, with electromagnetical relays. Nowadays, most systems installed are based on integrated electronics and there is a tendency to use microcomputers.

The current (electrical) systems used for emergency shut-down purposes are hard wired and each family makes use of a certain principle of inherently fail safe logic. The functionality of an ESD system is directly implemented in hardware out of building blocks for the Boolean operators and delays by interconnecting them with wires. These building blocks are fail safe, i.e., any internal failure causes the outputs to assume the logically false state. Unless implemented wrongly, this results in a logically false system output, which in turn causes a shut-down. Thus, any failure of the ESD system itself will lead to a safe state of the process (generally a process shut-down). This technology, used successfully for decades now, has some very strong advantages. The simplicity of the design makes the hardware very reliable. The one-to-one mapping of the client's specification expressed as functional logic diagrams (FLD) to hardware modules renders implementation mistakes virtually impossible. The "programming" consists of connecting basic modules by means of wires, stressing the static nature of such systems. Finally, the fail safe character of hard wired systems is a very strong advantage. But there are also disadvantages that gave rise to the work reported here.

Economical considerations impose stringent boundary conditions on the development and utilisation of technical systems. This holds for safety related systems as well. Since manpower is becoming increasingly expensive, also safety related systems need to be highly flexible, in order to be able to adjust them to changing requirements at low costs within short times. In other words, safety related systems such as ESDs must be program controlled in order to relinquish hard wired logic from taking care of safety functions in industrial processes. Owing to their simplicity, the most promising alternative to hard wired logic in ESD systems are *programmable logic controllers* (PLC), which can provide the same functionality. However, although a reasonable hardware reliability can be obtained by redundancy, constructing dependable software constitutes a serious, still unsolved problem.

There is already a number of established methods and guidelines, which have proven their usefulness for the development of highly dependable software employed for the control of safety critical technical processes. Prior to its application, such software is further subjected to appropriate measures for its verification and validation. However, according to the present state of the art, these measures cannot guarantee the correctness of larger programs with mathematical rigour. Prevailing legal requirements demand that object code must be considered for the correctness proofs of software, since compilers are themselves far too complex software systems, as that their correct operation could be verified. Depending on national legislation and practice, the licensing authorities are still very reluctant or even refuse to approve safety related systems, whose behaviour is exclusively program controlled.

In order to provide a remedy for this unsatisfactory situation, it was the purpose of the work reported here to develop a special — and necessarily simple — computer system in the form of a programmable logic controller, which can carry out safety related functions as required in emergency shut-down systems. The leading idea followed throughout this design was to combine already existing software engineering and verification methods with novel architectural support. Thus, the semantic gap between software requirements and hardware capabilities is closed, relinquishing the need for not safety licensable compilers and operating systems. By keeping the complexity of each component in the system as low as possible, the safety licensing of the hardware in combination with application software is enabled on the basis of well established and proven techniques.

2   The Software Building Blocks

All emergency shut-down systems can be constructed from a set of function modules containing just four elements, viz., the three Boolean operators *And*, *Or*, *Not* and a *timer*. For reasons of simplicity we restrict the number of inputs for both *And* and *Or* to two. The functionality is not effected, since any multiple input function can be described with a finite number of the two input gates. It is also sufficient to use only one type of timer. All other forms of timers used in hard wired logic can be implemented by, if need be, adding inverters. The timer has one Boolean input, I, one Boolean output, O, an adjustable delay time, t, and an internal state, d, with $0 \leq d \leq t$. Its functionality can be informally described as follows:

- Initially, the input is *false*, the output is *false*, and the internal counter, d, has assumed its maximum (as set), so $d = t$.

- As the input becomes *true*, the output remains *false* and the counter, d, decreases, i.e., the timer starts counting down.

Counter	Input	Output
$d = t$	true	false
$0 < d < t$	true	false
$d = 0$	true	true
$d = t$	false	false
$0 < d < t$	false	false
$d = 0$	false	false

Table 1: The timer output as function of input and counter value

- As soon as the counter becomes zero and the input is still *true*, the output turns *true*.

- If the input is *false*, after having been *true* for less than the preset delay time t, then the timer is reset. That is, the output becomes *false* and the delay time assumes its initial (maximum) value.

- If the input becomes *false* after d is 0 and, thus, the output has become *true*, also a reset operation is performed.

We observe that there are two values that may have to be changed. First, obviously, the logical output could change as a function of the input and the internal state. Secondly, the internal state may need updating, depending on both the logical input and the internal state.

Although the number of internal states of the timer is numerous, three interesting ones can be extracted, viz., $d = t$, $d = 0$ and $0 < d < t$. They are displayed in Table 1.

The functionality of the timer can be represented by a simple Boolean expression for its output:

$$O \equiv (d = t) \wedge I$$

What the module still lacks is a realisation of time. Hence, we define *time*, with $t_0 \leq time < \infty$. In a system *time* can be implemented in both hardware or software. For accuracy reasons, we have chosen a hardware solution: *time* is implemented in form of a counter triggered by a quartz stabilised time base.

An implementation of the four functions modules discussed above has been proven correct using predicate calculus [1]. This was trivial in the case of the Boolean functions. The correctness proof of the timer was straightforward, but took a few pages. The interested reader is referred to [2], because size restrictions prohibit to include the proof into this article.

3   The Software Engineering Paradigm

The analysis of functional logic diagrams suggests to introduce a new programming paradigm, viz., to compose software out of high level user oriented building blocks instead out of low level machine oriented ones. Whereas a single machine instruction taken out of a program context does not reveal its purpose, the occurrence of a certain function module instance usually gives already a clue about the problem, its solution, and the module's rôle in it.

The development of ESD software is carried out by process engineers in the traditional way of drawing FLDs. The latter describe the mapping from Boolean inputs to Boolean outputs as functions of time such as, e.g.,

> *if* a pressure is too high *then* a valve should be opened *and* an indicator should light up *after* 5 seconds.

In Figure 1 an example of a FLD is given. The FLD describing the functionality of an average ESD system contains thousands of blocks, laid out over many drawing sheets.

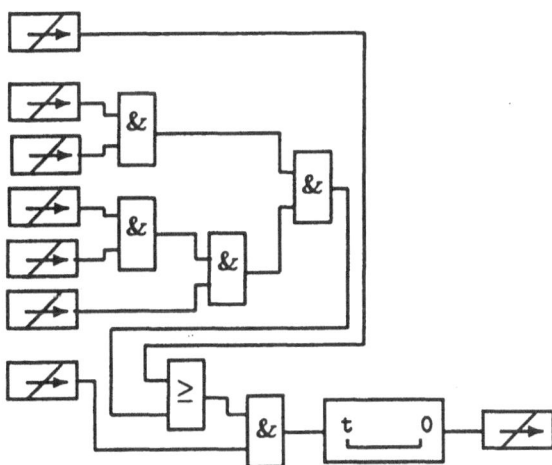

Figure 1: An example of a FLD (with dyadic Boolean operators only)

This specification level programming method consists of graphically interconnecting instances of the above mentioned four basic function modules with each other by lines, i.e., single basic functions are invoked one after the other and, in the course of this, they pass parameters. The interconnections between function blocks have to meet just one restriction: each input must be connected to exactly one output. Besides the provision of constants as external input parameters, the basic functions' instances and the parameter flows between them are the only language elements required by this programming paradigma.

A compiler transforms the graphically represented program logic into object code. Owing to the simple structure, this logic is only able to assume, the generated programs contain no other features than sequences of procedure calls and some internal moves of data. The verification of the compiler transforming the graphical software representation into object code is still impossible — but also not necessary, because for FLD software only the module interconnections need to be verified. As outlined below, for this task the architecturally supported method of back translation is employed.

4   The Architectural Concept

In order to facilitate the conceivability of the implemented software and of its execution process, we design an architecture for an ESD oriented programmable logic controller with, conceptually, two processors:

- a control flow processor (master) and

- a basic function block processor (slave).

Thus, we achieve a clear and physical separation of concerns: execution of the basic function modules in the slave processor and all other tasks, i.e., execution control, sequential function chart processing, and function module invocation, assigned to the master. This concept implies that the application code is restricted to the control flow processor, on which the project specific safety licensing can concentrate. Special architectural support for the cyclic operating mode of programmable logic controllers is implemented in the master processor. To enable the detection of faults in the hardware, a dual channel configuration has been chosen, which supports diversity in form of different master processors and different slave processors.

At least one of the master processors should have the most simple organisation possible for the considered application requiring only two instructions, one of which is MOVE. The other one implements a special architectural support for the cyclic operating mode of PLCs. Since only one step is active at any given time, a memory protection mechanism prevents the erroneous access to the program code of the inactive steps. The STEP instruction is the only means to perform a branch. It solely allows to return to the initial address of the active step's program code if the corresponding transition condition is not fulfilled.

The capabilities of the slave need to be somewhat more complex and are implied by the operations of the four basic function modules. The objective of a PLC for ESDs suggested to employ the VIPER [3] chip in the slave, because it is the only available microprocessor whose design has been formally proven correct. The slave processor performs all data manipulations and takes care of the communication with the environment. It has no program RAM, but only executes the basic function modules whose code is contained in firmware ROMs.

To recognise hardware faults, all processing is simultaneously performed on two master/slave pairs. A number of comparators checking the outputs from the master processors before they reach the slaves and vice versa completes a fault detecting two-channel configuration. The master and slave processors communicate with each other through two FIFO-queues. They execute programs in co-ordination with each other as follows. The master processor lets the slave execute a function block by sending the latter's identification and the corresponding parameters and, if need be, also the block's internal state values via one of the FIFO-queues to the slave processor. Here the object program implementing the function block is performed and the generated results and new internal states are sent to the master processor through the other FIFO-queue. The elaboration of the function block ends with fetching these data from the output FIFO-queue and storing them in the master's memory. To avoid faults during operation, the function modules' object code is put in the slave's read-only program memory, after the correctness of the code has been established. The master/slave configuration has been chosen to physically separate two system parts from one another: one whose software only needs to be verified once, and the other one performing the application specific part of the software. Needless to say, that the latter requires indvidual safety licensing. This concept implies that FLDs are solely mapped onto the control flow processor, to which project specific safety licensing can be restricted. Figure 2 gives a conceptual diagram of the master/slave PLC architecture.

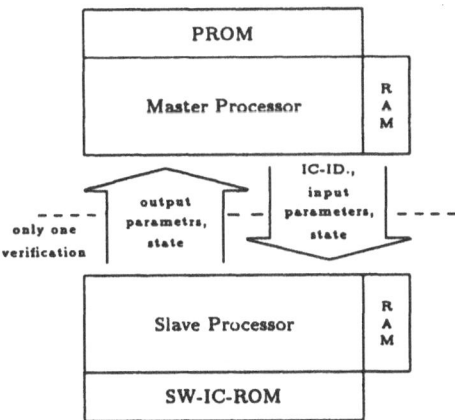

Figure 2: Configuration of a PLC with master/slave processors

In available PLCs, the execution time for a step generally varies from one cycle to the next depending upon the program logic performed and the external conditions evaluated each time. Therefore, the measurement of external signals and the output of values to the process is usually not carried out at equidistantly spaced points in time, although this may be intended in the control software. To achieve full determinism of the time behaviour of programmable logic controllers, a basic cycle is introduced. The length of the cycle is selected in a way as to accommodate during its duration the execution of the most time-consuming step occurring in an application. It is supervised that the execution time of a step does not exceed this cycle period by awaiting, at the end of the step's program processing and after the evaluation of the corresponding transition condition(s), the occurrence of a clock signal, which marks the begin of the next cycle. An overload situation or a run time error, respectively, is encountered when the clock signal interrupts an active application program. In this case a suitable error handling has to be carried through. Although the introduction of the basic cycle exactly determines a priori the cyclic execution of the single steps, the processing instants of the various operations within a cycle, however, may still vary and, thus, remain undetermined. Since a precisely predictable timing behaviour is only important for input and output operations, temporal predictability is achieved as follows. All inputs occurring in a step are performed en bloc at the beginning of the cycle and the thus obtained data are buffered until they will be processed. Likewise, all output data are first buffered and finally sent out together at the end of the cycle.

5 Safety Licensing

With the implementations of all four basic function blocks employed in FLDs having been proven correct and, as parts of the architecture, being invisible from the application programming point of view, for any new ESD project only the proper mapping of a particular interconnection pattern of invoked function block instances on object code needs to be verified. For this purpose we subject the object code loaded into the master processor to back translation, a safety licensing method [4] which was developed in the course of the Halden nuclear power plant project and which is — although rigorous — essentially informal, easily conceivable, and immediately applicable without any training. Thus, it is extremely well suited to be used on the application programming level by people with the most heterogeneous educational backgrounds. The ease of understanding and use inherently fosters error free application of the method. It consists of reading machine programs out of computer memory and giving them to a number of teams working without any mutual contact. All by hand, these teams disassemble and decompile the code, from which they finally try to regain the specification. The software is granted a safety license if the original specification agrees with the inversely obtained re-specifications. Of course, in most circumstances the method is extremely cumbersome, time consuming, and expensive. This is due to the semantic gap between a specification formulated in terms of user functions and the usual machine instructions carrying them out. Applying the programming paradigm of basic function modules, however, the specification is directly mapped onto sequences of module invocations. The object code consists of just these calls and parameter passing. The implementation details of the function modules are part of our architecture. Thus, they are invisible from the application programming point of view and do not require safety licensing in this context. Consequently, back translation can lead, in one easy step, from machine code back to the problem specification, which is given in the form of FLDs. For our architecture, the effort required to utilise the method of back translation is by several orders of magnitude lower than for the von Neumann architecture.

Back translation is a verification method to the carried out with diverse redundancy. Originally, this called for different teams of human inspectors. Since in the case considered here there is only one rather simple inverse analysis step, we are optimistic that the licensing authorities will eventually accept the following procedure. Verification by inverse analysis is carried out by a number of different programs, which should be proven in practice but do not need to be formally verified. Such programs are to yield graphical outputs. An official licensor performs the inverse documentation as well, compares his results with the ones of the verification programs and with the original graphical application program under inspection and, upon coincidence, issues a safety license. Such a procedure is in line with the dependability requirements for diversely redundant programs demanded by the licensing authorities and necessitates only the minimum of highly expensive human involvement, viz., one licensor, who is always indispensable to take the legal responsibility for issuing a safety license.

In order to prevent any modification by a malfunction, in our safety oriented architecture all programs must be provided in ROMs. For practical reasons, generally there are two types of these memories. The code of the basic function modules resides in mask programmed ROMs. On the other hand, the code representing FLDs is written into (E)PROMs by the user. This part of the software is subject to project specific verification to be performed by the licensing authorities, which finally still need to install and seal the (E)PROMs in the target PLCs.

6   Conclusion

In our society there is a growing concern for safety (which goes hand in hand with the increasing awareness for the environment). This has important consequences for the assessment of program controlled systems. One has begun to realise the inherent safety problems associated with software. Since it appears unrealistic to abandon the use of computers for safety critical control purposes — on the contrary, there is no doubt that their utilisation in such applications is going to increase considerably — the problem of software dependability is exacerbating.

62

In a constructive way, and using presently available methods and hardware technology only, in this paper for the first time an architecture was defined, which enables the safety licensing of a complete programmable electronic system including the software. The measures to achieve this objective were:

- using hardware as much as possible, but not necessarily in the most (hardware-) cost efficient way, since now there is cheap hardware in abundance (the additional hardware costs are equivalent to the cost of a software engineers for about half a day),

- utilisation of a high level, graphical software engineering method,

- closing of the semantic gap between architecture and user programming by basing the software development on a set of function blocks with application specific semantics,

- removal of compilers from the chain of items requiring safety licensing,

- avoiding the need for a complex operating system, and

- by providing a feasible application level and architectural support for the software licensing method of back translation.

Employing VIPER microprocessors, we have built a prototype of the PLC architecture described. Its utilisation in practice showed that implementing the functionality of a hard wired ESD system with our PLC architecture is feasible, and that the programming paradigm based on formally verified function modules can render error free software. We hope that the concept presented here will lead to the replacement of hard wired systems safeguarding industrial processes by programmable ones executing safety licensed and, thus, highly dependable software.

References

1. Dijkstra EW, Feijen WHJ. Een methode van programmeren. Academic service cop., 1984

2. Scheepstra J. PLC-Implementation of Emergency Shut-Down Systems. Master's thesis, University of Groningen, 1992

3. Kershaw J. The VIPER Microprocessor. Report No. 87014, Royal Signal and Radar Establishment, Malvern, England, 1987

4. Krebs H, Haspel U. Ein Verfahren zur Software-Verifikation. Regelungstechnische Praxis 1984; 28: 73 — 78

The Fast Cost Effective Design and Concurrent Certification of the Safe Computer for a Real Time Train Control Application

G. A. Mutone

AEG Transportation Systems, Inc.,
Pittsburgh, PA, USA

M. Rothfelder

Institute for Software, Electronics,
Railroad Technology (ISEB) of TÜV Rheinland,
Cologne, Germany

Abstract

This paper starts with a general description of the AEG Transportation Systems, Inc. Automatic People Mover System. Subsequently, the specific safety requirements of the ATP, and the consequent design features to meet these requirements are described. Following this introduction, details of the relationship between designer and certifier, the utilization of embedded rules-based systems, the concurrence of the design and certification process, and the de-coupling of the safety functions from the hardware are given. It is described how the dramatic improvements in the traditional large costs and long schedules normally associated with both the design and certification of safe computer systems are made possible.

1 General Description of the Automatic People Mover

1.1 Overview

The AEG Automatic People Mover System consists of driverless operated trains which usually run on a guideway with a concrete surface. The trains are guided by an I-Beam and may be configured as consists of one, two or more vehicles. The access of passengers to the guideway is prohibited by automatically operated station doors, which are normally closed. They open in synchronism with corresponding vehicle doors for passenger exchange only when the train is in the station and the doors on both sides (vehicle and station) are properly aligned. The operation of the Automatic People Mover System is controlled and supervised by the *Automatic Train Control System* (ATCS). At Frankfurt Airport the headway is designed to be down to 90 seconds. The Frankfurt Airport Passenger Transfer System will be the first application of the new Automatic Train Protection system design which is described in this paper.

1.2 Automatic Train Control System

1.2.1 General Overview

The Automatic Train Control System consists of three major computer-based subcomponents with different functions. The *Automatic Train Protection* (ATP) ensures the safety of operation such as the safety of moving trains or passenger exchange in stations. The ATP does not allow unsafe system states. The *Automatic Train Operation* (ATO) provides operational control of train speeds, programmed station stopping, station and vehicle door operation, as well as passenger audio and visual information. The third function, the *Automatic Train Supervision* (ATS), is responsible for the supervision of the Automatic People Mover systems, route and headway control, and reporting of alarms.

The Automatic Train Protection system consists of two major sub-systems, the Wayside ATP and the Vehicle ATP. Each is based on vital dual channel cross-checked computers and other vital I/O hardware. The safety functions of these two subsystems satisfy specific safety requirements on the ATP as identified in the Safety Requirements Catalog (see chapter 2 Safety Requirements). Below, the allocation of safety functions to either Wayside or Vehicle ATP is given.

1.2.2 Wayside ATP

The Wayside ATP fulfils the following major safety functions:
- Detection of Trains
- Provision of Safe Speed Codes (Selection and Transmission)
- Safe Switch Operation
- Safe Station Door Operation
- Vital Inputs and Indications for the Central Control Operator
- Protection of Maintenance Vehicle

1.2.3 Vehicle ATP

The Vehicle ATP serves, among others, the following purposes:
- Reception of Speed Commands
- Supervision of Actual Train Speed
- Supervision of Safe Travel Direction
- Safe Vehicle Door Operation
- Safe Station Stopping
- Safe Reaction on Unintentional Train Separation

2 Safety Requirements

2.1 Analyses of Potential Hazards and their Causes

As comprehensive as possible all potential hazards and their causes are identified by means of a Safety Analysis. In the early phases, this analysis comprises the *High Level Fault Tree Analysis* (HLFTA) and the *Preliminary Hazards Analysis* (PHA). Hazards such as collisions with switches or other trains, end-of-line run-through, passengers gaining access to the guideway, or overspeed in curves are considered carefully in the HLFTA and the PHA. These analyses provide the designers of the ATP with very detailed information on the possible causes of potential hazards which, in turn, provide all necessary means and measures to protect against these hazards. The necessity of these means and measures is documented in the *Safety Requirements Catalog* (SFRC). This document states all requirements for safety functions which need to be fulfilled by the ATP in order to ensure safe operation of the Automatic People Mover System. The Safety Requirements Catalog is the basis for all further ATP-related development and certification steps.

Correcting the problems early in the design and development process usually is much less expensive than fixing them later in the process. As the hazards are identified very early in the design and development process by the above-mentioned Safety Analysis, the requirements are determined very early and hence additional costs in order to fix problems later are minimized.

2.2 Results of Safety Analysis and Consequent Design Principles

Most of the potential hazards identified during the safety analysis might lead to injury or death of persons or damage to equipment. Generally speaking, the worse the potential consequences of the hazards are the more rigorous the measures to avoid them must be.

One might have the idea to differentiate between each safety relevant function according to its potential damages in order to engage different means and measures to protect against the respective risk. But many of the functions are implemented by the use of software. This makes it difficult to ensure that functions of a lower integrity level do not have any unsafe impact on those functions which protect against higher risks.

Due to these difficulties, each function of the ATP is considered to be of high safety relevance and hence is designed in a vital fashion. Functions which have no safety relevance are allocated to the Automatic Train Operation System (ATO). For instance, while the ATO controls the doors such that station and vehicle doors open synchronously, the ATP ensures safety by not allowing the vehicle to move if either the station or the vehicle doors are not closed and locked. Both computers are separated such that the non-vital ATO can not interfere with the vital ATP. This strict separation of vital and non-vital functions minimizes the verification, validation and certification effort necessary for the vital ATP and for the entire Automatic Train Control System.

The very fundamental regulations for the certification of the Frankfurt Airport Automatic People Mover are BOStrab [1] and DIN VDE 0831 [2]. As far as DIN VDE 0831–the German standard for electrical railway signalling systems–is applicable, all ATP functions are designed to be signal-safe. In other cases, where neither BOStrab nor DIN VDE 0831 are immediately applicable–in case of software for instance–other German or international standards which represent the current state of the technology are applied. This is required by BOStrab. For instance, DIN V 19250 [3], DIN V VDE 0801 [4] or MÜ 8004 [5] are applied. According to DIN V 19250 [3], the ATP is categorized as of *Anwendungsklasse* (*Integrity Level*) 7. Consequently, applicable recommendations of DIN V VDE 0801 are followed.

All the components of the PTS are designed with an inherent high level of reliability so as to deliver an overall system availability of 99.65% or higher. This level, demonstrated quantitatively, is established by contract and consistent with actual levels reached in numerous operating installations of the AEG people movers.

2.3 Design of Vital Hardware–Examples

The safety of a computer-based train control system depends directly on the reliability of the underlying hardware. The hardware must fulfil its specified functions correctly and safely. Failures must not lead to unsafe states [2]. Hence, failures of hardware components need to be considered carefully.

Some failures can be excluded by the application of vital design properties. Other failures cannot be excluded and need to be detected to ensure that the system goes to a safe state in case of such failures. The following sections describe these vital design principles by means of examples taken from the Automatic Train Protection system.

2.3.1 Dual Channel Cross-Checked ATP Computers

The ATP computers which have to ensure the safety of operation are designed as dual channel cross-checked computers. Both computers cross-check each other continually. Each single computer channel compares the inputs and outputs to and from the actual process with the equivalent data from the other channel before it allows the transition from one safe state to another state. If one computer detects a mismatch of cross-checked data, immediately appropriate actions are taken to transfer the system to a safe state, which is the shut-down of the concerned guideway portion. All vehicles in this section will come to a stop and any further movement is prohibited by the ATP until personnel has fixed the problem and the ATP checked and confirmed that safe operation is restored.

2.3.2 Occupancy Detection and Speed Code Distribution (TX/RX System)

The *Occupancy Detection* of the Automatic People Mover System is based on *Track Circuits*. A Track Circuit is occupied through 'short-circuit' by multiple shunts on each vehicle. The Wayside ATP sends the appropriate speed commands to each vehicle via the Track Circuits and the *Transmit/Receive System* is used to transmit the speed codes to the Track Circuits. Each Track Circuit is fed with speed commands even when there is no train currently occupying it. If there is no train, the sent speed codes are read back by the Wayside ATP. They are then compared against the transmitted codes. If they do not match, a failure in the Transmit/Receive System is assumed and a shutdown is initiated. This ensures a periodic test and failure detection of the Wayside ATP components engaged in the Speed Code Distribution and Occupancy Detection. Here, the fulfilment of Failure Detection Period Requirements is designed into the system.

2.3.3 Failure Detection Periods

Other components may not be guaranteed to be tested within their specific failure detection period by the above-mentioned approach. Here, other measures need to be taken to ensure the necessary periodic tests.

One approach to test these components, for instance, is that the ATP performs periodic checks of their operation. Since the respective outputs are verified by a readback, the component can be considered tested if it is operated and no failures are encountered. If that is not the case, the ATP alarms this fact to the Central Control Operator in order to take appropriate action.

2.3.4 Failure Exclusions and Fail-Safe Design

In other cases, failures are excluded by the use of special design properties, such as German signal relays and so-called gravity drop-out relays. Here, the correct function of the relay is ensured by the chosen mechanical design, weights and gravity. Other components are designed in such a way that all failures which are to be assumed will always lead to a safe state.

2.4 Development of Safety Relevant Software

The safety relevant ATP software is developed using the high-level language Pascal. The compiler used to generate the object code is validated and proven. It supports a module concept. Hence, the ATP software is designed according to industry-accepted design principles of *Information Hiding*, *Data Abstractions*, *High Cohesion,* and *Low Coupling*. This ensures low complexity and highly comprehensible modules. Consequently, generation and verification of the source code is less error-prone. Additionally, these design principles improve the maintainability which in turn makes changes less error-prone and results in less effort. The verification and validation of the ATP software involves rigorous methods which are described in Section 4.4.

3 Rules-Based Interlocking Engine

Conventional interlocking systems usually are designed for specific applications. In the past, different guideway layouts or extensions to existing systems led to considerable effort to customize and certify each separate configuration of the wayside train protection system. Furthermore–in case of extensions to existing systems–this approach led to additional system outages during the exchange of components. The *Rules-Based Interlocking Engine* (RBIE) [6] has been designed to circumvent these problems.

3.1 Overview

The basic principle of the *Rules-Based Interlocking Engine* is to make the implementation of safety functions independent from the underlying hardware and core software of the ATP. The ATP becomes a generic means to realize safety functions. This is accomplished by the following way. A description of the actual system with *application-specific* information is supplied to the *application-independent* Wayside ATP. To achieve this, a description of the system's layout and other details is created. This description is called the *Guideway Definition File*. The Guideway Definition File is placed in the ATP (See Figure 1). The information in the Guideway Definition File is then transformed (parsed) into an internal representation suitable for interpretation by the Rules-Based Interlocking Engine software. During runtime the vital decisions of the Vital ATP Computer are based on generic interlocking rules. These rules are evaluated by the Rules-Based Interlocking Engine regarding the application specific relations as defined in the Guideway Definition File.

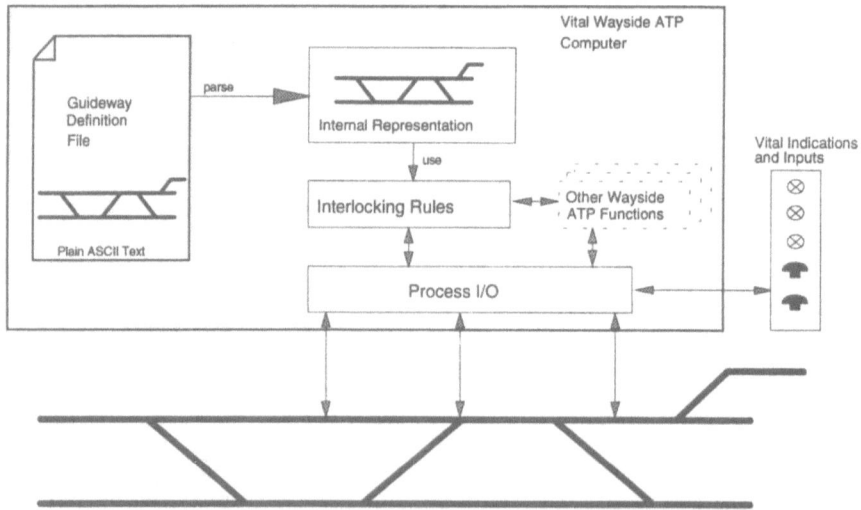

Figure 1: Simplified Architecture of the ATP Computer with focus on the RBIE

Because of this strict separation of application-specific information in the Guideway Definition File from the generic Rules-Based Interlocking Engine embedded in the Wayside ATP, the generic Rules-Based Interlocking Engine is certifiable as a type. Once the Rules-Based Interlocking Engine is certified according to the applicable standards and regulations, the effort for the development and certification of the interlocking of new systems is reduced to efforts related to the generation and certification of a valid guideway definition file.

3.2 Guideway Definition File

As mentioned above, the application specific information on the people mover application which the ATP is intended for is laid down in the Guideway Definition File in a comprehensible and easy-to-read manner. Among others, the Guideway Definition File gives information on the guideway layout, arrangement of stations, allowable speeds and application-specific obstacles which might violate the clearance profiles such as facade washers.

3.3 Validation of Guideway Definition File

The Guideway Definition File is a well-structured plain text and easy-to-read ASCII file. The language is especially designed to be comprehensible to people mover experts without any special software knowledge. Hence the validation of the Guideway Definition File may concentrate on the railroad aspects. Validation may be supported through the graphical exercise of the information contained in the Guideway Definition File.

3.4 Advantages

The approach of the Rules-Based Interlocking Engine leads to improvements with respect to safety as well as economics. The validation of application-specific design characteristics is much less error-prone compared to conventional approaches. Besides the safety aspects this approach is very cost-effective. Once the ATP with its Rules-Based Interlocking Engine is type certified, time and efforts spent in the configuration and certification of other ATP applications are reduced dramatically.

4 Project-Accompanying Safety Certification

4.1 Relevant Regulations and Standards

The first application of the described Automatic Train Protection system is the Frankfurt Airport Passenger Transfer System (PTS). The system has been designed according to German regulations, like the BOStrab [1], DIN VDE 0831 [2].

In cases where these regulations do not provide sufficient detail regarding the safety requirements other standards are engaged that represent the state of technology such as DIN V VDE 0801 [3] and MÜ 8004 [5].

4.2 Certification Process

The certification process conducted by the certifier Institute for Software, Electronics and Railroad Technology (ISEB) within TÜV Rheinland Sicherheit und

Umweltschutz GmbH, consists of five major steps which concur with the manufacturers development process:
- Set-up of a Certification Plan
- Review of manufacturer's Quality Assurance System
- Adaptation of the certifier's configuration and documentation management system to the manufacturer's documentation structure
- Verification and Validation
- Safety Trials
- Compilation of a Final Report on the system's safety

In close cooperation with the manufacturer the Certification Plan defines kind and scope of certification steps to be conducted by the certifier such as reviews, inspections, audits, analyses, and tests. The depth and thoroughness of the certifier's activities depend immediately on those activities which are planned and performed by the manufacturer and on the insight and understanding which the certifier gains from these activities.

Firstly, the basic steps of the certification process are defined in principle. Concurrent to the certification process these basic steps are refined as the development phases emerge. This process leads to a common understanding of the set of relevant items which are subject to certification such as documents, processes, software, and hardware components. For each relevant item, it is determined which activities need to be performed by the certifier. The certifier then adjusts his configuration and documentation management system to the set of relevant documents. Furthermore, a schedule is agreed upon which defines the dates of the submittal of certification relevant items as well as the duration of each certification activity.

The more rigorous and comprehensive the QA measures of the manufacturer are and the better these measures are documented, the more the certifier may limit his efforts to minimal measures to gain confidence in the manufacturer's QA measures. For this reason the manufacturer's intended QA system is reviewed in the early phases of the certification in close cooperation between the manufacturer and the certifier. This is done based on appropriate documentation, e.g. *Software Quality Assurance Plans* and *Software Verification and Validation Plans* taken from former projects. Guidelines for documentation, design methods, coding standards, and project management as well as the structure of test plans are established in these early phases. The actual certification activities then consist of the following main steps which are often similar even in different projects:
- Review of Concept Descriptions
- Validation of System Requirements
- Validation, Verification and Certification of Hardware and Software
- Integration Tests
- Safety Trials

In each phase of the certification process a close contact between certifier and developers makes it possible to communicate emerging problems as soon as possible and hence fix them as early as possible.

The following sections describe the methods engaged during the development of the Automatic Train Protection system. This only provides an impression of the certification process. A detailed description would have exceeded the boundaries of this paper.

4.3 Hardware Certification

The hardware certification consists of the review of documentation pertinent to all ATP hardware components. Among others, Lower Level Fault Tree Analyses, Environmental Analyses, Descriptions and Specifications, Detailed Hazards Analyses, FMEA, Worst Case Analyses, Test Plans and Test Results for each component are inspected and, when necessary, supplemented. Boards and other components are inspected thoroughly. For instance, the creepage distances are examined carefully. Some components such as relays are tested thoroughly by the certifier. The execution of tests of other components by the manufacturer are witnessed.

4.4 Software Validation and Verification

4.4.1 Software Life Cycle and QA System

The software is developed by the manufacturer according to a software life cycle model consisting of a *Concept Phase, Requirements Phase, Design Phase, Implementation Phase, Test Phase (Unit Tests through System Integration Tests), Installation and Checkout Phase, Operations and Maintenance Phase.* For each phase, the *Software Quality Assurance Plan* and *Software Verification and Validation Plan* exactly define the specific tasks of design, verification and validation. The entity (development or V&V) responsible for performing a particular task, as well as the means for documenting the task results, are defined. A group of engineers is assigned by the manufacturer to be responsible for all V&V efforts. These engineers are independent from the development engineers. This leads to a high quality of items submitted to the certifier and consequently less certification effort.

Besides the definition of V&V tasks, the manufacturer's Quality Assurance System takes into account further regulations such as Coding Standards, Documentation Guidelines, or Software Requirements Specification Procedures. Each component of the Quality Assurance System is assessed by the certifier.

4.4.2 Validation and Verification Methods

All documents have been validated and verified thoroughly by the certifier with methods like walk through, reviews, inspections, and static analysis. The manufacturer conducts comprehensive tests on units, during software integration, and during system integration. These tests are witnessed by the certifier. Prior to the actual test execution, the certifier reviews the test plans generated by the manufacturer's V&V Group in order to ensure that all relevant system states are tested. When necessary the certifier supplements these test cases. In order to achieve a comprehensive set of test cases the Cause-Effect Graph Method is used by the manufacturer's V&V group to define the requirements-based Black Box test cases. Black Box tests are executed first and branch coverage achieved by these cause-effect graph based test cases is measured. If this coverage does not reach 100% branch coverage, additional White Box tests are conducted. This systematic test approach leads to comprehensive tests of the ATP software.

4.4.3 Type Certification

As soon as certain components (sub-components, boards, computers, software components like the Rules-Based Interlocking Engine) are type certified they may be regarded as building blocks. The certification of a specific application may then focus on the correct application of these building blocks and the validation of application-specific configuration data.

5 Conclusions

The chosen approach leads to improvements in costs, scheduling and safety. These improvements are made possible mainly by the flexibility and reusability of the Rules-Based Interlocking Engine, the flexible and effective concurrent certification process, and the manufacturer's development-independent Quality Assurance activities. In turn, as shown above, the rigorous verification and validation measures ensure the safety of the Automatic Train Control system with the above-mentioned significant cost improvements.

References

1. Verordnung über den Bau und Betrieb der Straßenbahnen (Straßenbahn-Bau- und Betriebsordnung - BOStrab) vom 11. Dezember 1987, Bundesgesetzblatt Teil I Nr. 58 vom 18. Dezember 1987
2. DIN VDE 0831/08.90 Elektrische Bahn-Signalanlagen
3. DIN V 19250/01.89 Messen-Steuern-Regeln: Grundlegende Sicherheitsbetrachtungen für MSR-Schutzeinrichtungen
4. DIN V VDE 0801/01.90 Grundsätze für Rechner in Systemen mit Sicherheitsaufgaben
5. Grundsätze zur technischen Zulassung in der Signal- und Nachrichtentechnik (Mü 8004), Deutsche Bundesbahn, Bundesbahn-Zentralamt München, 01.02.1993
6. G. A. Mutone, J. Daubner, Vital Automatic Control of Guided Transports with Real-Time Expert Systems, ITTG '93, Lille, France, September 1993

Design and Analysis of a Failsafe Algorithm for Solving Boolean Equations

Harvey E. Rhody, Ph. D.
RIT Research Corporation
Rochester Institute of Technology
Rochester, NY (USA) 14623

Vittorio Manoni
SASIB Signalamento Ferroviario
40128 Bologna, Italy

James R. Hoelscher
General Railway Signal Corporation
Rochester, NY (USA) 14620

1 Introduction

Since the installation of the first mechanical interlocking in 1856, railway signal engineers have developed a set of rules which define the essential requirements for safe train movement. In the majority of cases this set of rules can be expressed as a closed set of boolean equations which, when implemented as written, yield a safe operating system. The boolean equation set will vary depending upon the particular requirements of each application. The set of general rules are imposed on the specific requirements of each application to yield a closed set of boolean equations which completely describe the safety and operational requirements of that application.

By implementing the set of equations with hardware elements which have known failure modes it is possible to not only create a safe operating system but also a failsafe operating system. For the past 50 years, this has been accomplished using the 'safety relay.' The 'safety relay' has a known set of failure modes. More importantly, it has by design eliminated some failure modes which are common in general purpose relays. This relay is designed, such that its front contacts will not be closed unless the relay's coil is properly energized. While this allows the signal engineer to implement a safe system, designing a relay to have a particular set of failure characteristics results in relays which are costly and physically large (GRS B1 relay is approximately 16 cm x 2.5 cm x 8.6 cm. and weights about 4 kg). Even a simple application may use in excess of 100 relays and have more than 1000 wire connections.

While relay based systems are very robust, they are also very inflexible. Changes to the control system requires the addition, deletion or modification of relays and interconnecting wires. This is labor intensive and in many cases it requires physical space that is not available. The microprocessor on the other hand offers extreme flexibility, small size, and low cost. However, it also offers hardware with a set of failure modes which cannot be totally defined, and, therefore, cannot be used to achieve a failsafe system *based on failure characteristics*. Thus, the challenge is to develop an *algorithm* that will allow the microprocessor to be used to solve boolean expressions in a way that insures that any and all failures which might affect the attainment of correct results are *revealed* and used to *force* the system to a known safe state.

One approach that has been implemented in safety critical applications is the use of multiple processors checking each other. This approach is implemented in various ways including: two or more identical processors with identical software, two or more identical processors with diverse software and two or more different processors with diverse software. In each case process results are checked for agreement, and the lack of agreement is used to force the system to a safe state. Each of these implementation methodologies has areas which must be thoroughly analyzed before they can be accepted as producing a failsafe system.

The identical processors using identical software approach requires that the initial software be proven to be completely error free. This technique is only effective in revealing independent hardware failures which cause the checked results to differ. It does not protect against software design errors. If both systems have the same embedded flaw then they will both act on it the same and the flaw will not be revealed. Proving that the software contains no embedded flaws or is error free is difficult if not impossible [1].

The use of identical processors with diverse software requires proof that the software is actually diverse. If the software is written to the same specification for the same processor, using the same command set the level of diversity of the final set of software appears questionable. Also, it appears to be difficult to prove that the software is actually diverse enough to reveal all embedded errors and/or hardware failures.

The approach of using diverse hardware and diverse software would appear to solve the problems of these other approaches, but it actually requires two complete developments, two complete sets of hardware and two complete sets of installation tests. This is costly, and hardware dependent. Also, in all of these cases the voting or checking algorithm must be developed and analyzed to prove that it provides the safe operation required.

These approaches, while not perfect, do offer the possibility that the microprocessor can be applied to safety critical systems. However, they suffer from serious practical problems related to the cumbersome analysis that must be done to validate the level of safety they provide in each installation. Much of the analysis must be repeated, in detail, for each system that is installed.

In order to use microprocessors in the rail industry we determined [2] that the solution must not only resolve the safety problem but it must also produce a system that is; 1) analyzable and verifiable to some minimum level of safety, 2) hardware independent, 3) application independent (such that it does not have to be safety verified for each new application), 4) easy to apply to different sets of boolean equations, 5) not based on proving the software to be error free and 6) cost effective. The developed solution has been named Numerically Integrated Safety Assurance Logic (NISAL) [3]. This solution is effective in meeting the stated goals and, because it is numerically based, it allows the upper bound of the probability of an unsafe event to be calculated [4]. This then allows the lower bound of the mean time between unsafe events (MTBUE) to be calculated. These calculations can be done once because they are an intrinsic part of the system design. They do not need to be redone for each application.

2 The Algorithm

The basic structure of a system that uses numerically integrated logic is shown in Figure 1. A set of sensors which have relay contacts that are either open or closed provides information on the state of the railroad system. These sensors are probed by the system to determine their values, and the TRUE or FALSE state is used in primordially safe boolean expressions to determine the proper settings for the output devices.

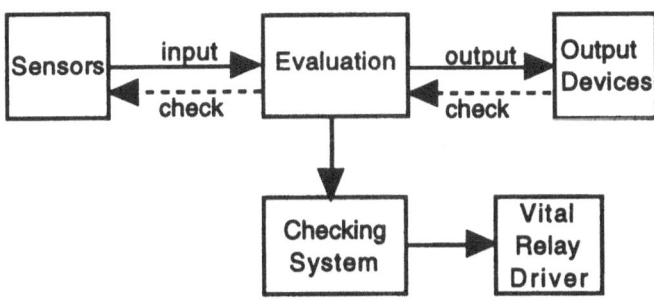

Figure 1: Simplified structure of the system.

The safety of the process is assured by causing it to generate checkwords that are used in a validation step by an independent agent, described below. Each process step – evaluating a boolean expresion, verifying the state of an input port, verifying the state of an output port setting, verifying that a section of computer memory was cleared, etc. – generates a checkword, and each checkword must satisfy the independent agent. In current implementations the checkwords are thirty-two bit binary numbers. The operations are so structured that the system generates each checkword by correctly doing the associated operation. If the operation is not done correctly then the checkword is effectively a random pattern with only a small chance of being correct. Thus, incorrect checkwords reveal inproperly done operations.

The independent agent consumes checkwords that are produced by the system and provides the energy to operate the output devices. If the agent is not satisfied with the sequence of checkwords it is receiving, it removes energy from the system and causes it shut down in a known safe state.

Neither the system nor the agent "know" the correct checkwords in the sense that they are stored in memory to be looked up. This would lead to the possibility of improper operation being unrevealed due to the correct checkword sequence being obtained from memory. As we have seen, the system must generate them by proper operations. Similarly, the agent responds very selectively to its input binary pattern. If the pattern is not correct then the response will not support the continued flow of system energy. The response characteristics are determined by the structure of the agent, which is matched to the proper checkword pattern in a manner similar to the correspondence of a matched filter to a particular signal.

The independent agent has two possible responses: (1) provide vital power, and (2) remove vital power. Providing vital power is a positive action that depends upon the correct input sequence. If the sequence is not correct, energy is removed from the system and it shuts down safely. The operation of providing or removing vital power can be accomplished in a number of ways, including the use of a vital power supply or a vital relay.

To illustrate the nature of the checking process, consider the simplest possible situation – one in which there is a single expression to be evaluated. Suppose that the states of input devices are represented by A, B, etc. and that the logical value of the expression is represented by x as in the equation

$$x = ABC + DE + FGHI + \cdots \tag{1}$$

x will have the value TRUE if and only if all of the variables in at least one of the product terms are TRUE. (Values which should be false for the equation to evaluate to TRUE would be complemented in the equation.)

Suppose that the TRUE and FALSE values of x are represented by N-bit binary patterns, and that the TRUE pattern is unknown by the evaluator. The proper pattern is produced by scanning the expression until the first product term that has proper values for a TRUE result is found. The parameters in that term are processed by the evaluator, in a manner described below, to produce an N-bit pattern for x. If no product term is found that has proper values for a TRUE result, then the N-bit FALSE pattern is inserted for x. The system logic is designed so that no FALSE result can cause an unsafe condition.

A large system will have many boolean expressions, but the principle is the same. Each expression is processed to produce a binary pattern, which is used as a checkword. By making the binary patterns sufficiently long, it is highly improbable that they can be produced by chance. It is even more unlikely that a system with a hidden failure will produce a correct sequence of checkwords by random selection.

The input sensors are read by transmitting particular digital patterns to the input test circuits and observing the returned patterns. A particular sensor is taken to have a TRUE reading only if the returned pattern is a codeword in a suitable error detecting code. This, however, does not produce the vital input sensor value. It only establishes a hypothesis about the setting of the input port. Any TRUE setting is then verified by transmitting another pattern through the port and using that pattern in the computations.

The settings of output devices are verified in a similar fashion. The state of the devices are set by sending commands based on the results of the boolean evaluations. The output states are then verified by transmitting state-dependent patterns through the devices and using those results together with the intended settings to produce the signal for the independent agent.

To guard against the possibility that some internal system failure could be unrevealed by a particular checking algorithm, the system may be designed to do a second set of checking operations that are informationally redundant but algorithmically diverse from the first set. The current implementation produces a pair of checkword results for each operation, say x and x', each created by a separate checkword algorithm implemented as a finite-state machine with a unique system matrix.

2.1   Structure of the Evaluation System

There are many ways to construct an evaluator which will produce a given output pattern from a sequence of input patterns. A basic form is a finite-state machine, which is started in an initial state and then stepped through a state sequence. Each parameter in a boolean product term is represented by a binary pattern, and those patterns serve as the machine inputs. Each pattern in the input sequence influences the next evaluator state. The final state is completely predictable from the initial state and the input sequence. The evaluator output, which is a function of the final system state, will be a proper binary pattern if it begins in the correct initial state and all of the input parameter values have their expected patterns.

A table of initial conditions, one for each product term of each boolean expression, is maintained. When a particular product term is to be evaluated, the initial condition for that product term is taken from memory and used as the initial state vector of the finite-state machine. The machine is then cycled with the parameter value patterns as inputs. When all of the parameters have been consumed the output is calculated as a function of the final state of the machine. An independent evaluation of each boolean expression is done with a second evaluator using parameters from the second input channel. The two results, x and x', form the diverse pair of binary patterns.

The state of a finite-state machine at a particular time $i+1$ is determined by its state at time i and the input at time i. If the inputs are represented by the sequence $\mathcal{U} = \{u_1, u_2, \ldots, u_n\}$ and if the states are represented by the sequence $\mathcal{V} = \{v_1, v_2, \ldots, v_n\}$, then

$$v_{i+1} = v_i T + u_i U \tag{2}$$

where T is the *system matrix* and U is the *input matrix*. If the system has r internal states and s inputs, then T is $r \times r$ and U is $s \times r$. The system behavior is determined almost completely by the properties of T.

If the system is initialized to be in state v_1, then the state v_{i+1} can be shown to be

$$v_{i+1} = v_1 T^i + \sum_{j=1}^{i} u_j U T^{i-j} \tag{3}$$

The state v_{i+1} is a function only of the initial state v_1 and the sequence of inputs $u_1, \ldots, u_i$.

The checkword for the evaluation of a boolean expression is computed by (1) finding the first product term that has proper parameter values to produce a TRUE result (or directly produce the FALSE checkword as the value for x if there is no such term); (2) selecting an initialization value for that product term (the value of v_1); (3) setting the initial value of the machine; (4) operating the machine with the product term i as input u_i; (5) after the n parameters of that product term have all been entered, using v_{n+1} as the system output. Thus, the output is determined by the initial state and the input sequence. For the algorithm to be useful for our purposes it must be possible to compute the initial condition to produce the target output value for each parameter set and it must be the case that an error in any bit of the initial condition or any parameter will cause the system to produce a different output pattern.

It is a relatively simple matter to compute the initial condition required to produce a desired target output t for a given input sequence $\mathcal{U}$. In (3) let $i = n$ and set $v_{n+1} = t$. Then sequentially step backwards through the states by doing the calculation

$$v_i = (v_{i+1} + u_i U) T^{-1} \tag{4}$$

Multiplying by T^{-1} is equivalent to stepping the machine backward. After n steps we reach the desired value for the initial condition, v_1.

The effect of an error in the initial condition or any of the inputs is related to the dynamics of the system when it is allowed to operate with zero input. Because the system described by (2) is linear, the effects of individual errors can be found by superposition. The effect of an error in the initial condition v_1 can be seen by allowing the initial condition to be

$$\hat{v}_1 = v_1 + v_1^e \tag{5}$$

where v_1 is the intended value, v_1^e is the error, and $\hat{v}_1$ is the initial condition actually used. The final state from this state is

$$\hat{v}_{n+1} = \hat{v}_1 T^n + \sum_{j=1}^{n} u_j U T^{n-j} \tag{6}$$

The difference between the actual final state and the intended final state can be found by subtracting (modulo 2 addition) (3) with $i = n$ from (6).

$$\mathbf{e}_{ic} = \mathbf{v}_{n+1} + \hat{\mathbf{v}}_{n+1} = \mathbf{v}_1^e \mathbf{T}^n \tag{7}$$

The above result shows that an error in the initial condition depends only on the error value $\mathbf{v}_1^e$ and the system matrix $\mathbf{T}$. It does not depend upon the intended initial condition or any input value.

Similarly, suppose that there is an error in a particular input value, say $\mathbf{u}_m$, $1 \le m \le n$. Then $\hat{\mathbf{u}}_m = \mathbf{u}_m + \mathbf{u}_m^e$ replaces $\mathbf{u}_m$ in the computation of the final state. The result is

$$\hat{\mathbf{v}}_n = \mathbf{v}_1 \mathbf{T}^n + \sum_{j=1}^{n} \mathbf{u}_j \mathbf{U} \mathbf{T}^{n-j} + \mathbf{u}_m^e \mathbf{U} \mathbf{T}^{n-m} \tag{8}$$

The effect of the input error is given by the last term in the expression.

$$\mathbf{e}_{in}^m = \mathbf{u}_m^e \mathbf{U} \mathbf{T}^{n-m} \tag{9}$$

Once again, the effect of the error does not depend upon the desired initial condition or the desired input values. It is simply a product of the input error, the input matrix, and the system matrix raised to the power $n - m$. The same result would be achieved by starting the system in state $\mathbf{u}_m^e \mathbf{U}$ and stepping it $n - m$ times with no other input. The mechanism for an individual error affecting the output is the system transient response, as determined by powers of $\mathbf{T}$.

Errors in other input parameters can be treated similarly. The total error in the final state is given by

$$\mathbf{e}_t = \mathbf{v}_1^e \mathbf{T}^n + \sum_{m=1}^{n} \mathbf{u}_m^e \mathbf{U} \mathbf{T}^{n-m} \tag{10}$$

Each error causes a system transient response to be reproduced. The total error is the sum of the error transients set up by the individual errors. This observation allows us to investigate the effects of errors on the system output by examining the transient response, or the autonomous system response.

2.2 Autonomous System Response

An autonomous system has all inputs set to zero. If it is started in a state $\mathbf{s}$ then the sequence of internal states is $\{\mathbf{s}, \mathbf{s}\mathbf{T}, \mathbf{s}\mathbf{T}^2, \ldots\}$. The powers of the matrix $\mathbf{T}$ must eventually begin to repeat, causing this to be a repeating state sequence. The fact that powers of $\mathbf{T}$ must repeat is a consequence of the Cayley-Hamilton theorem, which says that a matrix must satisfy its own characteristic equation.

$$\Phi(X) = |\mathbf{T} + X\mathbf{I}| \tag{11}$$

The characteristic equation is a polynomial in X. If $\mathbf{T}$ is $r \times r$ then the highest power of X in $\Phi(X)$ is X^r.

$$\Phi(X) = a_0 + a_1 X + \ldots + a_{r-1} X^{r-1} + a_r x^r \tag{12}$$

where the coefficients are from the binary number field $\{0, 1\}$ and, in particular, $a_r = 1$. Since $\mathbf{T}$ must satisfy this equation,

$$\mathbf{T}^r = a_0 \mathbf{I} + a_1 \mathbf{T} + \ldots + a_{r-1} \mathbf{T}^{r-1} \tag{13}$$

Every power of $\mathbf{T}$ can be represented as a polynomial in $\mathbf{T}$ with maximum degree $r - 1$.

$$\mathbf{T}^k = b_0 \mathbf{I} + b_1 \mathbf{T} + \ldots + b_{r-1} \mathbf{T}^{r-1} \tag{14}$$

where the coefficients are binary numbers. There at most $2^r - 1$ distinct powers of $\mathbf{T}$ (excluding $\mathbf{T} = 0$), so that the state sequence has a maximum length of $2^r - 1$ before repeating. This maximum length will be achieved for linear fininte-state machines in which the characteristic equation is a primitive polynomial of degree r.

A maximum-length LFS machine of order r can be constructed by building a linear feedback shift register with feedback connections determined by the coefficients of a primitive polynomial $\Phi(X)$ of degree r. There are at least two primitive polynomials of every degree, so this is always possible. A good discussion of the shift-register implementation of finite-state machines and their properties is contained in Peterson and Weldon [5].

2.3 Error Effects

The independent agent "expects" a checkword sequence $\{z_i\}$. Its design is such that it will fail to provide system energy if this sequence is incorrect. In the current implementation each z_i is a thirty-two bit binary number that is generated by the operation of the LFS machines as described above. To find the probability that the independent agent will fail to shut the system down in the event of a computational error, we need to compute the probability of getting the proper value of z_i with an erroneous evaluation.

The effect of an individual error depends upon both its value and the time of occurrence. The effects of multiple errors combine by superposition, as shown by (10). A combination of errors will be invisible only if the effects sum to 0. This can occur only when there are two or more errors because no single error can produce the 0 state.

As an example, let us look at the requirements of a pair of errors, say at times $t = j$ and $t = k$ with $k > j$, such that their combined effect is 0 at some later time, say $t = n$. The requirement is

$$\mathbf{u}_j^e \mathbf{U} \mathbf{T}^{n-j} + \mathbf{u}_k^e \mathbf{U} \mathbf{T}^{n-k} = 0 \tag{15}$$

$$\mathbf{u}_k^e \mathbf{U} = \mathbf{u}_j^e \mathbf{U} \mathbf{T}^{k-j} \tag{16}$$

The right-hand side can be any nonzero pattern. In the current system in which $r = 32$, there are $2^r - 1 = 2^{32} - 1 = 4,294,967,295 \approx 4.3 \cdot 10^9$ distinct patterns. The left-hand side must match the pattern that is chosen by the right-hand side. Starting at a random initial state, produced for instance by a random error pattern combining with the shift-register contents at time j, the system is equally likely to be in any of the nonzero states $k - j$ steps later. The probability of the error at time k just canceling out the random pattern caused by the error at time j is one in $2^r - 1$, which can be made as small as desired by the choice of r. In the current implementation this probability is about $2.33 \cdot 10^{-10}$.

An equivalent analysis applies to the case of more than two errors. The chance that several errors will combine to produce a 0 result is approximately 2^{-r}.

Recall that the system uses two channels for diversity in error checking. The probability that both channels produce a 0 response to two or more errors is the product of the individual probabilities, or about $5.43 \cdot 10^{-20}$. In a system in which there is an independent error opportunity about once every 100 milliseconds, the expected time between such occurrences (MBTUE) is $5.8 \cdot 10^{10}$ years.

3 Applications

This algorithm has been applied in the Vital Processor Interlocking (VPI), the Microcabmatic[1] and the Apparato Statico con Calcolatore Vitale (ASCV[2]) products. The VPI and ASCV products are used to control signals and switches at interlockings (locations were multiple tracks are connected together to allow various train movements) and the Microcabmatic product is used on board vehicles (locomotives and transit vehicles) to insure safe speed enforcement. These products have been applied in more than 350 installations and to date have achieved over 3,000,000 hours of safe operation. There has not been a failure of the algorithm to insure safe operation. Furthermore, the systems have proven to be reliable and have not suffered from unnecessary shut downs. For example, the average VPI system has demonstrated a mean time between system shutdowns (assuming no systems utilize backup or standby redundancy) of over 50,000 hours due to hardware failures including those in the input and output interface circuitry.

The algorithm has been reviewed and analyzed by rail and/or transit authorities in at least 6 different countries, and all have found it to be acceptable for their use. Since the algorithm is not application dependent, this equipment has found use in other rail applications were boolean equations are used, including speed limit selection control and highway crossing gate control. Also, since the algorithm is application independent it is easily used by countries whose signalling philosophies differ from that used in the USA.

---

[1]VPI and Microcabmatic are registered trademarks of GRS Corporation.
[2]ASCV is a registered trademark of SASIB Signalamento Ferroviorio.

The power of the microprocessor has allowed these products to include embedded diagnostics. These diagnostics have been instrumental in system maintenance and in reducing the time needed to restore a system after a failure. A computer aided applications package has also been developed for these products which allows the application process to be partially automated and allows the systems to be simulated before they are installed. The computer applications package and simulator in conjunction with the flexibility of this scheme has been instrumental in applications which progress through multiple stages as a result of changing track work.

4 Conclusion

The NISAL algorithm has proven to be both robust and flexible. It has provided a system which is analyzable and allows the maximum probability of an unsafe event to be calculated independent of the hardware and the application. It is easily applied and adapted to the different signalling philosophies found in different countries. It does not require a proof that the software is error free. Its acceptance worldwide has allowed the power and advantages of microprocessors to be applied to the rail industry while maintaining high safety standards.

References

[1] Butler, R. W. and Finelli, G. B. The Infeasibility of Quantifying the Reliability of Life-Critical Real-Time Software. *IEEE Trans. on Software Engineering* 1993; 19:3-12.

[2] Hoelscher, J. R. and Balliet, J. B. Microprocessor-Based Interlocking Control – Concept to Application. 1986 APTA Rapid Transit Conference, June 4, 1986.

[3] Rutherford, D. B., Jr. Failsafe Microprocessor Interlocking – An Application of Numerically Integrated Safety Assurance Logic. Proc. Institute of Railway Signal Engineers. London. Sept. 1984.

[4] Rutherford, D. B., Jr. A Vital Digital Control System with a Calculable Probability of an Unsafe Failure. IEEE CH2830-8/90/0000-0001. August, 1990.

[5] Peterson, W. W. and Weldon, E. J., Jr. Error-Correcting Codes. The MIT Press, Cambridge, 1972.

Session 3

SAFETY ASSESSMENT

Chair: G. Rabe
TUEV
Norddeutschland, D

Programmable Electronic System Analysis Technique in Safety Critical Applications

M.J.P. van der Meulen
Department of Industrial Safety
Institute for Environmental and Energy Research
The Netherlands Organization for Applied Scientific Research TNO
Apeldoorn, The Netherlands

T. Stålhane
SINTEF DELAB
Trondheim, Norway

B. Cole
Software Metrics Laboratory
Glasgow Caledonian University
Glasgow, Scotland

Abstract

The PESANTE[1] project intends to arrive at an integral approach towards PES[2] assessment. Elicitation of knowledge on relations between PES characteristic and functioning is the first step. Here categorical analysis plays a major role. The results of this phase will be used to tune a Bayesian inference network. This network is able to assess PESs given an amount of information on the PES characteristics. The techniques chosen are able to cope with heterogeneous and missing data. PESANTE will cover software and hardware aspects, as well as the human factor. Also, it can indicate the value of information to be procured next; this makes sure a balanced assessment is being made.

---

[1] PES = Programmable Electronic System

[2] PESANTE = Programmable Electronic System ANalysis TEchnique

1 Introduction

Before installing safety critical programmable electronic systems in industry, one wants to assess the dependability aspects of the system. This is done with the purpose of:
- Choosing the right PES in the right application; balance of costs and performance.
- Balancing the overall dependability of a total system; the dependability requirements for the PES are related to those of the total system.
- Numerical assessment of plant dependability performance.

The problem with the available techniques is that they tend to be unbalanced, not using as much data as would be possible, and arriving at unusable measures.
- They are unbalanced. Most techniques only address hardware performance. Some address software performance and only few address the human factor. Experience has shown that all three of these are important to arrive at dependable systems. Almost no technique addresses the combination of the three of them.
- They ignore information. Each technique only includes data that fits in its mathematical framework. From these, most techniques available at the moment disregard much information. More specific: they only include numerical data on only a few aspects of the system's characteristics.
- They result in unusable measures. Unusability of measures happens in two ways: too much information on a small part of a system (e.g. MIL-HDBK-217 calculation of hardware failure rate [1]), or information that is hard to handle (e.g. the number of remaining bugs in a program).

TNO has defined the PESANTE [2] project, in cooperation with SINTEF and Norsk Hydro, Dow Europe, and Glasgow Caledonian University's Software Metrics Laboratory. The aim of the PESANTE is to measure dependability characteristics of programmable electronic systems in process industry. The method is to be used for highly reliable PESs in safety applications resulting in a balanced, complete and usable assessment.

2 Baseline of the Project

Dependability assessment of safety related systems is getting more crucial every day. More and more, PESs control and guard critical processes. Nowadays the insight has grown that dependability of electronics depends on three factors:
- Hardware.
- Software.
- Human factor.
All three of these have to be examined in order to arrive at a sound assessment of a system. The following sections shortly describe the state of the art in these fields relevant to PESANTE.

2.1 Hardware Dependability

As all dependability engineering, hardware dependability assessment can be done qualitatively and quantitatively. The first methods used were essentially qualitative (dependability by design). Then the quantitative approach emerged, amongst others initiated by the U.S. Defence organizations: the MIL-HDBK-217 [1]. This approach has gained much popularity and is used for almost all PESs at the moment. In spite of its popularity, much has been said against the applicability of the handbook [3]. By this, and the notion that safety has to be an integral part of the design in highly dependable systems (making things impossible is better than making them very improbable) the qualitative approach gains field again.

2.2 Software Dependability

In software dependability engineering, the most widely applied methods are quantitative. Most methods use data on debugging and failure times for estimating failure rates [4]. There is much discussion about the validity of the approach and its correlation to reality. Case studies have shown good results, but the question remains whether the methods give valuable clues for the individual case.

2.3 The Human Factor

The notion that human dependability is a major factor in the functioning of all systems is rapidly gaining field. Human factors play a role in all stages of a design: specification, implementation, debugging, maintenance, upgrading, and last but not least use. The emphasis on quality assurance at all stages is one of the examples of the effects of this notion. The other way around, the incorporation of data on the human factor in an assessment of a system is in development. Numerous publications [5] discuss the matter and try to find solutions to the problem. The results are promising. It has become clear that without assessment of the human factor dependability estimation of a system is incomplete.

2.4 Integral Approach Towards Dependability Assessment

All three of the aspects mentioned above play a role in dependability estimation of industrial electronics. This poses the following problems to the assessor:
- Finding the right balance in assessment of the three aspects.
- Combining findings on the three aspects into one assessment. The findings may be of quite different nature. Also some information on an aspect may be lacking. Two examples may clarify this: sometimes information has to be discarded because the tool used cannot handle the kind of information, also sometimes analysis methods demand data for arriving at an answer, even if the data is not available. Essentially, most information one can gather on a system is heterogeneous: having different units, scales, and importance. This poses hard demands on the assessor and also on his tools.

- Assessing the value of information. If a certain amount of information on a system is available and one is not yet satisfied with the results: what information should be gathered next? What information is most valuable to the assessor in order to improve his ordeal. E.g. if one has a lot of information on hardware dependability, extra information on hardware dependability might not be very valuable as opposed to information on the human factor. How can an assessor choose which information he should gather?

This tendency to try to balance can be found everywhere now. It is a type of thinking towards a goal, rather than thinking from tools [6]. The goal is a valid assessment of safety performance of a system. PESANTE integrates some of the suggested approaches and is described in the following paragraphs. Based on the insights gained in the course of the project tools will be developed.

3 Technical Description

The PESANTE project is a combination of mathematical and knowledge elicitation techniques. The project is a synthesis of approaches from various disciplines: traditional software, hardware and human dependability analysis, categorical analysis, elicitation and Bayesian inference techniques.

To implement the technique, first knowledge elicitation will take place, using proven techniques as the Delphi method. This will concern both objective and subjective information on systems. The information originates from users and all others familiar with the performance of the PESs. These people are asked questions illuminating the quantities under study from different viewpoints. The idea is that these people all have information, but cannot combine this data into the metrics needed. PESANTE will systematically help them to do that.

Also input to the categorical analysis are the results of standard assessment techniques as MIL-HDBK-217 calculations, software dependability measures as MTTF estimations, and human factor assessment techniques. The technique proposed is to be a step beyond these methods: it is to combine all results into one assessment.

3.1 Categorical Analysis

The technique to analyze the information is a categorical analysis [7]. This will quantitatively reveal correlations between (combinations of) characteristics. If by example a certain test strategy is necessary to realize a certain maintainability, this correlation will show up. Later these relations will be used the other way around: to asses the system from its characteristics [8,9].

In PESANTE categorical analysis will be used to elicit relations between heterogeneous data and dependability aspects. Categorical analysis is available now for some time, and mainly used in psychology and sociology, where complex relationships are common (as well as the need to unravel them). The technique is used to reveal those relationships.

Characteristic for categorical analysis is its ability to cope with heterogeneous and missing data. This enables PESANTE to include all sorts of data available on PESs in safety applications.

3.2 Bayes' Reasoning

For assessment the relations between PES characteristics and dependability aspects as found using categorical analysis will be implemented using Bayes' reasoning [10]. This tool will use Bayes' reasoning to conclude on dependability aspects based on given input. This input does not need to be complete, the tool will be able to cope with missing data. As categorical analysis, the technique used is not new. It is used with success in other applications like fault diagnosis. Only few attempts for using artificial intelligence in the field of dependability engineering have been done until now. The advantage of using Bayes' reasoning are:
- It can cope with missing data.
- It can give estimations of the value of data.
- It can give an impression of the reasoning behind a result: how did the system come to this conclusion.

3.3 Combination of Categorical Analysis and Bayes' Reasoning

In short, the techniques mentioned above will be combined in PESANTE as follows.

Procurement of information on PES functioning will be done using techniques readily available as the Delphi method and other expert opinion collection techniques.

Categorical analysis will be used to elicit relations between heterogeneous data and dependability aspects.

After elicitation of the relations, a model based on Bayes' reasoning will be tuned accordingly. The resulting tool will assess future PESs.

4 Conclusions

The PESANTE project will deliver an integral approach to dependability assessment of Programmable Electronic Systems. Valuable are the following characteristics:

- It will be able to combine heterogeneous data on the PES under consideration in order to arrive at a sound assessment.

- It will be able to determine the value of data: what data is most valuable to arrive at a better assessment. This makes sure that a balanced assessment is being made.

- It will be able to cope with missing data. In most cases, data on a PES are incomplete. Most methods demand input, available or not. PESANTE will base the assessment on the data available.

References

1.   MIL-HDBK-217, Military Handbook, Reliability Prediction of Electronic Equipment, Rome Air Development Center, Griffiss Air Force Base, New York.

2.   Stålhane, T., M.J.P. van der Meulen & B. Cole, Reliability Assessment for Programmable Electronic Systems using Subjective and Objective Categorical Data, Proceedings PCPI '93, p101-8, Düsseldorf, Germany.

3.   O'Connor, P.D.T., Reliability Prediction: Help or Hoax? Solid State Technology, p59-61, August 1990.

4.   Bendell, A. & P. Mellor, Software Reliability; State of the Art Report, Pergamon Infotech Ltd., 1986.

5.   A Resource Guide for the Process Safety Code of Management Practices, Responsible Care, Chemical Manufacturers Association, October 1990.

6.   Majone, G. & E.S. Quade (ed.), Pitfalls of Analysis, John Wiley & Sons, New York, 1980. ISBN 0-471-27746-0.

7.   Gifi, A., Nonlinear Multivariate Analysis, John Wiley & Sons, New York, 1980, ISBN 0-471-92620-5.

8.   Neil, M., Multivariate Assessment of Software Products. In: The Journal of Software Testing, Verification and Reliability, Vol. 1., (4) 17-37, 1992.

9.   Bendell T., The Use of Exploratory Data Analysis Techniques for Software Reliability Assessment and Prediction.

10.   Horvitz, E.J., J.S. Breese & M. Henrion, Decision Theory in Expert Systems and Artificial Intelligence, Knowledge Systems Laboratory, Technical Report No. KSL-88-13, Stanford University, California, July 1988.

Safety Assessment - The Critical System Suppliers View

Author
C J Goring
August Systems Limited (UK)

Abstract

This paper firstly analyses the decision making of a supplier in applying for certification. The process of safety assessment as seen from the supplier side is described, in particular the experience of being assessed by TUV to Safety Requirements Class 5/6 of DIN V VDE 0801.

By describing in detail the various phases of the assessment it is hoped that by this paper an aid to other would be applicants is given in easing their assessment path. The paper firstly reviews other national and international standards and schemes and looks forward to the possibility of a unified ISO standard and assessment procedure.

In conclusion the paper briefly analyses the cost/benefit of assessment and certification.

1. Introduction

As a supplier of safety critical systems to both the Nuclear and Petrochemical industry, August Systems begun in early 1990 to review the case for independent safety certification. From this review it was decided to proceed with safety certification for a Triple Modular Redundant Software Implemented Fault Tolerant Safety System. This paper analyses the reasons for this decision and seeks to provide aid and guidance to other suppliers in the path of safety certification.

The path of independent safety certification is a costly and time consuming process and no vendor should set out on this path without the resolve and financial resources to see the process through to the end.

2. Assessor and Certification Choice

Worldwide there are a small number of authorities that will provide safety certification to various national and application directed standard. Currently there are no international standards for safety systems to be assessed too and therefore the first decision that any vendor must make relates to the choice of assessor and the choice of standard to be assessed against.

The market to which the relevant safety system is targeted will often influence the choice of assessor and standard. For nuclear installation each major country has its own standards and regulatory organisation, eg. the National Nuclear Inspectorate (NNI) in the UK. This reliance on application and national standards limits the commercial viability of a more generalised safety systems approval.

Other industrial safety conscience markets such as Petrochemicals rely on standards of codes and practice such as the PES 1 and 2 guidelines issued by the Health and Safety Executive (HSE) of the UK and the Engineering Equipment and Materials Users Association (EEMUA) guidelines, however more recently for programmable systems an acceptance of TUV Certification to Class 5 and 6 has been considered appropriate by a number of major operators in this field.

Industry specific standards and certifying authorities such as the Federal Aviation Authority (FAA) in the USA, and the Civil Aviation Authority (CAA) in the UK, have obtained international acceptance on a wide scale for flight safety. To a lessor extend the national certification authorities for railway signalling have also achieved some international recognition.

As it can be seen the choice is wide and is normally governed by a combination of both the targeted industry and the national authority. In the international market of petrochemicals August Systems chose TUV as the certifying authority and DIN V VDE 0801 as the standard for safety systems. The choice was governed by a combination of major customer acceptance and international acceptability. The class of certification 5/6 was chosen to match the market for the majority of safety systems for the petrochemical market. Higher classification systems such as 7/8 may be readily obtained by a combination of diverse systems in highly critical plant areas (the references provide more information on high Integrity Protection System - HIPS where class 7/8 safety is required).

The DIN V VDE 0801 specification provides a risk graph aid to the user to determine the class of safety for the application, this shown in Figure 1.

Extent of damage
S1   minor injury to one person
S2   serious permanent injury to
        one or more persons, death of
        one person
S3   death of several persons
S4   catastrophic consequence

Frequency of exposure to hazard
A1   seldom to quite often
A2   permanent exposure

Avoiding the hazards
G1   possible under specific conditions
G2   almost impossible

Probability of unwanted event
W1   very low
W2   low
W3   relatively high

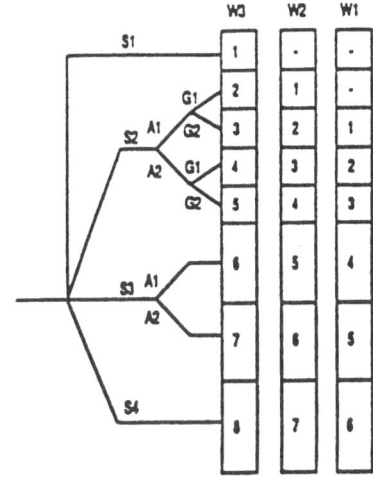

figure 1: Risk graph, requirement classes 1-8

3.   Assessment to DIN V VDE 0801

The assessment covers three basic phases which are:

a)   The Design Analysis
b)   The Hardware and Software Inspection and Testing
c)   The Certification

Each of the three phases has its own pre-requisite and imposes different work loads on both the vendor and the assessor.

3.1   The Design Analysis

The design analysis phase for DIN V VDE 0801 is called the concept review and consists of three distinct parts; requirements class selection, documentation inspection and analysis of the safety concepts of the system.

3.1.1 Requirement Class Selection

By holding informal reviews with the vendor, basically enabling the TUV staff to understand the general concepts of the system that is being offered for certification, the suitable certification class is selected and agreed. From this point onwards all measurements and results will be interpreted in relation to the selected requirements class. The levels of the efficiency of the measures taken for safety and the types of failure caused are simply defined by Table 1 which is given in DIN V VDE 0801.

Table of Required Efficiency of Measures against failures
Depending on the Requirement Classes (according to DIN V VDE 0801)

cause of failure	frequency of failure	location of failure	measures against failures	1	2	3	4	5	6	7	8
systematic		hardware	avoidance	basic measures	basic measures	basic measures				high	high
systematic		hardware	control								high
systematic		software	avoidance	basic measures	basic measures					high	high
systematic		software	control								high
random	single	hardware	control					high	high	high	high
random	multiple by accumulation	hardware	control							high	high
handling operation			avoidance	basic measures							high
handling operation			control								high
external and environmental influences			avoidance	basic measures							high
external and environmental influences			control								high

required efficiency of measures against failures

| none | simple | middle | high |

table 1

3.1.2 Documentation Inspection

For a system to be capable of being accredited with a safety classification it must be thoroughly and accurately documented. The certifying authority uses this phase to both validate the documentation and to give a further insight into the operation of the submitted system.

For the vendor this period will almost certainly result in documentation corrections being instigated and if the system is not thoroughly documented, additional documents will need to be produced to provide a complete documentation set.

The results of the section will confirm or otherwise that the system is capable of proceeding to the next phase and being accredited to the appropriate class.

3.1.3   Analysis of Safety Concepts

The vendor will be required to submit a Safety Concepts Review Document that completely describes all of the safety aspects and concepts of the submitted system. This document will almost certainly not exist within the vendor organisation, however the information to produce this document would normally be readily available from the vendors standard documentation packages.

This document plus all of the standard software and hardware documentation will allow the certifying authority to produce a system level failure mode effects analysis and will also provide an indication of the effectiveness of the measures taken to prevent system failure or fail to danger.

When the system has successfully passed through this first phase the indepth testing, inspection and validation can proceed.

3.2   Hardware and Software Inspection and Testing

This phase of the certification is probably the most intense for the certifying authority, it consists of Hardware Inspection and Validation, Software Inspection and Validation, Systems Integration and Safety Concepts Validation.

3.2.1   Hardware Inspection

The certifying authority now completes a detailed inspection of every hardware assembly. This inspection includes track thickness and track gaps on printed circuit boards as well as a complete analysis from documentation to final product.

Detailed failure mode effects analysis are carried out on all circuits where safety or fault tolerance are seen to be critical, this analysis encompasses all critical component tolerance analysis using CAE tools.

The Mean Time Between Failure calculation provided by the vendor are checked and finally the system is type tested to the vendor specification. This final exercise is normally completed shortly before certification as by experience TUV have found that it is better to ensure full certification is achievable before type testing takes place.

3.2.2   Software Inspection

It is at this stage that any short comings in the design approach of the software to be accredited will become evident. For the rigorous designers, there will be little to do but perhaps provide additional structure guide documentation.

If however the source code and structures are not well documented then there will be a significant amount of back documentation to be completed before the assessing authority can proceed. It must be remembered that the assessors will have little or no background knowledge about the design to be assessed.

The process of assessment indicates a thorough inspection of the documentation, data structures and data flows against quality criteria, programming codes of practice and consistency against the specification.

The software is analysed for measures taken to guard against faults and errors, such techniques as defensive programming, on-line software test routines and error traps are confirmed to be present or otherwise, and this enables TUV to determine the category that the software design belongs to.

The software is also subjected to a computer aided white-box test which provides analysis of coverage, run-time, data range and control flow.

3.2.3   Integration and System Test

By this time in the evaluation the authority will now have an intimate knowledge of the detailed workings of the hardware and software system under review. This will have been obtained not just by review of the thorough documentation but also by a series of meetings held with the vendors engineers. This acquired knowledge plus the experience of the examiners enables the certification authority to complete a large number of fault injection tests both on hardware and software.

The purpose of fault injections is to confirm the theory of operation for fault detection, time of fault detection, fault tolerance, fail safety and reparability. Injected faults include simulated hardware failures, actual software corruption and simulated fail to danger scenarios are carried out and each result logged and analysed.

4.   Application Specific Criteria and Final Report

With the results of the test complete it is now possible for the evaluating authority to generate a final report which defines the process types that can be protected (eg. processors that have fail safe states), the main time that faults may be in the system undetected and any special requirements of configuration.

5. Accreditation Standards

Currently there are no internationally accepted standards to which accreditation can be achieved. In the authors opinion it will be several years before a fully accepted international standard is available. Two national groups and one international group have been working towards producing an acceptable standard and the following provides a review of the current status of those standards.

5.1 Instrument Society of America ISA SP84

The ISA SP84 group is a process industry group that has been working on generating standards for safety control applications in the process industries particularly for programmable systems. The committee consists primarily of users and a few selected vendors and the standards are in advanced state of production with perhaps the exception of software criteria. Because of its advanced status it is possible that these standards will be submitted for international recognition in the coming two years, they are however currently limited to process applications.

5.2 DIN V VDE 0801

This specification is a German National standard which is not industry specific. The standard has been evolving over the last decade and its current status gives it acceptance both in germany and by certain international companies. I would expect the German standard groups to push for DIN V VDE 0801 to be accepted as the international standard for safety accreditation. The standard certainly does provide the basic coverage that an international standard will need, however in the authors opinion it does require further development, specifically in the area of safety metrics.

5.3 IEC 65A Working Groups 9 and 10

These European/International working groups have been operating for a number of years in an attempt to produce an acceptable ISO standard. The two working groups have been contributing to different aspects of safety system WG10 provide a generic system approach and WG9 concentrates on safety software.

It is hoped by the author that these standards will eventually become accepted by ISO although my expectation is that at least 2-3 years of further work is required and probably a degree of amalgamation with both the SP84 and the VDE 0801 standard are required.

6.   Conclusion

With no one international standard to accredit against, the drive to invest in accreditation by a safety system supplier will often be governed by his customers. If a sufficient number of customers continue to push for accreditation by a recognised authority then the supplier will need to make a commercial decision as to whether the investment in accreditation is necessary.

The cost of accreditation is of course dependent on the size and complexity of the equipment submitted, but will almost certainly fall within the range of $50,000 - $250,000.

Additionally there is the ongoing cost to ensure that modifications and updates are re-accredited.

From the suppliers viewpoint one internationally accepted standard with accreditation authorities in all of the major industrial countries would be the preferred outcome.   Until this is achieved we will have to continue to make commercial decisions on when and where accreditation is achieved.

References

4.   EEMUA Publication No 160, Safety Related Instrument Systems for the Process Industries

5.   ISA SP84, Electrical/Electronic/Programmable Electronic Systems For Use in Safety Applications (Draft Standard)

6.   Qualification of Programmed process Control Systems for Safety Requirements, H Gall and W Hammerschmidt (TUV)

7.   Safety First - Applying Computer Based Safety Systems in the Offshore Environment, C J Goring (August Systems)

Hazard Analysis Using HAZOP: A Case Study[1]

Morris Chudleigh
Cambridge Consultants Limited
Cambridge, England

1   Introduction

There is an increasing use of computing in safety related applications and ensuring that such systems are conceived, designed and produced with appropriate attention to safety is not easy. The process of identifying undesirable events and their consequences is known as hazard analysis. Carrying out a hazard analysis at varying stages during the development process is now being mandated in emerging standards produced by the International Electrotechnical Commission (IEC) [1] and the U.K. Ministry of Defence [2].

With computer based systems there is the particular problem that design faults tend to dominate random hardware faults and so identifying potential hazards early in the design process is crucial. There are well established methods for carrying out hazard analysis in many domains of engineering such as petro-chemical but it is not clear that such methods are readily applicable to computing based systems. Further, there is little advice available to assist with hazard analyses using computer technology.

To address the above problem we have modified the Hazard And OPerability (HAZOP) technique [3] and used it successfully in the computing domain. Our evolution of the HAZOP technique has taken place over a number of years and most applications have been Commercial in Confidence. However, a medical imaging application is being studied as a collaborative research project and serves as an appropriate case study.

The remainder of this paper explains our conclusion that a modified HAZOP is an effective way to carry out hazard analysis on computer based systems. We find that:

- The HAZOP method is well established and fits well into overall safety management
- The method can be extended to the computer software field
- The experience to date is that the approach is powerful and we illustrate this with the medical case study.

1.The case study reported here is part of a collaborative project between The Centre for Software Engineering Limited, Cambridge Consultants Limited and The Human Genetics Unit of The Medical Research Council.

To put the results in context the next section gives a description of the experimental medical system. Subsequent sections deal with our efforts to develop a modified HAZOP and a case study of the results obtained through its application to the experimental medical system.

2 The Medical Diagnostics Application

There are many laboratories worldwide which carry out cervical screening by expert manual inspection of slides prepared from smear samples. The purpose of screening is to recognise the presence of abnormal cells in a sample which may contain few such cells, and is likely to contain possibly confusing other matter. Usually, the review is manual and includes an assessment of the degree of abnormality, leading to diagnosis and treatment. For screening of healthy patients on a regular basis, there are likely to be few abnormal cells but for diagnostic screening of a sick patient, abnormal cells will be expected.

The U.K. Medical Research Council Human Genetics Unit in Edinburgh (HGU) has developed over a number of years a semi-automated screening system for cervical smears [4]. The basis of the automated system is the computer analysis of a slide prepared from the smear sample. The first version of the system relied on custom hardware to capture and process images and the HGU are now in the process of re-engineering the system to run on a modern computing platform without the use of custom hardware. The collaborative project of which the work reported here is part involves the re-implementation of part of the new system currently under development.

The system relies on efficient image processing and classification algorithms to find the abnormal cells. An initial search at high speed is carried out at low resolution to identify suspicious objects (an object is something that appears to the software to have cell-like characteristics but may not in fact be a cell). A second pass is then done at higher resolution so that a more detailed analysis can filter and rank the objects. The top-ranking objects are classified and a decision made to class them as normal, ask for a skilled human review, or pass for a full conventional review.

Outside the core imaging system, there are various preparatory and post-analysis processes to be taken into consideration. The former include administrative tasks to handle samples and accompanying paper-work sent from General Practitioners and clinical agencies. This is then followed by slide preparation and the transfer of slides to the automated imaging system. Configuration may then be necessary to initialise certain batch processing control parameters before the automated process itself can be started. Once this has been completed, there follows a certain amount of tidying-up before results are passed for diagnostic review, signing out or some form of quality control check. Results of the analysis must finally be returned to the clinic which supplied the sample and the analyzed slide placed in a long term archive.

The core image processing part of this application forms the case study for our application of HAZOP to a software based system.

3 HAZOP Studies are an Effective Part of Safety Management

This section describes the process of ensuring that safety is considered during all phases of the life of a system, the safety lifecycle. It then gives an overview of one of the established methods for identifying hazards, the HAZOP, followed by some of the advantages of the method.

3.1 The Safety Lifecycle

In a number of established industries, particularly petrochemical, the possibility of failures having an adverse impact on safety has been recognised for many years. However, it was not until a number of accidents and near accidents had taken place that it was recognised that a systematic approach to the management of safety was required. The approach taken is independent of the industry. To clarify the following, two definitions are given, both taken from Ref [1]:

A hazard is a physical situation with a potential for human injury.
Risk is the combination of the frequency, or probability, and the consequence of a specified hazardous event.

The basic steps in the overall safety lifecycle are:

- System definition generating an overall description of the system under review
- Hazard analysis to identify the potential for hazardous events
- Risk analysis to judge the safety risk of the defined system. This quantifies the potential for hazardous events and evaluates their consequences.
- Judgement of the acceptability of the risk
- Activities, if necessary to reduce the risk to an acceptable level. This might be by modifying the architecture of the system, including extra measures to avoid or contain safety failures, or ensuring that the system is built to standards that are appropriate to the level of risk.
- Implementing the system to the required standards, followed by effective operation and maintenance

In recent years there has been a huge growth in the use of computers and, as we are all aware, all computer systems are liable to contain design mistakes. The fact of the fallibility of computer systems when used in safety critical situations is now being addressed formally in standards work. Two new standards have been drafted: one from the International Electrotechnical Commission addresses functional safety of computer based systems [1] and the other from the U.K. Ministry of Defence, addresses hazard analysis for computer based systems [2].

Both the standards use the above safety lifecycle and stress the importance of carrying out hazard analysis. The traditional industries have developed a number of structured methods to help ensure that the hazard analysis is complete and thorough, such methods include 'what if' analysis, failure modes and effects analysis (FMEA), and HAZOP. The next section introduces one of the most effective hazard analysis methods, HAZOP.

3.2 The Traditional HAZOP

The full name of HAZOP is Hazard and Operability, and this gives pointers to the two facets of its purpose. The HAZOP ensures both that features that could lead to undesirable outcomes (ie hazards) are avoided and that necessary features are incorporated into the design for safe operation. The method was developed in the U.K. by ICI in the late 1960's and is well established in the petro-chemical sector. A reference for its use in that industry is given in [3].

In these industries, the plant design is normally described by piping and instrumentation diagrams (P&IDs). The HAZOP study is carried out by a team of knowledgeable engineers who carry out a systematic examination of the design. They postulate, for each element of the system design in the P&IDs, deviations from the normal operating mode and then assess the consequences of those deviations with respect to any safety or operability problems.

For each deviation, the team asks 'can it happen?' and, if it can, is this likely to lead to a hazard. The team will take into consideration any mitigating features, such as control valves, alarms etc, which might control the hazard. To formalise the process, a series of guidewords are used to define particular deviations and these are applied to each relevant parameter for each process component.

Thus, for fluid flowing in a pipe, relevant parameters might be **flow, pressure, temperature** and deviations examined would include **high, low, no, reverse, as well as.** In theory, each guideword should be applied to each relevant parameter for each part of the process description. In practice, this is very time-consuming, and an experienced HAZOP leader will use judgement to control the correct detail of questioning in each area.

The results can be presented in a number of ways and we have found a software package developed by our parent company, Arthur D Little, running on a personal computer to be effective [5]. Results are presented under the following headings:

Item number - a simple count of items logged from the beginning of the HAZOP
Equipment item - a description of the area of plant for which a deviation has been found
Parameter - such as temperature, pressure, flow etc.
Guideword for Deviation - such as high, low etc.
Cause - the circumstances that could give rise to the deviation
Consequence - the effect on the plant that the deviation might lead to
Indication/Protection - any feature that will either identify the deviation or mitigate it's effects (eg an alarm signal or a pressure relief valve).
Question/Recommendation - questions arise from items considered a potential hazard which cannot be resolved by the meeting. Recommendations are generally for changes to the design or particular actions to be taken during operation.

Answers/Comments - This allows the later insertion of answers to questions raised or notes which the team consider relevant to the design but are not questions or recommendations.

The HAZOP team is normally small, four to eight people, comprising a leader, a team secretary, people who understand the design intent, people who are experienced in the operation of similar plant and specific technical experts as necessary.

3.3 Advantages of HAZOP

There are three main advantages of the HAZOP approach to doing hazard analysis: it is carried out by a team; it deals with the design in a systematic top-down manner; and it is capable of being applied during all phases of a system lifecycle.

The team approach allows a variety of expertise and viewpoints to be applied to the system. Our experience is that many problems are caused by interactions between parts of the system and by differing understandings of designers and operators. The team approach allows such issues to be explored and there is less impact to a mistake by one team member. Having key personnel on the team means that any problem areas are brought immediately to their attention.

Dealing with the system in a top-down manner allows concentration on the key issues arising from the potential hazards and allows system-wide implications to be assessed. Having a clear view of the whole system allows the team to explore non-obvious interactions between parts of the system. This is in contrast with bottom-up approaches such as FMEA which must analyse every component to a similar level of detail which is time-consuming and error prone. We have found that the HAZOP is useful in identifying which are the most critical areas to concentrate on in any later FMEA.

The approach may be used during all phases of the life of a system. We have used it successfully both at the conceptual design stage and for safety assessment on completed systems.

4   HAZOP Has Been Modified

The positive experience of Arthur D Little in using HAZOP made it their technique of choice for hazard analysis in the petro-chemical industry. In recent years they have worked with their subsidiary company, Cambridge Consultants, to explore whether the HAZOP method could be extended to other domains, particularly those where electronics and computers were involved. It became clear that work was needed in two different areas; first to find appropriate representations and, second, to see what were an appropriate set of parameters and deviation guidewords.

4.1   An Effective Representation Method is Needed

With computer based systems there is not, obviously, a P&ID to work from. For electronic designs there are clear parallels with block diagrams and circuit diagrams and there is much uniformity of representation. Transfer of the approach was thus straightforward. This was not the case with software. There are many representational methods which involve diagrams and some, using mathematics, which have no pictorial representation. We have found that the HAZOP approach is tractable with a variety of pictorial methods but that the dataflow representation of structured design is the most natural to work with.

With our exemplar diagnostic image processing system, little in the way of design documentation was available and so we worked with the system designers to build a dataflow representation of the system using a CASE tool called Software through Pictures [6]. Figure 1 shows the first level of decomposition from the top level context diagram.

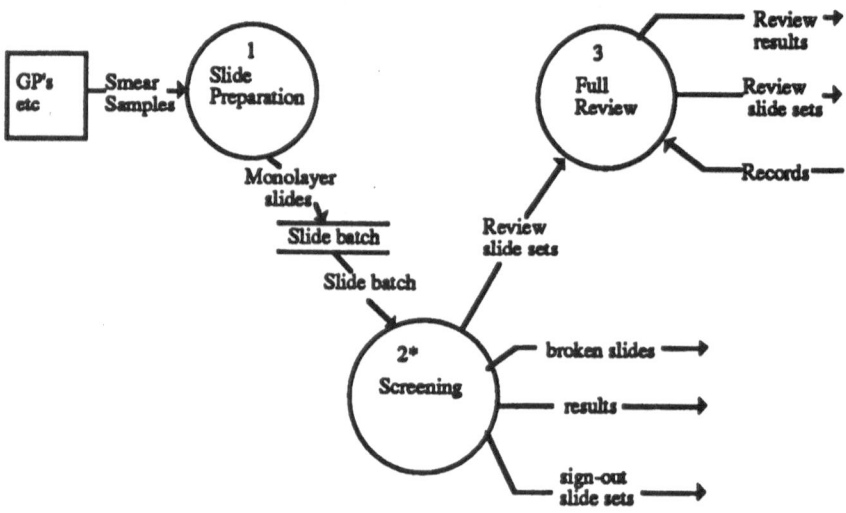

Figure 1: Cervical Screening System

The asterisk in process 2 indicates that a lower level decomposition has been generated. The three processes show the activities of:

- Slide preparation - preparing slides from smear samples
- Slide screening - the computer based imaging and diagnostic process
- Full review - manual expert checking of slides recognised as suspicious by the computer system

Figure 2 shows a simplified version of part of a lower level process, the process of carrying out a scan of the slide at low resolution. One of the sub processes, Bright Field, is used below to illustrate the HAZOP method so its function will be described here. Four or five frames are captured by using a microscope, digital camera and a special framestore supporting direct memory access of reduced scale images (typically every 5th pixel). The process performs a logical OR operation on the captured frames to determine maximum light intensity on a pixel-by-pixel basis. This data can then be used to correct object images for optical density (that is, making them invariant to lamp brightness and variation in field illumination)

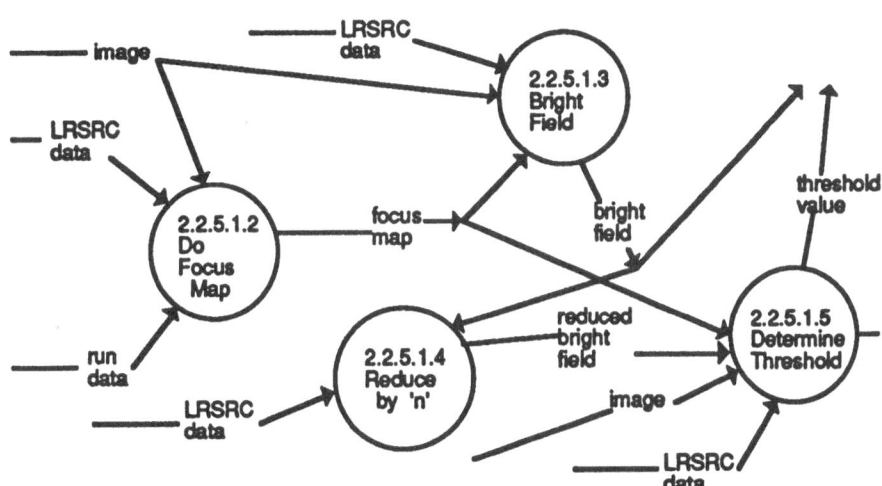

Figure 2: Part of Low Resolution Search Scan

We found that the use of the dataflow paradigm and a CASE tool enabled an accurate and easy to understand representation of the design to be produced quickly. One of the main advantages of the dataflow approach we have found over a number of projects is that the design is readily understandable by all interested parties, even those without any computing background.

4.2   New Guidewords Have Been Derived

Since the software and image-processing domains differ significantly from petro-chemical engineering, we found that the standard HAZOP guidewords did not provide a rich enough set. Over a number of projects [7] we have evolved a set which is generally applicable to software based systems and examples of these are shown in Table 1 below. The full vocabulary of guidewords was larger and included words that were only used on a few occasions. These arose through the complex mix of technologies in the experimental imaging system and the need to express a particular deviation. The set is evolving as our experience increases and we believe that improvements might arise if we differentiated between different data types.

PARAMETER	GUIDEWORD
data value control value	no incomplete incorrect stale (unchanging)
algorithm function	wrong sequence wrong specification wrong algorithmic derivation
data rate	too high too low no
comms protocol	corrupted transmission wrong prefix wrong handshake
timing (of actions)	too early too late too slow too fast out of sequence

Table 1: Parameters and their Deviation Guidewords

5 The HAZOP Process Worked Well in Practice

5.1 Results

The HAZOP was carried out by a team of four people including an experienced leader and one of the designers of the re-engineered diagnostic system. The HAZOP was of the full computer based image processing system and some 100 items were noted during the work.

The HAZOP was applied in a top-down manner so that each process was taken in the numeric order in which it occurred in the dataflow decomposition: first, deviations of the input dataflows were explored, followed by examination of the process itself which naturally defined the possible deviations and their consequences for the process outputs. Many data items were used as inputs to more than one process and of course they only needed to be addressed in detail once during the HAZOP. Thus, as each new level was examined, the data passed down from a higher level was marked and only reviewed briefly to check that possible corruptions or transformations had not been overlooked. Some generic conclusions from the results will be given and then some of the specific items from the process Capture

Bright Field described above.

The generic conclusions may be divided into three themes: First, efficiency led gains in processing speed have to be balanced against the potential for hazards due to incomplete sampling. To achieve the benefits of automatic screening the system throughput must be high and the necessary use of complex image processing algorithms and a series of filters means that not all a sample may be scanned.

Second, recognition of abnormal cells can arise from hardware faults and current software implementation practice. Hazards may arise through data corruption either leading to overlooking of abnormal cells or through recognition of normal cells as abnormal. The experimental approach to date has concentrated on algorithm research using the most convenient tools and methods and has given less attention to achieving high levels of safety integrity.

Thirdly, key data items for set-up and operation must maintain their integrity. In particular, processing histories are kept for a number of days on a rolling basis and classifier learning data is used by key algorithms: these and similar data must maintain their integrity so that consistent system operation and correct tracking of patient details are ensured.

The process **Bright Field** shown in Fig 2 is used to illustrate the HAZOP.

Recommendation 59 *Bright field* high cell density could lead to incorrect OD calculation, effect could be a 'blind spot'. Carry out sensibility check.

Question 60 *Bright field* drift of lamp intensity could lead to missed objects. How can this be checked for?

Question 61 *Bright field* Depends on light maxima which could be too high if slide sample acts as lens, forming ring on image. Can other algorithms be used?

Question 62 *Bright field* if value too high, too low or randomly incorrect, subsequent OD corrected images will be corrupted. Can this be checked for?

It should be noted that this was a HAZOP of an experimental system, developed to show feasibility and it is not surprising that we found the system lacking in a number of areas. The value for the design team is that they believe team members are now well placed to address hazard concerns in order to mitigate areas of potential risk.

6 The Approach Has Advantages and Limitations

The advantages of using a HAZOP described in Section 3.3 above have been found to be independent of the application domain and of the technologies involved. For the computing domain described in this paper the application of the method was novel to the HGU personnel involved. The combination of the dataflow representation and the systematic approach to examining deviations that could lead to hazards worked effectively. Once the team became used to the method (our experience on this and other projects is that familiarisation takes about half a day) productivity was high and the full HAZOP of the system took only a few days. This high productivity and early focusing on important issues is, we believe, a major advantage of the HAZOP as contrasted with bottom-up approaches.

There are, however, limitations to the approach. In the process industries guidewords for deviations from design intent are well established and application is reasonably straightforward. With software based systems we have found that application is less easy :

- The expertise of the HAZOP leader is crucial in focusing the discussion on potential hazards and not to exploring 'interesting' areas.
- Slavish adherence to the guidewords is not sufficient. We have found that significant flexibility and multi-disciplinary design experience by the team is necessary to explore unusual interactions.
- Independent technical experts, experienced in safety critical computer system design, are necessary to gain full benefit.

Thus we have found that the HAZOP approach provides an effective way to carry out a systematic and cost-effective hazard analysis but that the use of technical expertise and design experience is still required. We postulate that this is because of the inherent complexity of software based systems and their propensity for design error.

References

1   Functional Safety of Electrical/Electronic/Programmable Systems. Generic Aspects. IEC 65A (Secretariat) 123. 1991

2   Interim DefStan 00-56. Hazard Analysis and Safety Classification of the Computer and Programmable Electronic System Elements of Defence Equipment. U K Ministry of Defence 1991.

3   A Guide to Hazard and Operability Studies. Chemical Industries Association Limited, 1987.

4   Husain O, Watts K, Lorriman F et al. Semi-Automated Cervical Smear pre-screening system: an evaluation of the Cytoscan-110. Analytical and Cellular Pathology, 5: 49-68, 1993.

5   HAZOPtimizer. Arthur D Little, Safety and Risk Management, Cambridge, Massachusetts.

6   Software through Pictures. Interactive Development Environments, California.

7   Chudleigh M, Catmur J. Safety Assessment of Computer Systems using HAZOP and Audit Techniques. In: Frey (ed) Safety of Computer Control Systems 1992 (Safecomp '92) pp 285-292.

Session 4

SAFETY ANALYSIS

Chair: F. Koornneef
Delft University of Technology, NL

Safety Analysis of Clinical Laboratory Systems

Authors
S.S. Dhanjal - Lloyd's Register
R. Fink - West Middlesex University Hospital

1.   Introduction

The Clinical Biochemistry Department (CBD) at the West Middlesex University Hospital (WMH) performs tests on constituents of body fluids to facilitate diagnosis, prognosis and monitoring of treatment. A rapid increase in the need for this service in recent years has led to extensive automation of the analysis and data handling operations within the department. The automation and data handling have been implemented by integrating a computerised Laboratory Information Management System (LIMS) into the operations of the Laboratory.

Quality assurance of the analytical processes is well established. However, in common with other safety related disciplines and applications, there is a concern about the reliability of the computerised data management system and the lack of generally accepted standards. These issues have been addressed as part of the DTI sponsored MORSE - (Methods for Object Reuse in Safety Critical Environments) project.

WMH is a member of the MORSE project consortium (Dowty Controls, Lloyd's Register, Transmitton Ltd, West Middlesex University Hospital and the University of Cambridge). The MORSE project has been inspired by recently proposed standards and guidelines [1,2,3,4] which are at various stages of development. These standards and guidelines apply to safety critical systems and bring together a range of existing procedures, methods and design practices which, in combination, are untried. These methods and guidelines include the application of safety analysis techniques at the system level and the use of formal specification methods for the development of software.   The operations of the CBD have been the subject of a case study within the MORSE project aimed at gaining experience of developing software according to the aforementioned standards and guidelines.

The recommendations resulting from the safety analysis on the CBD related to the software are to be reimplemented using the RAISE [5,6] formal specification method.

This paper presents the experience gained in defining the CBD system and the application of hazard identification techniques within the overall safety analysis.

2. Operations of the Clinical Biochemistry Department

The CBD provides clinical and laboratory services for a wide spectrum of on-site and off-site users. A block diagram of the functions performed within the department is presented in Figure 1. Patient samples are received in one of two reception areas where they are given a unique reference number before being prepared for analysis. Details of the tests required and the patient demographics are entered into one of three terminals which are connected to a file server running the database and network software programs.

The file server is connected to a number of work-stations, which are bilaterally interfaced to computers which in turn control large capacity (7000 tests/hr) analyzers. Patient details and test request information is transferred electronically from the work-stations to the analyzers which perform the analyses required along with quality control checks before returning validated results via the work stations to the file server hard disk. Further quality and validity checks are performed before test results are printed on hard copy for dispatch to clinical staff. The data are finally archived.

3. Safety Analysis within the Morse Project

An improvement in the safety of the overall laboratory operation was an important objective of the MORSE project. To this end, it was necessary to look at the laboratory as a complete system made up of hardware, software and manual operations.

Safety analysis at the system level can typically be carried out in the following stages,
- system definition,
- hazard identification,
- hazard analysis,
- risk analysis and assessment.

The exact requirements and degree of detail to which the analysis is conducted is likely to be affected by factors such as the safety criticality of the system being considered, the financial and human resources available, the stage of development and the timescale of the project.

The need for improvements to the system design and operation are usually identified during all stages of the safety analysis.

The experience gained in the system definition and hazard identification stages of the project are discussed below.

4.   System Definition

The first stage of the safety analysis of the CBD was to produce a clear definition of the system under consideration, its boundaries and its intended mode of operation. Past experience of applying safety analysis techniques in other industries indicates that the techniques can be best applied if the system is described in terms of a number of related modules (hardware or functional blocks) on a flow diagram. The existing documentation at the laboratory did not describe the system in this form and therefore a new representation of the system was produced. This representation, aimed at capturing the main activities of the hardware installed, sample handling procedures etc. (human - computer, computer - computer, computer - black box, black box refers to a computer hardware and software package sold to the laboratory by an external supplier).

A description of the CBD system was therefore produced in the form of,
i)   serum sample flow diagrams,
ii)   data flow diagrams,
iii)   functional block diagrams,
iv)   hardware interconnections,
v)   descriptive text (purpose of each module, inputs and outputs, description of its function).

Example representations of the system are presented in Figures 1,2.

It was then necessary to investigate possible failures and potential consequences in a systematic manner. This was carried out through a Failure Modes, Effects and Criticality Analysis (FMECA), and Hazard and Operability (HAZOP) Study [7,8]. The development and use of each of these techniques is discussed below.

5.   Failure Modes, Effects and Criticality Analysis (FMECA)

FMECA is a technique that can be applied to a system which can be broken down into individual components. The components can be hardware blocks or functional blocks. The methodology requires the assessor to have a clear understanding of the function of each component along with all the inputs to and outputs from it.

The failure modes of each component can then be investigated in a systematic and rigorous manner to establish the causes and the effects of the failure. This information is recorded on a form which is designed to collect information to establish,
-       how each component can fail,
-       what the causes of failure are,
-       what the effects of failure are,
-       how critical the effects are,

- how often the failure occurs.

5.1 Experience of Applying FMECA To the CBD System

Making a judgement about the criticality of an identified failure mode has been the subject of further development within the project in order to ensure that the team performing the FMECA had a consistent approach. A criticality rating system was developed to score the attributes of data in a representative way. It was thought that "units of data" in the laboratory environment comprised integrity, flowrate and the effects of these on the system as a whole. Account was not taken of factors external to the laboratory such as the state of the patient and whether the doctor had made a correct diagnosis.

The attributes of data were scored as follows,

Integrity Rating (A)	Degree to which integrity is lost for individual unit of data (categories 0,1,2),
Flowrate rating (B)	Delay to flow of data through the component being investigated caused by the failure of that component (Category 0,1,2).
System Effects (C)	Likely effect on data leaving the overall system taking into account any recovery mechanism (Categories 0,1,2)
Failure rate (D)	Frequency with which the failure is likely to occur (Category 1,2,3)

The criteria for scoring 0,1, or 2 - was set such that any score represented approximately equal importance within each of the attributes A,B or C.

All the above aspects have been combined together in the following manner to establish a total criticality rating for the failure mode identified.

$$(A + B + C) D$$

The appropriateness of the rating system can only be assessed through the use of engineering judgement.

The criticality rating thus established was then used to prioritise the hazards identified and the recommendations made for improving the design and operation of the system.

5.2 Results From FMECA

Table 1 is an extract from the full FMECA that was undertaken and only shows the details of the criticality analysis. It refers to one of two external hard disks attached to the file server, for which three failure modes are shown. The second entry is described as follows: if one hard disk fails, due to hardware error (failure cause), the affected drive cannot store data (local consequences) and since the disk is "mirrored" by an identical disk to which it is paired, information from the backup disk is used automatically. Consequently, there is no data corruption (integrity rating 0) and no impairment of data flow (flow rate rating 0). There are no system consequences and this failure occurs less than once per month (failure rate 1), giving a total rating of 0.

In example 2, sample tubes receive bar-code labels bearing critical information about the identity of the patient from whom the sample was drawn. Four failure modes are shown only the first of which will be explained in some detail as follows. If the wrong label is attached (failure mode) due to human error (failure cause), the sample will be wrongly identified (local consequences). The data is corrupted (integrity rating 2) but data flow through the component is unimpaired (flow-rate rating 0). The error may be identified at a later stage (system rating 1) and such problems occur approximately once a week (failure rate 3) resulting in a total rating of 9.

These examples appear to reflect engineering intuition about the relative criticality of the failure modes discussed.

6. Hazard and Operability (HAZOP) Study

The HAZOP study technique was initially developed in the 1960's and 1970's within the Mond Division of ICI for application within the process industries. The technique consists of a critical and systematic review of the system under consideration by a multi-disciplinary team. The review is coordinated by a chairman who leads the investigation into sections of the system with the aim of establishing their design intent and the ways in which that section can deviate from the defined design intent. The deviations from design intent are investigated in a systematic manner by application of a number of guide words such as more, less, no. These are described in more fully in [8].

6.1 Experience of Applying HAZOP to the CBD System

Past experience in applying the HAZOP technique is primarily in the process industry where the guide words are used to investigate deviations in parameters such as temperature, pressure and flow. Clearly these parameters are not relevant to the operations of the CBD. It was therefore necessary to apply the basic guide words to activities (e.g. input password) or functions (e.g. create Print file from Day file) being performed.

6.2 Results From HAZOP

Table 2 is an extract from the HAZOP study and presents an example of a hazard identified within the electronic data transfer operations. This situation refers to the transfer of test requests from the file server (Figure 2) to one of the analyzer workstations. A program, called from DE1, copies records relevant to the analyzer to a print file. If the hard copy print-out is considered satisfactory by the analyzer operator, a further transactional file is created from which the records are downloaded to the appropriate analyzer workstation. Once the records have been downloaded, all records pertaining to this analyzer are flagged in order to prevent downloading of duplicate records.

In the example presented in Table 2 the activity being considered is the flagging of fields within the dayfile after the work has been accepted by the operator. The purpose of this activity is to prevent the same work requests being sent forward with future batch transfers.

A scenario has been identified where more records are flagged than should be when the operator accepts the printout of work. The cause of this is additional work for the analyzer in question being added to the dayfile from the other data entry terminal (DE2) after the request for the printout has been initiated. This will mean that some samples will remain untested.

The software to carry out the above operation can be easily modified to avoid the loss of work requests caused by this scenario.

Example 2 is an extract taken from the HAZOP on the LIMS hardware which is shown in Figure 3. All node terminals are connected by cables to a junction box (multi-function access unit) which is in turn connected to the file server computer. The file server is a stand alone PC with external hard disks which are mirrored by Novell networking software.

In this example the deviations from design intent of the junction box have been considered. A scenario has been identified under the guide work "no" where the failure of data flow through this section of the system would result in the whole network becoming inoperable. Action has been recommended to investigate contingency measures in place to recover from such a failure.

7. Discussion

It was found that the degree of scrutiny of the system possible as part a hazard identification exercise depended on the amount of detail included in the system definition. A number of revisions of the system definition were necessary to enable assessment to an adequate level of detail so that useful results could be obtained from the study.

The importance of system definition is stressed particularly since the application of safety analysis in other such laboratories will require a considerable amount of effort to define the system in a form that will facilitate such analysis.

The application of FMECA and HAZOP to the operations of the CBD was generally regarded to be a successful exercise. Each of the techniques has its advantages and disadvantages which are discussed further below.

The main advantage of the FMECA as compared with the HAZOP was that it enabled a criticality rating to be applied to the hazards identified and hence eased prioratisation of the recommendations. In the case of the CBD the analysis was carried out by a number of people working on their own. It was found that the progress of the analysis and degree of scrutiny depended significantly on each person's understanding of the system. It may be possible to improve the progress rate and the degree of detail by employing small teams of personnel according to their knowledge of the components being investigated.

The HAZOP study was performed by a team of people with varied knowledge about safety analysis techniques and the CBD system. In this case it was found that the written system definition could be considerably enhanced by the knowledge of the team members during the study. The HAZOP approach was found to allow much more free thought about potential failures and hence was better at picking up failures resulting from combinations of events.

The role of the chairman in stimulating thought within the HAZOP team was seen as critical in the application of the technique. The lack of previous experience of applying HAZOP in a clinical laboratory environment resulted in a number of teething problems to begin with. These were primarily caused by the use of inappropriate guide words.

The approach of using the basic set of guide words applied to activities and functions appeared to work well. This approach can be recommended as a good starting point when applying HAZOP in an industry where its use is untried.

8.   Where Next

The application of the safety analysis techniques has been very successful on the CBD. It is however recognised that a considerable amount of effort went into the preparation stages where the system was defined in a form that would easily facilitate safety analysis.

With this constraint in mind it is felt that the use of safety analysis techniques can be more readily accommodated in new developments where the system can be defined in the required form at the outset as part of the overall documentation of the project.

It is thought that the use of HAZOP is more conducive to existing systems where the system definition can be enhanced during the study by the knowledge of the HAZOP team members.

9. References

1. Ministry Of Defence. Hazard Analysis and Safety Classification of the Computer and Programmable Electronic System Elements of Defence Equipment - Interim Defence Standard 00-56/1. UK Ministry of Defence, April 1991.

2. Ministry Of Defence. The Procurement of Safety Critical Software in Defence Equipment - Interim Defence Standard 00-55/1. Ministry of Defence, April 1991.1

3. IEC/TC65A (Secretariat) 122. Software for Computers in the Application of Industrial Safety-Related Systems. Geneva 1992.

4. IEC/T665A (Secretariat) 123. Function Safety of Electrical/Electronic/Programmable Electronic Systems: Generic Aspects. Part 1: General Requirements. Geneva 1992.

5. S. Brock and C.W. George. RAISE method manual. Technical Report LACOS/CRI/DOC/3/v1, CRI, 1990.

6. K.E. Erikson and S. Preh. RAISE overview. Technical Report LACOS/CRI/DOC/9/v2, CRI,1990.

7. BSI. BS5760 : Reliability of Systems, equipment and components, Part 5. Guide to failure modes, effects and criticality analysis (FMEA and FMECA). British Standards Institute, 1991.

8. A Guide to Hazard and Operability Reviews by the Chemical Industries Association (CIA).

Failure Mode	Integrity rating	Flowrate rating	System rating	Failure rate rating	Total rating
Example 1.					
Bad sectors.	0	0	0	1	0
Mechanical failure	0	0	0	1	0
Driver board failure.	0	1	1	1	2
Example 2.					
Wrong label attached.	2	0	1	3	9
Bar code printer failure.	0	1	1	1	2
No label attached. No number on tube.	2	0	1	2	6
No label attached. Number written on tube.	2	0	1	2	6

Table 1 — Examples of Hazards Identified by FMECA on the CBD.

Item Description	Guide Word	Deviation	Causes	Consequences	Action Required
Example 1.					
Flag records in DAYFILE	More	More flagging in DAYFILE	Records entered after print sequence started but before flagging stage.	Some works requests will not be transferred forward to the analysers.	Consider means of preventing this.
Example 2.					
Junction box.	No	No flow	Failure of star topology junction box.	No communication between nodes and file server and with each other.	Review procedures for removing connections from failed box to the other junction box that has spare ports. Review redundancy of junction box as it stands.

Table 2 — Examples of Hazards Identified by HAZOP on the CBD.

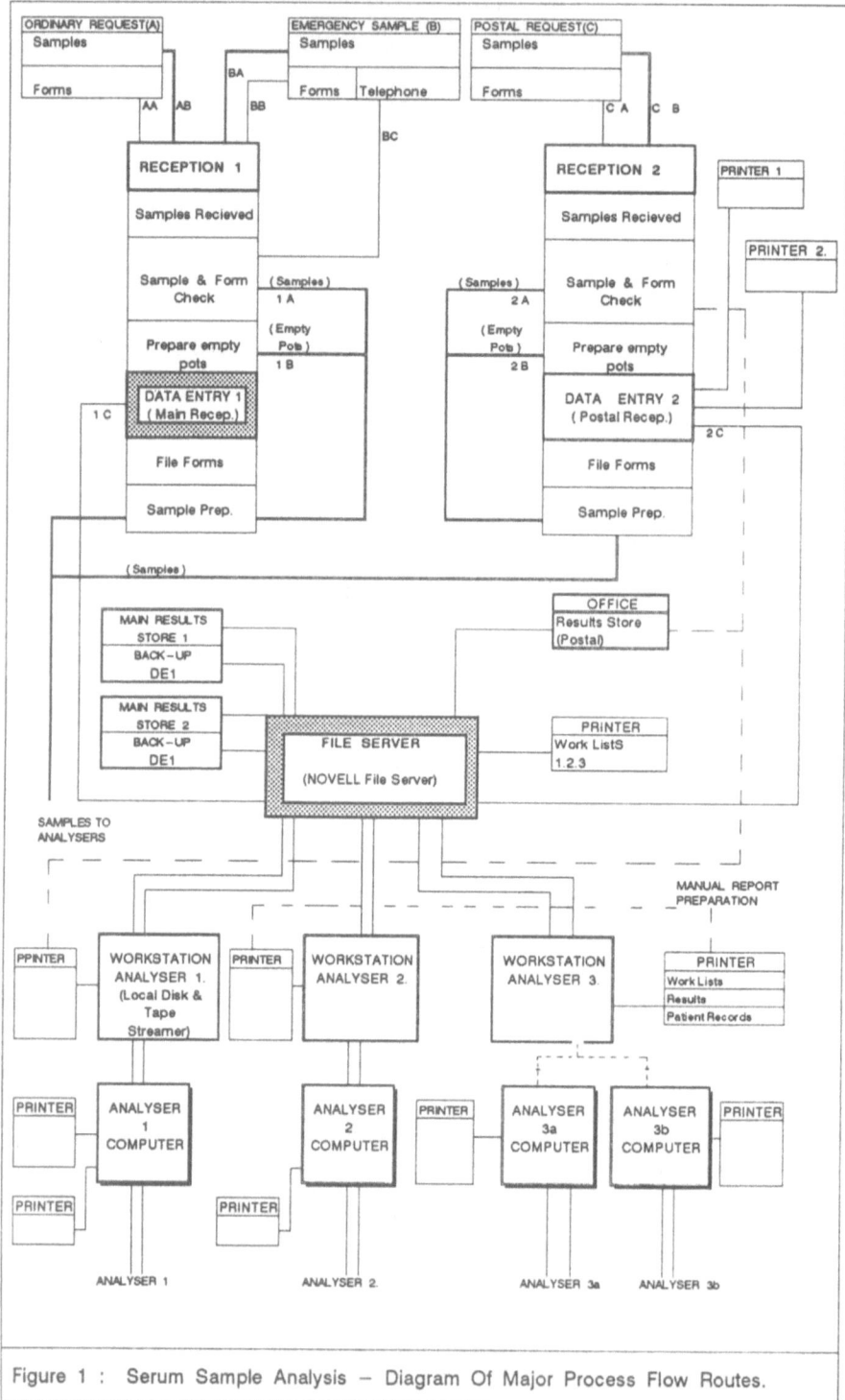

Figure 1 :   Serum Sample Analysis — Diagram Of Major Process Flow Routes.

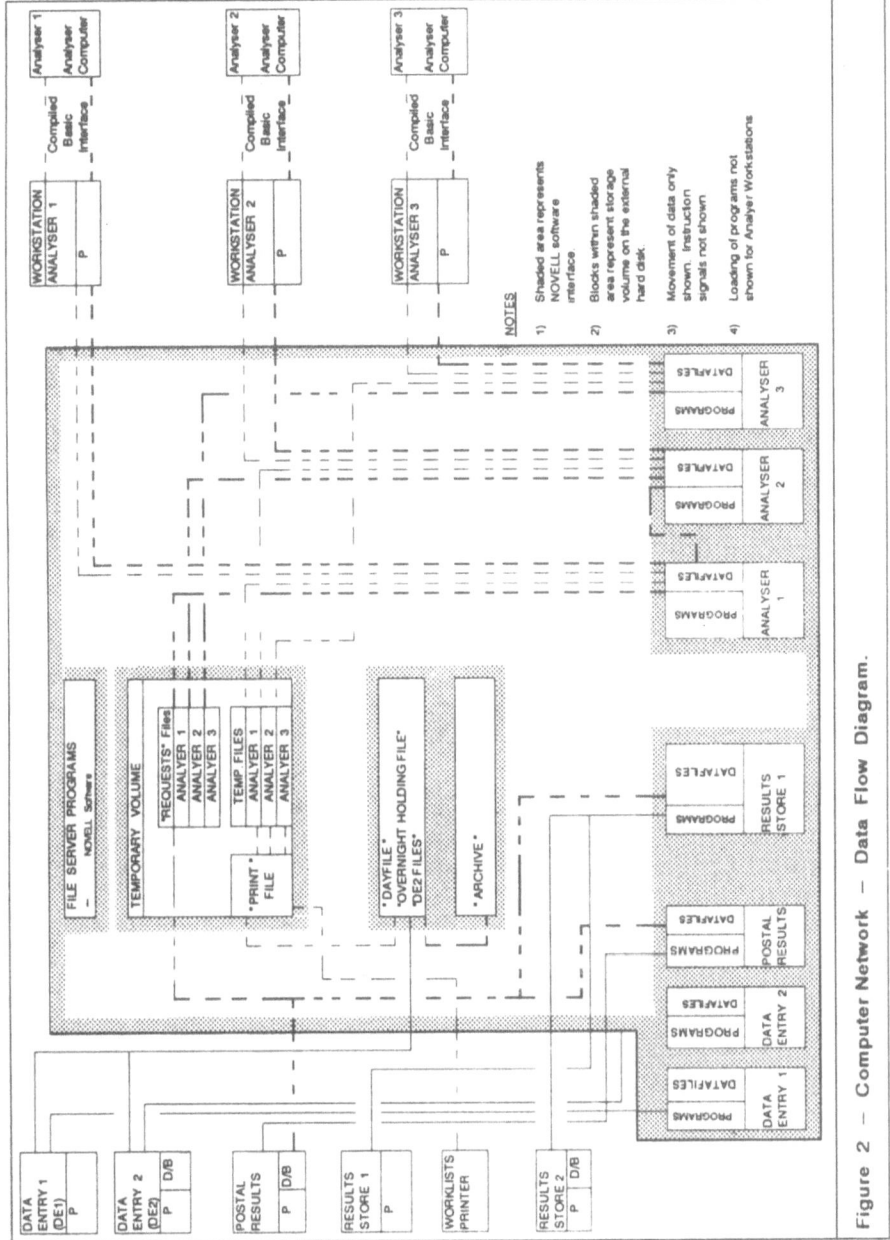

Figure 2 — Computer Network — Data Flow Diagram.

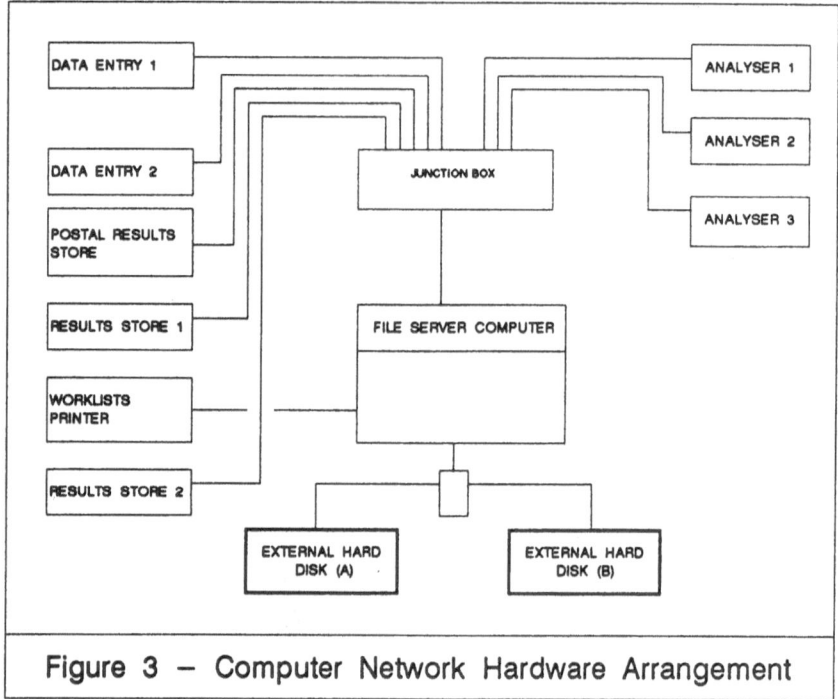

Figure 3 – Computer Network Hardware Arrangement

The Benefits of SUSI: Safety Analysis of User System Interaction

M F Chudleigh and J N Clare
Cambridge Consultants Limited
Cambridge, England

1 Introduction

The use of computer based systems has many advantages, including increased functionality, increased flexibility and, hopefully, ease of use. Because of these advantages their use is increasing dramatically, including applications where failure could have an adverse impact on safety. It is important to remember that most such systems have contact with human users and are used in organisations where there are set procedures and ways of working.

For industries such as petro-chemical, rail transport and air transport the consequences of a failure could be catastrophic and they have realised that evaluating and controlling the risk to humans arising from failures of such systems is vital. The process of evaluating the possible safety failures is known as hazard analysis. One particular method of carrying out hazard analysis is the Hazard and Operability Study (HAZOP) [1].

In recent years there has been an increasing realisation that considering the place of the human in a system (human factors) should be an important part of the system design. The critical issue is that as systems become more complex the human operators are increasingly prone to 'error'. In this case error is used to describe behaviours that were not the designer's intent. In part, such behaviours occur because the operators are not able to comprehend their role with the system. However, a key part is that the operators choose to do something different to the designer's intent because of new working procedures, or because the system did not perform as the designers intended. However, once an incident occurs then we normally identify human error as the root cause [2].

When we consider the design and implementation of such computer based systems we find that at least three specialists are involved: those who design functionality into systems (application designers); those who design the user interface (human computer interaction (HCI) specialists); and those human factors specialists who examine how particular tasks are carried out.

In order to build effective, safe systems it is clear that all three specialists must be able to communicate with each other. However, the reality on most systems is that the three areas all have their own specialist vocabulary and models of the system, and they tend to work independently of each other. In addition, industries which have an established record of building safety critical systems are likely to have specialists in safety. These personnel, again, have their own particular jargon and often may not be familiar with the techniques of developing computer based systems. This lack of co-ordinated coverage during system development has the potential to lead to hazardous situations.

We have developed an approach that we believe shows promise in dealing with the above problem. The approach is called SUSI, standing for Safety analysis of User System Interaction. SUSI comprises two parts:

- A common representation of all entities in a systems so that   communication between specialists is enabled; coupled with

- a structured, hazard analysis procedure which addresses features that are particular to human-machine interactions

The remainder of this paper describes the approach and gives examples of its use in two different applications at different points of the system lifecycle. One is of a new medical system design and the other is of an operational maritime system.

2   SUSI: A Common Representation

In describing our work in developing a common representation for systems we address three main areas. First, the key realisation that a common representation is both necessary and possible: it underpins the majority of our work in user system interaction. Second, the applicability and limitations of the chosen approach.

2.1   A Common Representation is Necessary and Possible

In building systems, a variety of expertise is necessary. Software application designers tend to treat the user as a separate entity from the system and leave an external source/sink labelled HCI. The building of user interfaces is then treated as a separate design activity given to HCI specialists with particular techniques. Human Factors specialists often work apart from system developers and concentrate on the activities carried out by humans (task analysis) with its own vocabulary [3]. This analysis may well be aimed at defining manning levels or training requirements.

In developing new systems, these specialists all tend to build their own models of the system which are not easy to correlate with the other models. In addition, the people who want the system, those who will use it and those who might have to judge its safety all need to have an understanding of the system. However, they all do need to communicate with each other to ensure they share a common understanding of the system. There needs to be a stable system view from each of the perspectives. Gaining that understanding is not easy with a plurality of system models and it is difficult to imagine the consequences to the rest of the system of changes in one representation.

An analysis of the features that need to be described in understanding human activities [4] and those used to describe a software system [5] show a commonality of entities:

        human task   software process

human information flow	software dataflow
human interactions	software control
human documentation	software database

With the existing widespread use of software dataflow analysis and description tools there would thus appear to be a basis for a common representation. In adopting a dataflow and process model for the human components of a system we can now generate an integrated representation of the overall system. Using this we can explore the consequences of failure in a consistent manner across the whole system. In Figure 1, part of a medical imaging system is shown to illustrate the convention used in this type of representation. The key components are circles to represent a process (human or machine), a solid line is a dataflow, a dashed line is a control flow (start, stop etc), and two parallel lines represent a data store (we show displays as data stores because data may be written to a screen, but there has to be an explicit human process to read the data).

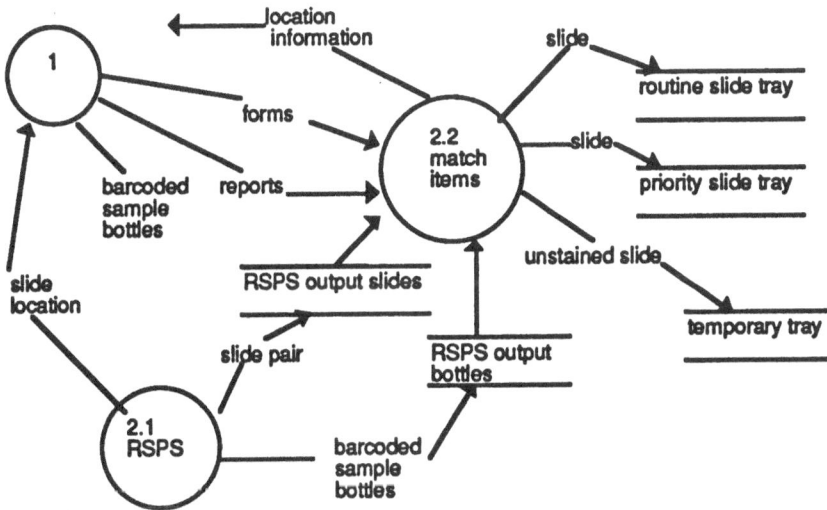

Figure 1: Part of Decomposition of Preparation and Scanning Process

2.2 The Scope of the Modelling Approach is Wide, but There are Limitations.

The approach works well as long as the human activity to be modelled is primarily information intensive. Thus activities such as decision taking, information transfer and classification/sorting are good matches, such as in the medical example above. The vast majority of human work falls into this, information intensive, class but there are limitations. The approach is not effective when the human processes have significant motor skills or introspective reasoning components. An example where motor skills take a prominent part might be the interaction between the driver of a Formula 1 racing car and the active computer systems used to manage many of the car mechanical functions.

We have found that a dataflow and process model for both the human and the computer parts of a system allows us to generate an integrated representation of the overall system. This model can then be used to explore various properties of the system. A major advantage we have found is that such models are easy to understand by non-computer specialists and that users of systems are able to comment on and critique designs at a very detailed level.

In the next section we show how the consequences of failure may be examined in a consistent manner across the whole system.

3 SUSI: Hazard Analysis Using Amended HAZOP

An approach to assuring safety of systems has been well established over many years. However, only in more recent years has the potential for computer systems to fail in ways that might impact safety, been recognised formally in standards work [6,7]. To facilitate the discussion two definitions from Reference [6] are given:

A hazard is a physical situation with a potential for human injury.

Risk is the combination of the frequency, or probability, and the consequences of a specified hazardous event.

Both emerging standards and existing practice in established industries uses the same basic lifecycle approach to addressing safety which may be summarised as follows:

* System definition generating a concise and complete description of the system under review
* Hazard analysis to identify the potential for hazardous events
* Risk analysis to judge the safety risk of the system as defined
* Risk acceptability: determination of whether the risks are acceptable
* Activities, if necessary, to modify the system definition or include additional measures in order to reduce the risk to an acceptable level

Hazard analysis is thus a key step in the process and a number of structured procedures have been developed to provide confidence that the hazard analysis is complete and thorough. This section introduces one of the main hazard analysis methods, HAZOP; explains how we have extended the HAZOP to address user system interaction; and describes some advantages and limitations of the approach.

3.1   What is a HAZOP?

The full name of HAZOP, Hazard and Operability Study, says a great deal about its purpose. It is to ensure both that necessary features are incorporated in a design to provide for safe operation and that features are avoided which could give undesirable outcomes (ie hazards). The technique was developed by ICI in the late 1960's and has grown to be well established in the petrochemical industries. An excellent introduction to the technique is given in [1].

In the process industries it is usual to describe plant designs in the form of Piping and Instrumentation Diagrams (P&IDs). The HAZOP is carried out by a small team with the following members: team leader; team secretary; personnel who have detailed knowledge of operation of similar systems; personnel who have detailed knowledge of the design intent of the system; specific technical specialists as necessary. The team work logically through the P & IDs, examining deviations from normal operation asking "can the deviation happen?" and if so, "would it cause a hazard?" (a hazard could be things such as a fire or release of toxic material). To guide the process a series of guidewords and potential deviations are used. Thus for liquid in a pipe, a relevant guideword is "flow" and potential deviations are "high, low, no, reverse". For fluids another guideword is "pressure" with deviations "high, low". In theory, each guideword/deviation should be applied to each process line and vessel . In practice, an experienced team leader will judge the correct detail of questioning for each area.

3.2 Modifying HAZOP for User System Interaction

A critical element of the HAZOP process is the choice of parameter keywords and guidewords for deviations. We believe the guidewords for petrochemical plant have limitations when addressing computer based systems and user system interaction. Other work by Cambridge Consultants and their parent company Arthur D Little has led to modifications of the HAZOP approach for computer based systems and this is reported in [8] and [9].

For our work with user system interaction we have developed a vocabulary of discrete entities which have associated deviations. These are shown below

Entity	Deviation	Comments
Process	Failure	Execution fails, data is used inappropriately
	Error	Process algorithms wrong or contain flaw(s)
	Wrong Process	Wrong process selected or human short cut
	Interrupted	Process not restarted appropriately
Data Flow	Corrupted	Data changed in transit
	None	Data does not exist
	Wrong source/sink	Data taken/sent from/to wrong place
Data Store	Corrupted	Data changed in store
	None	Data not stored or not found
Control Flow	Corrupted	Wrong control signal
	None	Control does not exist
	Wrong source/sink	Sent to/received from wrong place

We have found that the traditional HAZOP team structure and general approach can be used without change. However, it has been found essential to have independent technical personnel who are experienced in system design and human factors as part of the HAZOP team.

3.3 Advantages and Limitations of the Modified HAZOP

The main advantages of the approach are: it is done by a team; it gives a top-down approach to the system; and it can be used both on new system designs and on existing systems.

The team approach brings a variety of expertise and viewpoints onto a common problem and concentrating on hazard identification leads to productive sessions. Also, the team, by providing a variety of viewpoints helps to avoid excessive investigation of non-credible hazards.

The top-down approach, examining the whole system first, allows the homing in to key issues based on the potential hazards and is very good at assessing system-wide implications. The approach of looking at deviations from design intent, then their causes and consequences encourages exploration of non-obvious interactions both of the user/operator with the automated system and of the automated system with its hardware environment. The use of a HAZOP provides guidance towards the most critical areas to concentrate on in any subsequent low level investigation.

The HAZOP fits naturally at all stages of the life of a system, from concept through to operation. In later sections we give an example of use during a medical system conceptual design stage and another example of analysing an existing operational maritime system.

There are, however, limitations to the HAZOP approach. We have found that straightforward application of the deviation guidewords is not sufficient: the process relies on the experience and intuition of the team members (especially of the independent technical experts). Further, the choice of an experienced HAZOP leader is key. It is the leader who controls the pace of the analysis and it takes significant experience to guide the team discussion to the most critical areas while still ensuring full coverage within usually tight time constraints.

4 The Use of SUSI in a New System Design: a Medical Laboratory System[1]

This section is divided into three parts: first a brief description of the system to partly automate screening of cervical specimens, then an outline of the dataflow description and finally some of the results from the HAZOP analysis.

4.1 The Medical Imaging System

The Human Genetics Unit (HGU) of the Medical Research Council in Edinburgh have produced an experimental version of a semi-automated system for screening of cervical smear samples to identify abnormalities which might lead to cancer. The

---

1.The work carried out here was part of a collaborative project between The Centre for Software Engineering Limited, Cambridge Consultants Limited and the Human Genetics Unit of the Medical Research Council.

HGU have been working closely with the Department of Pathology of the University of Edinburgh to carry out trials on the system and both parties were closely involved in the work presented here.

Within a cytology laboratory the equipment would consist of two major parts, a robot slide preparation system (RSPS) and a slide scanning system (SSS). The RSPS takes in sample bottles submitted by clinics and transfers a part of the material as a monolayer sample onto slides.

The SSS is an image processing system which inspects objects on the slide and classifies them into various categories. Where abnormal objects are identified, the system stores digitised images for subsequent human inspection. Both the above systems are supported by a computer based system providing overall administration and interfaces to the laboratory main computer which stores patient records.

4.2   Development of the Dataflow Description

The full system description is far too long to be included here: we give simplified versions of two levels of the description to illustrate the method. The system context identifies the complete system under consideration and its principle interactions with the external world. Here there are two external entities; the clinics or surgeries which collect samples and have reports returned and, within the laboratory, the archives where reports and samples are stored. Note that the total screening system includes the human administrators, technicians and medical personnel who interact with the machine sub-systems.

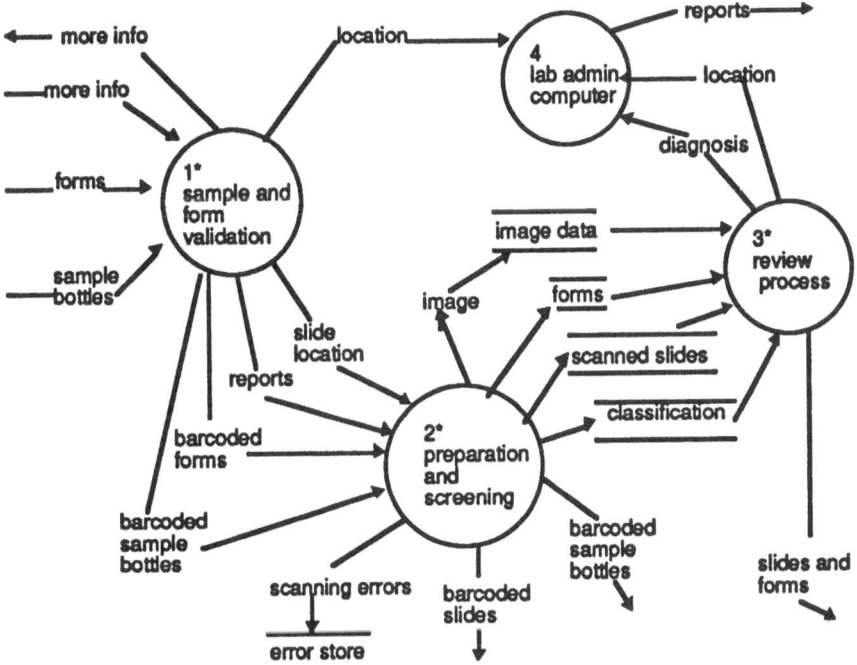

Figure 2: Top Level Decomposition of Medical System

The top level system is then decomposed whilst preserving the external data flows. Figure 2 shows the division into four processes: sample and form validation, preparation and scanning, review process, and laboratory computer data management. Figure 1 (shown earlier) gives the decomposition of the preparation and scanning process. In discussing the design of the system it became clear that process 2.2 Match items was a key and human intensive process. It was necessary to include an explicit process because the automated part of the system had few opportunities to cross check for errors in the pairing of patient forms and patient sample bottles. As the design proceeded it was also recognised that there were other quality control and sorting processes that could also be carried out at this stage of the overall process. As a result the Match Items process is an intensive human process which achieves the following: cross-checking, visual quality assessment of prepared slides, attachment of label to each slide and assignment of samples for priority processing where there are clinical indications. To aid the human carry out the process a screen is used to display status information and a barcode reader pen is used to cross-check barcoded items against patient details.

The dataflow design was produced by a team of a consultant (one of the authors), a cytopathologist and a member of the design team of an experimental automated system. The team approach ensured that the various viewpoints on the design were captured in an effective manner and partitioned activities between human and machine in a way that was readily understandable to all those involved. This level of common understanding had not been achieved previously despite regular and detailed liaison.

4.3   Application of the modified HAZOP

The HAZOP team included representatives of both the design team and the intended users of the system. The HAZOP leader role was taken by a member of Cambridge Consultants who had no direct involvement in generation of the dataflow design but was aware of the overall goals of the project. The design consultant provided the technical expertise in system design and human factors.

The HAZOP process focused on two types of issue. One of these related to design aspects which would enhance operability and safety, the other was from the experience of current staff who were able to identify practical and procedural constraints which would impact the viability or safety of the system. Below we give some of the observations recorded during analysis of the Match Items process.

Recommendation 1. Bar coded form wrong through mismatch of names. Wrong patient gets matched to slide. Do double check.
Recommendation 3. No report/label because detached. Leads to delay. Print report and label on same form.
Recommendation 4. Wrong slide pair because of incorrect location of slide. Wrong patient could be matched. Check barcode and accession number match.

5 The use of SUSI in an Existing System: Maritime Control

A hazard analysis has been carried out on the navigation systems for a vessel. This was part of a series of investigations into hazards associated with coastal traffic operating in crowded waters. The object was to identify areas where hazards could be expected to have significant effects. The SUSI methodology was used as the basis for the analysis. The first stage was the production of dataflow descriptions of the activities of watch keeping officers. This was done on a series of sea voyages during which the dataflow models were constructed and reviewed by experienced, seaman officers. Figure 3 shows the first level decomposition of processes, most external data flows have been omitted to aid intelligibility. As can be seen the principal processes are navigation, ensuring that you are where you should be; collision monitoring to identify other mobile hazards; and conning, giving orders for changes to ship speed and course. Each of these processes were decomposed to lower levels for the actual analysis.

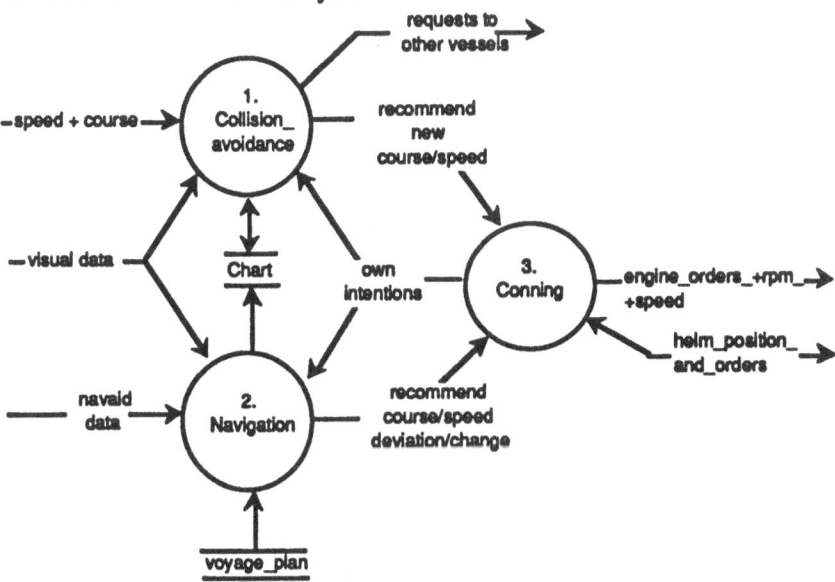

Figure 3 Decomposition of Ship Navigation

On completion of the data flow analysis a HAZOP team was assembled consisting of a HAZOP leader, secretary, dataflow/human factors expert and experienced navigating officers. At the lowest level of each branch of the decomposition the guide words were used for every data flow, process, data store and control flow. The output was recorded using a HAZOP database tool. The HAZOP output was revised post team session in order to ensure consistency of responses and to cover areas that the HAZOP team were unable to detail at this time (ie lack of detailed knowledge of the operating procedures).

The next stage was to develop fault trees with top events of collision, grounding etc. The fault trees were constructed using available data on navigation aid failures and human error rates. Human error rates were factored by stress level, complexity and experience. The human error rates used were for single events and it was noted that normal navigational procedures require multiple independent sightings. If errors are suspected then new sightings are taken. For example a visual fix requires the bearing at regular times of three fixed points. All three have to coincide for an error free fix to be recorded. In many cases this would then be cross checked with a radar fix or a navigational beacon.

The SUSI approach was both necessary and valuable for this task as there were no simple descriptions of the navigation tasks or use of navigational aids. There were a series of operating procedure manuals, the training materials for ships navigation officers and various instructions to mariners issued by regulating bodies. In order for the HAZOP team to operate this mass of data had to be compiled into an easily digestible form which could also be validated against actual operational practice. The development of the dataflow description provided a clear and easily understood view of these activities (one training officer has decided to use them as navigational training material).

The overall process of hazard analysis was completed in four weeks, from boarding the first ship to delivering the final report.

References

1   A Guide to Hazard and Operability Studies. Chemical Industries Association Limited, 1987.
2   Westrum R. Technologies and Society. Wadsworth Inc. 1991
3   Stammers R, Carey M, Astley J: Task Analysis. In: Wilson J, Corlett E (eds) Evaluation of Human Work. Taylor & Francis 1990, pp 134-160.
4   Wilson J, Corlett E (eds) Evaluation of Human Work. Taylor & Francis 1990
5   Hatley D, Pirbhai I . Strategies for real-time system specification. Dorset House, 1988.
6   Functional Safety of Electrical/Electronic/Programmable Systems. Generic Aspects. IEC 65A (Secretariat) 123. 1991
7   Interim DefStan 00-56. Hazard Analysis and Safety Classification of the Computer and Programmable Electronic System Elements of Defence
8   Chudleigh M, Catmur J. Safety Assessment of Computer Systems using HAZOP and Audit Techniques. In: Frey (ed) Safety of Computer Control Systems 1992 (Safecomp '92) pp 285-292.
9   Chudleigh M. Hazard Analysis Using HAZOP: A Case Study. In these proceedings.

Panel Session

TECHNOLOGY TRANSFER BETWEEN ACADEMIA AND INDUSTRY

Moderator: F. Redmill
Safety Critical Systems Club, UK

Issues Affecting Technology Transfer
and
Experience with a Community Club

Felix Redmill

Redmill Consultancy
and Co-ordinator of the Safety-Critical Systems Club
22 Onslow Gardens, London, N10 3JU, UK

1   INTRODUCTION

Achieving technology transfer is a perennial problem. Yet, without it, industry's problems can remain unsolved, and solutions to one problem are not generalised to more or broader applications.

There are two objectives of the paper. The first is to raise a number of issues concerning the transfer of technology from academia to industry. The second is to show how a community club is helping to facilitate technology transfer.

2   ISSUES IN TECHNOLOGY TRANSFER

Some of the typical problems affecting the transfer of technology from academia to industry are listed, with brief notes.

2.1   Traditionally there is a delay between the development of a technology in academia and its implementation in industry. Some say that this can be as long as ten years. Many industries have no formal contacts in academia and they do not know how to find out what new technologies are available. Moreover, academia is not noted for effective marketing of its products.

2.2   The terms in which academics typically present their findings are those understandable to other academics, and not to industrialists. This contributes to the delay in the recognition and implementation by industry of new technologies.

2.3   Even when transfer does occur, it is usually to a small sector of industry, and the majority of those who need the technology do not receive its benefit.

2.4   Feedback from industry to academia is not standardised. It is seldom good and almost always slow. Thus, potentially useful but flawed or unready technologies are rarely corrected.

2.5   Communication between the users of a technology, particularly across industrial sectors, is at best poor and often non-existent. Thus, the lessons learned by one company, or industry, are not communicated, and others must undergo the learning curve for themselves. When the technology is flawed, finding the problems and suffering their consequences must also be repeated.

2.6   When there is effective marketing of new technology, for example by consultants, it is not uncommon for inappropriate technologies to be transferred. Consultants often transfer their pet technologies rather than those most suitable to the problem in hand.

2.7   Typically, technologies are developed for specific applications in specific industry sectors. Often, however, they are proposed as being effective over a much wider range of applications. When they are found to be ineffective, those who have used them may be left suspicious of academic innovations.

2.8   Much technological development in academia is directed towards obtaining degrees and preparing publications rather than to getting the technologies right. The resulting technologies may then not be properly refined. With luck, such technologies do not get transferred. When they are transferred, they are often not suitable for use, and industry wastes time discovering this.

2.9   In the normal course of events, a great deal of technology transfer is of proven technologies to new environments rather than of the introduction of new technologies.

3   THE SAFETY-CRITICAL SYSTEMS DOMAIN

Safety-critical computer systems are a relatively new phenomenon, but the domain is expanding rapidly. Moreover, it is not growing out of nothing, but is the coming together of a number of other fields. The two main components are safety engineering which, though not new, has traditionally been restricted to a small number of industry sectors, and software and systems engineering, which is itself a new field. In addition, safety-critical systems are expanding into almost every sector of industry, and demanding input from other specialisms, such as human factors and quality management. This leads to the following observations on the need for technology transfer in the safety-critical systems domain.

3.1   There is an urgent requirement for the development and transfer of new technologies to meet the particular safety demands of safety-critical systems.

3.2   Software and systems engineers are typically not familiar with safety engineering, and there is a need for the transfer to them of existing safety technologies.

3.3   Similarly, there is a need for safety engineers to become familiar with computer systems and software technologies and practices.

3.4   The knowledge and technologies of the human factors domain, on such issues as human dependability and human-computer interaction, urgently need to be assimilated in the safety-critical systems community.

3.5   The need is urgent for an improved awareness, in all sectors of industry, of the application of safety-critical systems.

3.6   There is the need for the transfer of existing technologies and practices between industries.

3.7   If the development and transfer of technologies is to keep pace with expansion in the domain, there needs to be focused research, and easily accessible communication links between industry and academia. Improved communication links would also allow transferred technologies to be improved and made fit for purpose.

3.8   Mechanisms are needed for communication across industry sectors, so that experiences of new technologies can rapidly be communicated.

4   SETTING THE SCENE FOR A COMMUNITY CLUB

Technology transfer may take place in a number of ways. For example, individual companies may make contact with one or more universities or research establishments; more and more, universities are encouraging the setting up of 'spin-off' companies for the purpose of marketing and selling their technologies; professors and other researchers, in their roles as consultants, are active in the transfer of technologies, and they carry considerable responsibility in choosing what they recommend to their clients; reports and publicity may catch the attention of industry. There is also the possibility of achieving technology transfer via a community club.

In 1991, in the UK, the British Computer Society (BCS) and the Institution of

Electrical Engineers (IEE) were contracted by the Department of Trade and Industry (DTI) to set up a community club for technology and information transfer, and for raising awareness, in the safety-critical systems domain. The two Societies were assisted by the Centre for Software Reliability at the University of Newcastle upon Tyne, who in turn engaged the current author to be the club's Co-ordinator. The club was launched in May 1991 and the high level of interest in it was demonstrated by the attendance of 255 delegates at the inaugural meeting in July of that year. By May 1993, there were 1682 members, of whom 130 were from outside the UK.

5   THE CLUB'S OBJECTIVES

The club exists to facilitate information and technology exchange, and to increase awareness, in the safety-critical systems domain. It is recognised that in order to be successful in this, the club must gain access not only to engineers and technicians but also to managers with decision-making responsibilities.

By facilitating communication across industry, the club's objectives are to:
- Increase the rate of dissemination of useful technologies;
- Prevent the spread of flawed technologies by the rapid communication of experience;
- Improve the industrial testing of new technologies;
- Bring industrialists together to plan feedback to academia and to coordinate the sponsorship of research.

By facilitating communication between academia and industry, the club's objectives are to:
- Improve the choice and application of technology;
- Accelerate the feedback to academia of experience in the use of technologies;
- Improve safety-critical computer systems which are supplied to industry;
- Facilitate the targetting of research;
- Accelerate the correction and improvement of flawed but useful technologies.

It is also the club's objective to provide a platform for reporting on research into new technologies and experience in their use in industry.

6   SUCCESS IN MEETING THE OBJECTIVES

6.1   Newsletter

A newsletter, of at least 10 pages, is published three times per year and distributed to all members. Typical contents are:
- Feature articles on safety-critical systems matters;
- A calender of events on safety-critical systems;
- A calender of events on related issues;
- Calls for papers for future conferences;
- Reports on new products;
- Reports on government studies or initiatives affecting safety-critical systems;
- Comments by members on safety-critical issues.

6.2   Seminars

By June 1993, the club had held nine seminars. Of these, seven were one-day and two were two-day events. The topics covered were:
- Inaugural meeting and introduction to safety-critical systems;
- Requirements for safety-critical systems;
- Education and training for safety-critical systems professionals;
- Safety-critical software and technology in the medical sector;
- Standards for safety-critical software;
- Human factors in safety-critical systems;
- Design for safety and reliability;
- Safety-critical systems in the nuclear sector;
- The safety case.

In the interest of bringing industrialists of all sectors together, the majority of the seminars are on topics which are of broad application. However, it has also been the club's policy to hold one sector-specific seminar each year. In 1992 and 1993, these were devoted to the medical and nuclear sectors respectively. In each of these cases, and particularly in the latter, many other industrialists and academics attended in order to learn the lessons of that industry. Thus, the objective of cross-fertilisation is being achieved.

The speakers at the first five seminars were invited to prepare chapters, based on their presentations, for a book. Twenty-two speakers responded, and the resulting book [1] was published by Chapman and Hall in 1993.

At the nine seminars held so far, the total attendance has been 1283. At each event, the delegates have been asked to complete questionnaires on the quality and value of the seminar and, without exception, the feedback has been positive.

By the end of 1993, two further one-day seminars will have been held, on

'Testing and validation of safety-critical systems' and 'Measurement of safety and reliability'.

6.3   Annual Symposium

The club has initiated an annual three-day symposium, the Safety-critical Systems Symposium (SSS), to be held in February of each year. The first, SSS '93, held in Bristol, attracted 190 delegates. Thus, the total attendance at the first ten club events was 1473 - an average attendance of 147.3 delegates.

The 19 papers presented at the symposium covered a broad spectrum, many reporting on research projects involving collaboration between industry and academia. One of the goals of the club is to provide a forum for the reporting of the results of these projects, and in the years to come the Safety-critical Systems Symposium will provide this platform. The proceedings of the symposium [2] were published by Springer-Verlag.

SSS '94 will be held in Birmingham in February 1994.

6.4   Ad Hoc Activities

The principle of the club's existence is cooperation rather than competition. Thus, the club has participated and assisted in activities not mentioned among its principal objectives. In this respect, it has co-sponsored events, assisted in the organisation of workshops, given advice on safety-critical issues, brought together potential participants of collaborative projects, and given publicity to safety-critical matters. The club continues to offer support whenever appropriate.

7   CONCLUSIONS

This short paper has listed a number of issues in the transfer of technology between academia and industry. It has also reported on the experience of how a community club can contribute to technology transfer, effectively and over a broad spectrum.

In two years of operation, the Safety-Critical Systems Club in the UK has attracted a large membership, staged ten successful events, published two books, and further facilitated technology transfer by the publication of a regular newsletter, the co-sponsorship of events, and the provision of advice. It provides a model which could be used in other parts of the world.

8   REFERENCES

[1]   Redmill F and Anderson T (Eds): *Safety-critical Systems - Current Issues, Techniques and Standards*. Chapman and Hall, London, 1993.

[2]   Redmill F and Anderson T (Eds): *Directions in Safety-critical Systems - the Proceedings of the First Safety-critical Systems Symposium*. Springer-Verlag, London, 1993.

Information on the Safety-Critical Systems Club may be obtained from Mrs J Atkinson, The Centre for Software Reliability, The University, 20 Windsor Terrace, Newcastle upon Tyne, UK.

Subsidiaries and start-up Spin-off companies of Inria

Panelist : Jean-Pierre Banâtre

Inria-Rennes/Irisa

Campus de Beaulieu - 35042 Rennes cedex, France

This document is a summary of Inria's policy for the creation of private companies as a privileged means for the transfer of basic research results.

1 Inria

The National Institute for Research in Computer Science and Control Automation (Inria) is a french scientific public institute under the responsibility of the ministry of research and technology and the ministry of industry. With its headquarters located at Rocquencourt near Versailles, Inria has five research centers located at Rocquencourt, Sophia-Antipolis, Rennes, Grenoble and Nancy respectively.

Inria brings together 1000 scientists, including 250 permanent research staff and more than 300 PhD students. Its budget in 1991 is of the order of 450 MF (75 M$). Inria's activities is information processing and control theory encompass basic and applied research, design of experimental systems, international scientific exchange, cooperative international programs and technology transfer. The latter is probably one of the most important ones in the context of information technology where changes happen rapidly.

2 Transfer of research is a must

Inria has, over the last 10 years, encouraged the dissemination of its research results and communicated them in the industrial and scientific communities. The transfer is organized through information seminars, research contracts, reception of researchers and engineers from industry in Inria's research teams and detachment of Inria's staff to industrial research centers.

This policy has been systematically accompanied by a diffusion, as large as possible, of results through publications and dissemination of software to partners (universities, public and industrial research centers, industrial departments of research, etc.). This diffusion allows for the evaluation of prototypes and the feedback by the users.

3 The creation of "high-tech" companies : a solution for transfer

Large companies are not always able to assume a direct transfer of research prototypes, especially for basic tools ; this situation may be due to the rigidity of large structures, the difficulties to manage the competition with internal teams, the difficulties to adapt new technologies to the internal strategy, the cost of knowledge transfer, but mainly, the transfer of products without the transfer of men.

Inria has been encouraging the creation of high tech companies by its own researchers ; these companies remain close to research for the main reason that their staff is composed almost exclusively of former researchers. Their business is moreover mainly product oriented.

In this context, fourteen companies have been created up to now in the environment of Inria. They are either subsidiaries or not, but their main goals are always the transfer, exploitation and commercialization of know-how and prototypes originating from Inria.

4 Subsidiary or spin-off ?

Inria's legal status (Public Institute for Science and Technology) allows the institute to share in the capital of private companies. In this context, Inria created three subsidiaries (Simulog in 1984, Ilog in 1987, O2 Technology in 1990) where it controls the majority of the capital, and Gipsi S.A. (coming from the groupment Gipsi SM 90) controlling a minority of the capital. The creation of such subsidiaries only occurs under the following conditions :

- a large investment in research,
- an advanced technology,
- a new but attractive market,
- a very competitive international context,
- the lack of motivated industrial partners.

During this time, researchers, engineers or former PhD students of Inria have created fifteen companies. They represent one of the most dynamic vectors for transfer. The majority of them industrialize and exploit products originating from Inria, under license. These licenses are negociated on a non-exclusive basis and royalties are determinated according to the rules as applied to usual partners.

One of the fifty licenses for prototypes active in 1990, nearly half of them involve spin-off companies and subsidiaries. More than 70% of the total amount of royalties are coming from these companies and the progression accelerates (only 25% in 1989).

Conversely, these companies help Inria to understand strategic information regarding market needs, helping Inria to take strategic decisions concerning research orientation.

Human Medium in Technology Transfer

W. Cellary

Franco-Polish School of New Information and Communication Technologies
Poznan, Poland

To analyze problems of technology transfer between academia and industry, first, it is important to realize expectations of both sides. In my opinion, academia expects money and problems to be solved, while industry expects problem solutions and people trained in research. A question is: what is the role of these people trained in research in technology transfer from academia to industry, and why they are expected by industry? One may believe that a medium of technology transfer is paper, i.e. written documents describing problem solutions. I claim that to transfer a new technology a human medium is required. In a new technology (I emphasize word *new*, meaning here *revolutionary*) developed in academia, i.e. mostly at the theoretical level, the main concept is established, but a lot of problems remains unsolved. I mean here the research problems, not just implementation ones. If the concept is revolutionary new, it is difficult to transfer it in its integrity to a foreign team. Still more difficult is to transfer an idea how to solve the related problems in such a way that the main concept is not violated or deviated. The most efficient way to transfer technology is to use humans, i.e. the researchers who developed it, as a medium. This makes, however, a painful hole in academia, and thus is successful if academic research teams are relatively big and may be split.

Let us now analyze a career of a young, over average talented researcher in computer engineering, who prepared his PhD in collaboration with a team developing a new technology. Assume that he is 28 years old when he gets his PhD, and that he is hired by industry to transfer technology and develop a product based on this new technology. Nowadays, mean life duration of a software product is around twelve years. If we add three years to develop the product, we get fifteen years. Assume that half of this time is creative, i.e., some scientific research needs to be performed, related with the product. We may call it an *offensive phase*. The second half, called the *defensive phase* is devoted mostly to maintenance and some development: moving to various platforms, integration with various software products available on the market, etc. In this phase, innovations are restricted, because of compatibility with previous product versions. Our researcher is 35 years old at the end of the offensive phase of the product. At this age he has a good potential of creativity and, moreover, he has good experience in development of industrial products, cooperative work in a team, etc. There is no reason to keep him to maintain the product during its defensive phase which is not creative. He is at a good point of his professional career to assume responsibility for a new advanced project. There are, however, two menaces: first, that he will continue the old project in a new frame; second, that he will create a new project, but his role will be reduced to the managerial aspects only. To avoid these menaces he has to be trained in new technologies developed in academia during the time when he was occupied with his first product. Seven years elapsed since his PhD is a very long period in so active scientific domain as computer engineering. During this period some new concepts and new research directions appeared that are more suitable and more promising for new products that have to defend themselves against other products even ten years after. A challenge for academia is to integrate and efficiently train such people, as a medium of advanced technology transfer. They need some special organizational solutions, because they cannot be simply mixed with postgraduate students. I think they need around two years of training: one year to study a new domain, and one year to start to produce original results in this domain. A good solution would be a kind of sabbatical. Its advantages are the following. For industry, it is an investment in future products using the most advanced technologies. For academia, it is the growth of research potential, financed by industry, and feedback from practice and applications. In the Franco-Polish School of New Information and Communication Technologies we are encouraging industry to apply this solution.

Technology Transfer - from Purpose to Practice

Bob Malcolm
Malcolm Associates Ltd.
Savoy Hill, London, UK

Abstract

It is submitted that technology transfer between academia and industry is not a matter of academia telling industry how to do things better, but of industry better understanding its own needs and being better able to evaluate academic work.

1 Introduction - the purpose

One of the recommendations of the IEE-BCS report on "Software in safety-related systems" [1] was that there should be a programme of awareness and dissemination of latest developments and of best practice, in parallel with a research programme. The research programme is now established, and there are now 35 projects, with over 130 industrial organizations, and over 30 academic institutions participating [2]. This present paper presents the considerations involved in establishing the related technology transfer programme.

2 Policy

The first step in any government-led initiative must be to identify the policy which both guides and constrains any action. In the present case the relevant policy was set out in a UK government policy paper concerned with innovation (a 'White Paper') from 1988 [3].

Frequently, technology transfer is discussed in the context of 'encouraging adoption of best practice'. This sounds very reasonable at first hearing, but, if interpreted too simplistically, it implies that someone knows what best practice is; that someone will pick technological winners which they will inflict upon everyone else. Would *you* trust any individual, or worse, a committee, to do this on your behalf?

An alternative approach was presented in the White Paper. This is that the role for any government-inspired technology transfer activity should be to make it easier for suppliers to select the most appropriate technology for their business. (Note that in this case 'suppliers' are *users* of technology.)

This policy is motivated by an economic argument. A government-led technology transfer programme is, in effect, intervention in the free market. The economic case for government support of any such action must be that there is some kind of failure of the free market to deliver an optimum product-price combination to end-users. However, a 'perfect' market requires 'perfect information'. It is not too difficult to make a case here that those operating in this business do not have 'perfect information' about the latest technological developments.

3 Theory

A study was performed of the principles of technology transfer, so as to inform those in government responsible for establishing any initiative [4]. The study identified some of the parameters to be considered in such an initiative, which were then used to guide the selection of technology transfer mechanisms.

3.1 The role of technology transfer in innovation

Technology transfer is implicitly part of technologically fuelled industrial innovation. Note - *innovation* - the putting to work of new developments, which is distinguished from their *invention*.

Taken literally, technology transfer is the actual transfer of the technology - the adoption by organizations of technology developed elsewhere, whether in academia or industry. It can be achieved either by organic technological innovation within the firm, or, for instance, by corporate acquisition of a company with new technological capability.

The innovation process varies from sector to sector, at least in detail, because of their different structures. There are also differences in the way in which technology is transferred across different groups of sectors - again due to different industrial structures [5]. However, in general, innovation is a consequence of *diffusion* of knowledge. Moreover, von Hippel proposes that a *"significant mechanism"* contributing to this diffusion is *"informal know-how trading"*, which is *"essentially a pattern of informal co-operative R&D. It involves routine and informal trading of proprietary information between engineers working at different firms - sometimes direct rivals."* [6]

Key players in this diffusion of knowledge are the industrial 'gate-keepers' [7]. These are the personnel in companies who have the external contacts with emerging and prospective technologies, and who are respected by the decision-makers and business managers, inside the company, who are able to use such technology commercially. Such individuals will be familiar with trends in their application sectors which are likely to influence the development of technology, either through market pressure for technological development or through constraints on the nature of such development coming from emerging regulatory changes or, again, the market. 'Gatekeepers' are able to filter ideas and information and promulgate them appropriately. They are important nodes in a communication network which is itself a vital part of technology transfer.

Technology transfer communication should be seen as two-way traffic. This is important because it appears that in many sectors *invention* as well as take-up innovation often originates with the industrial technology users [6]. Indeed, recent research puts it more strongly: *Successful technological change is more likely to be market-led than science-led."* [8]

3.2 Leaders & followers, small & large

Companies which embrace innovation - the 'leaders' - see it as a necessary investment, while aware of the risks. And it is an ongoing investment, rather than a once-off transfer of instances of presently available technology. *"Companies who get there first do not benefit from a lasting loyalty from customers. Successful innovation means more than initiating a product or technique and bringing it to fruition. It means constant improvement, ..."* [8]

However, many companies - the 'followers' - are not geared up to innovation. They do not positively adopt this as their route to future wealth [9]. The view they take is that it is sufficient, and even best for the shareholders, to come second - to let someone else take the risks [10].

It is interesting that the present political climate is one in which emphasis is placed on encouraging the economic growth of smaller companies. Perhaps surprisingly, the evidence indicates that small companies are not noticeably more innovative, on balance [11]. It is the large firms who are more likely to have their own R&D activity, who are both easier to target and best equipped to take-up new technology, and who can make a bigger dent in the overall industry culture. As they are also a source of innovations ripe for diffusion elsewhere, they are an important contributor to the technology transfer communication network. So, it would not make sense to exclude such firms from any initiative.

3.3 Force or facilitate?

Any technology transfer *initiative* must be distinguished from the actual transfer itself. Now, the role of a technology transfer initiative *could* have as its purpose either the transfer itself, or the *facilitation* of the transfer.

At its most extreme, the former takes us back to the 'technology-push' approach, in which somebody decides what should be transferred. What must *not* happen is that some central decision-making committee should pick winners from the technology-push supply side (even under the guise of 'consultation') and make everyone use their preferred technology. After all, who would believe that a central committee could be both sufficiently competent and unbiassed by either lobbying or by having fixed ideas? And, anyway, however well it is done, any such prescription will ultimately stifle innovation.

In this context, we should address the potential role of regulation as a means of accelerating take-up of new technology. This is certainly possible, but it must be handled very carefully and intelligently. The intention, once again, is *not* to enforce the take-up of any particular technology. Such prescription is anti-competitive, restricts trade, and stultifies competition, innovation, and technological development - just the opposite effect, in the long term, to the effect which is sought.

Where regulation is necessary, Rothwell and Zegfeld refer to *"the desirability, where possible, of formulating regulations that allow maximum freedom in developing technology for compliance."* [7]. It is not too difficult to avoid constraining compliance to particular technologies. It is at least as important, and usually much more difficult, to frame regulation which does not presume certain *classes* of technological solution.

On the other hand, if it can be done properly, then tough, technology-free, targets can stimulate the development of a range of innovative solutions. It is important that the targets are indeed tough, else the inclination is to adopt 'best practice' targets, based on existing technology, which tend to be lowest common-denominator targets, thereby, once more, inhibiting competition and innovation rather than accelerating it.

Returning to the alternative of *facilitation*, any facilitation of technology transfer might either directly support the explicit adoption of new technology (while not prescribing what that should be, of course) or it can more indirectly attempt to overcome - or undermine - some of the barriers to innovation.

The two major barriers are, it seems, a lack of awareness of new developments and practice elsewhere - despite the availability of information for those that positively look for it - and an inability to assimilate change, primarily because of shortcomings in *"the strategic ability of management to integrate externally acquired technology into an overall business plan"* [9]. It is fairly clear that a technology transfer initiative could address the first of these. It is less clear that problems with the strategic ability of managements lie within the scope of technology transfer. Indeed, in the UK there is now a much more broadly-based attempt to inspire an innovatory culture ([12], for example).

However, technology transfer actions of the right kind can help. It appears that 'perceived performance gap', compared with fairly close competitors, is a major motivator for innovation. *"The technological strategy [of business units within firms] is to achieve the full potential of the product or process [around which they are organised] in a way that at least matches the performance of rivals"* [8] So information about what is happening elsewhere in industry is at least as important as information about technology. A technology transfer initiative can address both of these.

3.4 The players and their parts

In essence, the very much simplified technology transfer model in the Buxton-Malcolm paper [4] comprises:
- awareness - coupled with 'interest'
- gatekeeping
- in-depth economic evaluation
- decision
- acquisition of technology and capability in its use

Again simplifying very much, we identify a number of types of individual in an organization. They play different roles in each of these stages, and require different types of information in order to perform properly:
- senior-managers - the 'decision-makers' as they are often called, except ..
- middle management - who might well put this year's bottom line before high-level highfalutin' ideas about change, and therefore, in reality, make the decisions by default
- gatekeepers
- engineers - 'the workers'
- researchers

Note the important requirement for 'interest' in any awareness activity. This is different from the salesmen's 'attention' (which is necessary but not sufficient). A 'perceived performance gap' referred to previously is one way to grab both the attention and interest of decision-makers and managers.

After awareness, assuming interest is aroused, we need to make sure that gatekeepers have access - and have preferably already had it - to the sort of technical and economic feasibility which *they* can believe, so that when asked, they do not block the introduction of technology simply because of a lack of information.

And then, assuming that interest is held, organisations will need convincing information, from a reliable source, of the advantages and disadvantages of alternative technologies.

Should the decision be to proceed, there will be a need for back-up; information on supply of tools, training, and so on. This must all be available well in advance of any decision though, since availability and supply will be considered during the earlier deliberations of the organisation.

Note that the order of these stages is not necessarily linear, and, in particular, awareness is often stimulated by gatekeepers.

4 The practice - mission & mechanisms

4.1 The mission statement

Having studied the theory, the mission statement for the proposed technology transfer programme was established. This is:

"To achieve, in the supply of safety-related programmable electronic systems, better informed application of safety engineering practices and better informed choice and application of appropriate software technology."

4.2 A choice of mechanisms

Having identified different classes of information required by different people in the diffusion of innovation, Table 1 was a first attempt to help to identify which mechanisms provide support for which of the different aspects, discussed previously. It does not purport to be complete, and the 'more-blobs-the-better' ratings are entirely subjective, as a starting point for discussion.

Note that some of these mechanisms perform dual roles. For instance, collaborative research, technology demonstrators, and 'application experiments' [13] are all means whereby technology can actually be transferred into participating organisations. But to others they are perceived as sources of information, accessed through the associated dissemination activities.

Contribution to:	Technical feasibility		Awareness				Economic feasibility			Capa-bility	Ability		
for *:	R	G	G	D	M	E	G	D	M	M	M	E	Cost
collaborative research	•••	•••	•••	•	••	••	•						£££
mailshots	••	••	••			••							£
press			•	•	•	••							£
gatekeeper clubs	•••	•••	••		•	••	••						££
technical feasibility demos.	••	•••											£££
economic feasibility demos.		••	•		•		•••	•	••				£££
technology centres					••	••		•	•	••			££
journeyman schemes			••		•		•••		••	••			£
case studies			•	••	••		•••	••	•••				£
sector clubs		•••	•••	••	•••	•	•••	••	•••	••			££
hazard analysis forum		•••	•••	•••	•••	•	•••	•••	•••				££
conferences	••	•	••		•	••	••		•				££
centres of excellence		••	•••				••			••	••		£££
standards					••	•							££
certification			•	••	••	•							£££
teaching company scheme			•		•	•							££
consultancy initiatives					•	•	•		•	•			££
management handbooks					•	•	•		•		•		£
manager forums					•••				••				££
internal communications			•	•	•	•							£
technology review								•••	•••				££
directories										•			£
technology survey reports		•	•		•		•		•	•			££
information centres								••					££
user-groups					•••			••	•••				££
summer schools										••	•	•	££
acquisition, recruitment												•••	?

* KEY: R: Researchers; G: Gatekeepers; D: Decision-makers; M: Managers; E: Engineers

Table 1. Appropriateness of various technology transfer mechanisms

4.3 The Safety-Critical Systems Club

On the basis of all these considerations, and given the existence of the collaborative research programme, we decided to combine the functions of gatekeeper clubs and sector clubs in a 'community club' - the Safety-Critical Systems Club.

The primary aim of the club is to facilitate the flow of information between practitioners within industry - both between peers and between 'leaders' and 'followers'. By enhancing awareness of current practices and of latest developments, the club accelerates agreement on what constitutes best practice, and enables evaluation of academic work.

Since one of the aims is to achieve greater consensus among both individuals and organisations, we must avoid proliferation of additional organisations, and of activities which overlap or compete with their activities. The club is therefore intended to undertake new activities only where necessary, and to encourage, stimulate, and perhaps facilitate activities of existing technology transfer organisations who may not be dealing with this subject at present, but who may be able to offer the most appropriate forum. Such existing organisations include trade associations, sectoral research associations, professional institutions, existing technology clubs, and publishing houses.

Such a club is, cheap, simple, informal, and, we believe, highly cost-effective. At the last count there were over 1600 members [14].

5 Effectiveness

The effectiveness of the club has yet to be proved. It is a requirement that all government supported initiatives of this type be evaluated. For evaluation we must develop some 'output measures' from the mission statement. This has yet to be done, but for this kind of technology transfer activity we will, for instance, be seeking evidence that organisations

- are better informed about safety engineering and software technology
- review their activities and technological needs more thoroughly
- are better able to judge whether changes in their practice are desirable and
- make such changes, when they feel they are appropriate.

References

1. Institution of Electrical Engineers. *Software in safety related systems.* IEE, London, 1989
2. Malcolm R. *The JFIT Safety Critical Systems Research Programme: Origins & Intentions.* In: Redmill F and Anderson T (eds). Safety-critical systems - Current Issues, Techniques and Standards. Chapman & Hall, London, 1993
3. Department of Trade & Industry. *DTI - the department for Enterprise.* HMSO, London, 1988
4. Buxton JN, Malcolm R. *Software technology transfer.* Software Engineering Journal 1991; pp 17-23
5. Pavitt K. *Patterns of Technological Change - Evidence, Theory and Policy Implications.* Science Policy Research Unit, University of Sussex, UK, 1983
6. von Hippel E. *Sources of innovation.* OUP, 1988
7. Rothwell R, Zegfeld W. *Reindustrialization & technology.* Longman, Harlow, 1985
8. Smith D. *Innovation: a review of recent ESRC research.* ESRC, Swindon, UK, 1992
9. Rothwell R, Beesley M. *The Importance of Technology Transfer.* In Barber J, Metcalfe JS, Porteous M. Barriers to Growth in Small Firms. Routledge, London, 1989
10. Private communication from the managing dirctor of a 'high-tech' company.
11. Rothwell R. *The Role of Small Firms in the Emergence of New Technologies* OMEGA Vol 12, No. 1 (1984)
12. Lilley P. *Innovation: Competition & Culture.* DTI, London, 1991
13. European Commission *ESSI: The European System & Software Initative - Briefing Package for Proposers.* CEC, Bruxelles, 1993
14. Redmill F. *Issues affecting Technology Transfer and Experience with a Community Club.* In these Proceedings.

INVITED PAPER

Dependability: from Concepts to Limits

Jean-Claude Laprie

LAAS-CNRS, Toulouse, France

Abstract

Our society is faced with an ever increasing dependence on computing systems, which lead to question ourselves about the limits of their dependability. In order to respond this question, a global conceptual and terminological framework is needed, which is first given. The analysis of the limits in dependability which is then conducted identifies design faults as the major limiting factor, a consequence of which is the concluding recommendation of applying a fault tolerance approach to the improvement of the production process.

Introduction

Our society has become increasingly dependent on computing systems and this dependency is especially felt upon the occurrence of failures. Recent examples of nation-wide computer-caused or -related failures are the 15 January 1990 telephone outage in the USA, or the 26-27 June 1993 credit card denial of authorization in France. The consequences of such events relate primarily to economics; however, some outages can lead to endangering human lives as second order effects, or even directly as in the London Ambulance Service failure of 26-27 November 1992. As a consequence of such events, which can only be termed as disasters, the consciousness of our vulnerability to computer failures is developing, as witnessed by the following quotation from the report *Computing the Future: A Broader Agenda for Computer Science and Engineering* [COM 92]: "Finally, computing has resulted in costs to society as well as benefits. Amidst growing concerns in some sectors of society with respect to issues such as unemployment, invasions of privacy, and reliance on fallible computer systems, the computer is no longer seen as an unalloyed positive force in the society".

Faced with this situation, a natural question is then "To which extent can we rely on computers?", or, more precisely, "What are the limits of computing systems dependability?". Responses to these questions need a conceptual and terminological framework for dependability, which in turns is influenced by the analysis of the

This work was partially supported by the ESPRIT Basic Resaerch Action PDCS (Predictably Computing Systems, project no. 6362)

limits in dependability. Such a framework can hardly be found in the many standardization efforts: as a consequence of their specialization (telecommunications, avionics, rail transportation, nuclear plant control, etc.), they usually do not consider all possible sources of failures which can affect computing systems, nor do they consider all attributes of dependability.

The considerations expressed in the above two paragraphs have guided the contents of the paper, which is composed of two sections. The first section is devoted to the main definitions relating to the dependability concept; those definitions elaborate on the definitions given in [Lap 92a]. The second section deals with the limits of dependability.

1 The Dependability Concept

1.1 Basic definitions

Dependability is defined as the trustworthiness of a computer system such that *reliance can justifiably be placed on the service* it delivers. The service delivered by a system is its behavior *as it is perceptible* by its user(s); a user is another system (human or physical) which *interacts* with the former.

Depending on the application(s) intended for the system, different emphasis may be put on different facets of dependability, i.e. dependability may be viewed according to different, but complementary, *properties*, which enable the *attributes* of dependability to be defined:

- the *readiness for usage* leads to **availability**,
- the *continuity of service* leads to **reliability**,
- the *avoidance of catastrophic consequences on the environment* leads to **safety**,
- the *avoidance of unauthorized disclosure of information* leads to **confidentiality**,
- the *avoidance of improper alterations of information* leads to **integrity**,
- the *ability to undergo repairs and evolutions* leads to **maintainability**.

A system **failure** occurs when the delivered service is not up to fulfilling the system's function. An **error** is that part of the system state which is liable to lead to subsequent failure: an error affecting the service is an indication that a failure occurs or has occurred. The adjudged or hypothesized cause of an error is a **fault**.

The development of a dependable computing system calls for the *combined* utilization of a set of methods and techniques which can be classed into:

- **fault prevention:** how to prevent fault occurrence or introduction,
- **fault tolerance:** how to ensure a service up to fulfilling the system's function in the presence of faults,
- **fault removal:** how to reduce the presence of faults,
- **fault forecasting:** how to estimate the present number, the future incidence, and the consequences of faults.

The notions introduced up to now can be grouped into three classes and are summarized by figure 1:

- the **impairments** to dependability: faults, errors, failures; they are undesired — but not in principle unexpected — circumstances causing or resulting from un-dependability (whose definition is very simply derived from the definition of dependability: reliance cannot, or will not any longer, be placed on the service);
- the **means** for dependability: fault prevention, fault tolerance, fault removal, fault forecasting; these are the methods and techniques enabling one a) to provide the ability to deliver a service on which reliance can be placed, and b) to reach confidence in this ability;
- the **attributes** of dependability: availability, reliability, safety, confidentiality, integrity, maintainability; these a) enable the properties which are expected from the system to be expressed, and b) allow the system quality resulting from the impairments and the means opposing to them to be assessed.

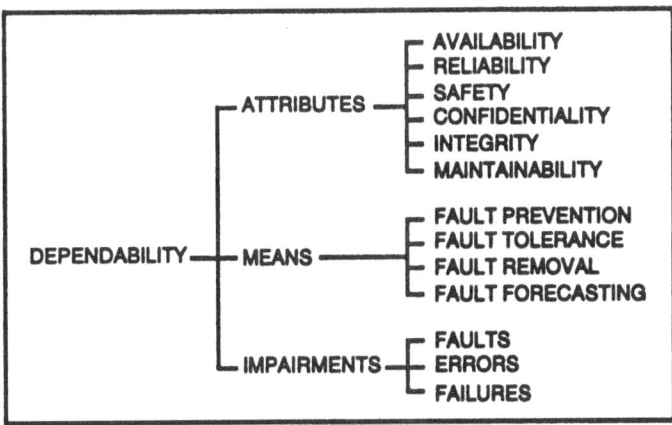

Figure 1 - The dependability tree

1.2 On the Attributes of Dependability

The definition given for integrity — the avoidance of improper alterations of information — generalizes the usual definitions (e.g., prevention of unauthorized amendment or deletion of information [EEC 91], or ensuring approved alteration of data [Jac 91]) which are directly related to a specific class of faults, i.e. intentional faults, that is deliberately malevolent actions. Our definition encompasses accidental faults as well (i.e. faults which appear or are created fortuitously), and the use of the word information is aimed at not being restricted to data strictly speaking: integrity of programs is also an essential concern; regarding accidental faults, error recovery is indeed aimed at restoring the system's integrity. It is also noteworthy that our definition can be interpreted by default in order to encompass subtle attacks against integrity, such as preventing suitable data updates. Integrity is a prerequisite for availability, reliability and safety, but it may not always be so for confidentiality, as in the case of passive attacks using covert channels or wire-taping.

Confidentiality, not security, has been introduced as a basic attribute of dependability. Security is usually defined (see e.g., [EEC 91]) as the combination of confidentiality, of integrity and of availability, where the latter are understood with respect to unauthorized actions, whereas availability is not usually restricted to such events, nor is integrity according to the above discussion; in addition, as noted in [Gas 88], confidentiality is the most distinctive characteristic of security. A definition of security gathering the three aspects of [EEC 91] is: the prevention of unauthorized access and/or handling of information; security issues are indeed dominated by intentional faults, but not restricted to them: an accidental (e.g., physical) fault can cause an unexpected leakage of information.

The definition given for maintainability — ability to undergo repairs and evolutions — deliberately goes beyond *corrective* maintenance, which relates to repairability only. Evolvability clearly relates to the two other forms of maintenance, i.e. a) *adaptive* maintenance, which adjusts the system to environmental changes, and b) *perfective* maintenance, which improves the system's function by responding to customer, and designer, defined changes. It is noteworthy that the frontier between repairability and evolvability is not always clear, for instance if the requested change is aimed fixing a specification fault [Ghe 91]. Maintainability actually *conditions* dependability when considering the whole operational life of a system: systems which do not undergo adaptive or perfective maintenance are likely to be exceptions.

From their definitions, availability and reliability emphasize the avoidance of failures, safety the avoidance of a specific class of failures (catastrophic failures), and security the prevention of what can be viewed as a specific class of faults (the prevention of the unauthorized access and/or handling of information). Reliability and availability are thus closer to each other than they are to safety on one hand, and to security on the other; reliability and availability can thus be grouped together [Lap 92b, Jon 92], and be collectively defined via the property of avoiding or minimizing the service outages. However, this remark should not lead to consider that reliability and availability do not depend on the system environment: it has long been recognized that a computing system reliability/availability is highly correlated to its utilization profile, be the failures due to physical faults or to design faults (see e.g., [Iye 82]).

The properties enabling the definition of the dependability attributes may be more or less emphasized depending on the application intended for the computer system under consideration. For instance, availability is always required, although to a varying degree depending on the application, whereas reliability, safety, confidentiality may or may not be required according to the application. The variations in the emphasis to be put on the attributes of dependability have a direct influence on the appropriate balance of the means to be employed in order that the resulting system be dependable. This is an all the more difficult problem as some of the attributes are antagonistic (e.g., availability and safety, availability and confidentiality), necessitating that tradeoffs be performed. Considering the three main design dimensions of a computer system, i.e. cost, performance and dependability, the problem is further exacerbated by the fact that the dependability dimension is less mastered than the cost-performance design space [Sie 92].

The discussion conducted in this section is summarized on figure 2.

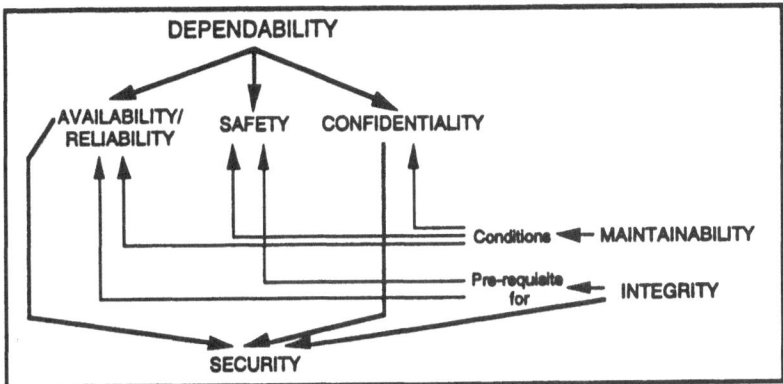

Figure 2 - Relationship between the dependability attributes

2 Limits in Dependability

Dependable computing systems are specified, developed, operated and maintained according to assumptions which are relative a) to the function(s) to be fulfilled, b) to the environment where the computing system is to be operated (load, perturbations from the physical environment, behavior of the operators and maintainers), and c) to the faults which are likely to manifest, in terms of their modes and frequencies. The achieved dependability is crucially depending upon the validation a) of the actual system with respect to these assumptions, b) of the assumptions themselves with respect to reality, and, recursively, c) of the assumptions of the validation itself (e.g., criteria according to which fault-removal is conducted). Limiting factors to dependability can thus originate from a variety of sources, due to inadequacies of the development-operation assumptions, or to imperfections in the validation of the system and of those assumptions.

Dealing in details with all the above issues is clearly out of reach, all the more as those issues are in fact closely related. In the sequel, we discuss the relationship between function and failure, we investigate the effectiveness of fault tolerance with respect to various types of faults, and we examine the distinction which has to be performed between the dependability which is actually achieved and the estimated dependability.

2.1 Function and failure

The function of a system is what the system *is intended for* [Kui 85]. The function is usually first described, or specified, in terms of what *should* be fulfilled regarding the system's primary aim(s) (e.g., performing transactions, controlling or monitoring a plant, piloting an airplane or a rocket). When considering safety- or security-related systems, this description needs to be supplemented with what *should not* happen (e.g., the hazardous states from which a catastrophe may ensue, or the disclosure of sensitive information). Such a description may in turn lead to specifying additional functions that the system *should* fulfill in order to reduce the likelihood of what should not happen (e.g., exhibiting a fail-safe behavior or authenticating a user and checking his/her access rights).

The description of a system's function(s) can be performed at various levels of details, according to several means of expression, from natural language to mathematics. A system may fail with respect to a given function, or with respect to a given level of detail, and still comply with the others.

Expressing a system's function(s) is an activity which is naturally conducted from the very first steps of a system's development. However, we all know that specifying a system's functions extends to the whole system's life, due to the inherent difficulty of eliciting a system's requirements. If the largest amount of effort devoted to this elicitation indeed takes place at the beginning of the system's development, it does not end with what is traditionally called "requirements analysis and specification" in life-cycle models such as the waterfall or the V models; this elicitation practically continues during all phases of a system's development, and during operational life as well: perfective maintenance is often performed in order to correct what are finally specification faults, i.e. oversights, mistakes or omissions. This remark is all the more important when considering failures: an unacceptable behavior can indeed be detected as such from its deviation from complying with the specification, but it may happen that it complies with the specification and is felt as unacceptable by users, thus in fact uncovering a specification fault; leaving aside the question of the consequences of such a situation, it can be said that in fact it helps eliciting what the real function of the system ought to be. This type of problem should not be minimized, for instance in advocating that systems generally exhibit more frequently failures due to the subsequent phases of their development than failures traceable to imperfections in the definition of their requirements (see e.g., [Gla 81]); however, besides the fact that severity has to be accounted for in addition to frequency, this frequency argument does not apply to safety-related systems: a consequence of the extreme care put into their design and realization is that specification imperfections generally constitute a significant source of the problems faced with in operation.

2.2 Effectiveness of fault tolerance

Before discussing the effectiveness of fault tolerance, let us stress that from our definition — ensuring a service up to fulfilling the system's function in the presence of faults, fault tolerance is a mean for providing a system with the ability to behave as expected, be it fail-safe or fail-operational.

The imperfections of fault tolerance, i.e. the lack of *fault tolerance coverage*, constitute a severe limitation to the increase in dependability which can be obtained. Such imperfections of fault tolerance are due either a) to design faults affecting the fault tolerance mechanisms with respect to the fault assumptions made during the design, the consequence of which is a lack of *error and fault handling coverage*, or b) to fault assumptions which differ from the faults really occurring in operation, resulting in a lack of *fault assumption coverage*, which can be in turn due to either i) failed component(s) not behaving as assumed, that is a lack of *failure mode coverage*, or ii) the occurrence of correlated failures, that is a lack of *failure independence coverage*. The influence of a lack of error and fault handling coverage [Bou 69, Arn 73] has been shown to be such that it not only drastically limits the dependability improvement, but that in some cases adding further redundancies can result in lowering dependability [Dug 89]. Similar effects can result from the lack of the other forms of fault tolerance coverage: conservative failure mode assumptions (e.g., inconsistent, or Byzantine failure modes) will result in a higher failure mode coverage, at the expense of necessitating an increase in the redundancy which can lead to an overall decrease in the system dependability [Pow 92]; correlated failures are known to defeat fault tolerance strategies based on the replication of identical items, either because of a) common design faults, or of b) identical sensitivity to workload or externally-induced perturbations [Hec 87].

Another significant source of limit in dependability are *temporary faults*. They have long been recognized as constituting the vast majority of hardware faults, and the progresses in hardware integration can only emphasize this tendency (see e.g., the field data from several sources in [Sie 92]). A direct consequence is that , although it is not always so [Geb 88], emphasis should be placed in the design of fault-tolerant systems on discriminating between temporary and permanent faults: the misinterpretation of a temporary fault as a permanent fault results in an unnecessary decrease in the available redundancies, thus in lowering dependability.

Temporary faults are not limited to hardware: the notion of temporary fault applies to software as well. Although such a notion has been introduced a long time ago [Elm 72], and more recent studies have shown that most of the software faults present during operational life are temporary faults [Gra 86], the very notion of temporary software fault is often felt as contradicting our perception of software. In fact, if it is not arguable that the ultimate cause of software faults are present as long as they are not fixed, it has to be recognized that most software faults manifesting in operation in large, complex, software are subtle enough in order that their activation conditions depend on equally subtle combinations of internal state and external solicitation, so that they can hardly be reproduced. Stated in other terms, the failure domain of temporary software faults can vary with the conditions of execution of the software, and be a null space under most operating conditions. Although it is generally recognized that software is the current bottleneck in terms of dependability [Gra 90, Cra 92], fault tolerance approaches aimed at temporary software faults have been paid little attention [Gra 86, Hua 93] when compared to the work devoted to tolerating permanent software faults, which necessitates design diversity [Ran 75, Che 78, Lap 90]. Because of the high cost of design diversity, software fault tolerance is currently mostly limited to some safety-critical applications [Vog 88] such as avionics, and, to a lesser extent, railway transportation or nuclear plant monitoring. The main limiting factor of the undeniable improvement in dependability brought about by design diversity is constituted by the unavoidable correlations between the software variants issued from diversified designs [Eck 91], which are a form of lack of failure independence coverage.

This brief overview of the factors limiting dependability in terms of fault classes would not be complete without mentioning the man-machine interaction faults, resulting from inappropriate interactions. Their importance is not new, and is not restricted to safety-critical systems [Toy 78, Dav 81], although they are generally felt as a major source of failure in those systems, for the same reason as already mentioning regarding specification faults, i.e. the extreme care put into their design and realization. Recognizing that most of interaction faults are in fact design faults of the system [Nor 83] should encourage the development of approaches for their tolerance [Max 86, Rou 87].

Fault tolerance is not restricted to accidental faults, which have been dealt so far in this section. Some mechanisms of error detection are directed towards both intentional and accidental faults (e.g. memory access protection techniques) and schemes have been proposed for the tolerance to both intrusions (i.e. intentional interaction faults) and physical faults [Rab 89, Des 91], as well as for tolerance to malicious logic (i.e. intentional design faults) [Jos 88]. However, fault tolerance is far from being recognized as a viable approach to security issues, in spite of the fact that the continuously increasing cost of failures attributable to intentional faults clearly show that the current approaches, mainly based on fault avoidance (i.e. fault prevention and fault removal), do not constitute a fully satisfactory answer.

2.3 Achieved dependability and estimated dependability

It is now widely admitted that performing dependability evaluation of hardware-fault-tolerant systems without accounting for the lack of fault tolerance coverage can only lead to grossly overestimated evaluations of dependability. However, it is noteworthy that if error and fault handling coverage has been devoted a large attention due to the ability of estimating it via fault-injection (see e.g. [Gun 89, Arl 90, Cho 92]), it is not so for assumption coverage as it needs experience to be accounted for in addition to fault-injection.

Regarding software reliability modeling and prediction, the vast majority of the published material focuses on the development and validation phases. Much less studies encompassing the operational phase appear in the open literature. As already mentioned in the previous section, a clear tendency in computing systems is to see software becoming the major source of failure; there is thus a growing need for performing, during their validation, reliability evaluations of software systems in order to forecast their (future) reliability in operation.

The current approaches to software reliability evaluation cannot satisfy this need. The main reason lies in the fact that they consider products in isolation from the process which produced them: the reliability predictions performed for a given software either a) from its successive times to failure for reliability growth modeling, or b) from its execution times without failures for statistical testing, can only be far below any reasonable reliability requirement (see e.g. [Par 90, The 91] for statistical testing). In addition, those predictions usually (and hopefully) are also far below what will be observed in operation (see e.g. [Kan 87, Gra 90]). Considering that software systems produced from scratch are exceptions, and that the current usual approach is rather to make evolutions from existing software, a logical consequence is to enhance the predictions performed for a given product from data relating to its validation with field data relative to previous, similar, software products [Lap 92c]. The corner stone of the approach is clearly the notion of similarity between the various generations of a software family; however, in addition to the usually mentioned negative dissimilarities resulting from added failure sources, we can expect positive dissimilarities to exist, resulting from progress in the development and validation methods and techniques.

3 Conclusion

The discussion conducted in the previous section clearly shows that design faults constitute the major limitation to dependability of computing systems, be they fault-tolerant or not. This is not at all surprising: a computing system is a human artifact and as such any fault in it or affecting it is ultimately human-made since it represents human inability to master all the phenomena which govern the behavior of a system. As a direct consequence, pushing forward the limits of dependability of computing systems can only be done via improving their production process.

Improvements in the production process can indeed be noticed, when looking at the evolutions between the successive editions of guidelines for the development and validation of computing systems (a noticeably good example is the recent new release of the guidelines for certifying airborne software, DO-178-B, as compared to the previous release, DO-178-A). This is however a feedback loop which takes a considerable amount of time in order to make profit of experience. Some shorter feedbacks are currently being explored, as the previously mentioned work we are carrying out on software reliability evaluation [Lap 92c], or the recently published work on employing accumulated knowledge in software testing [Wil 92]. Both examples relate to validation: they are aimed at improving fault removal and fault forecasting, but they are not directly and explicitly concerned with the progressive reduction of the faults created during the development and inserted in the developed system before fault removal takes place. This is where the most effective feedback loop should exist, which would be nothing else than applying to the production process a fault tolerance approach: detecting the errors in the development, diagnosing and removing the causes, i.e. the faults in the production process.

We have to be conscious that such an approach questions directly the very organization of the producers: carrying out, and making profit of, the necessary experience does not appear compatible with the project-oriented organizations which currently prevail. It can only be hoped that the statistics on the continuously growing costs of computer failures can be a strong enough incentive to moving the organizational inertia. Let us just mention that the yearly computer failure cost is now exceeding more than 10 Billions of Francs in France, both accidental and intentional faults included (and has been constantly larger than the profit of the whole computing industry in France, including construction, distribution and services), and is close to 4 Billions of Dollars in the USA for accidental faults.

Acknowledgements

I would like to thank the conference organizers for giving me the opportunity to present the ideas contained in the paper. Thanks also to Alain Costes and Yves Deswarte for their helpful comments and discussions.

References

Arl 90   J. Arlat, M. Aguera, L. Amat, Y. Crouzet, J.C. Fabre, J.C. Laprie, E. Martins, D. Powell, "Fault injection for dependability validation: a methodology and some applications", *IEEE Transactions on Software Engineering*, Special Issue on Experimental Computer Science, vol. 16, no. 2, Feb. 1990, pp. 166-182,

Arn 73   T.F. Arnold, "The concept of coverage and its effect on the reliability model of repairable systems", *IEEE Trans. on Computers*, vol. C-22, June 1973, pp. 251-254.

Bou 69   W.G. Bouricius, W.C. Carter, P.R. Schneider, "Reliability Modeling Techniques for Self-Repairing Computer Systems", *Proc. 24th ACM National Conf., 1969*, pp. 295-309.

COM 92   "Computing the Future", Report of the Committee to Asses the Scope and Direction of Computer Science and Technology of the National Research Council, *Communications of ACM*, vol. 35, no. 11, Nov. 1992, pp. 30-40.

Che 78   L. Chen, A. Avizienis, "N-version programming: a fault-tolerance approach to reliability of software operation", *Proc. 8th IEEE Int. Symp. on Fault Tolerant Computing (FTCS-8)*, Toulouse, France, June 1978, pp. 3-9.

Cho 92   G.S. Choi, R.K. Iyer, "FOCUS: an experimental environment for fault sensitivity analysis", *IEEE Trans. on Computers*, vol. 41, no. 12, Dec. 1992, pp. 1515-1526.

Cra 92   R. Cramp, M.A. Vouk, W. Jones, "On operational availability of a large software-based telecommunications system", *Proc. 3rd Int. Symp. on Software Reliability Engineering*, Research Triangle Park, North Carolina, Oct. 1992, pp. 358-366.

Dav 81   E.A. Davis, P.K. Giloth, "No 4 ESS: performance objectives and service experience", *The Bell System Technical Journal*, vol. 60, no. 6, July-Aug. 1981, pp. 1203-1224.

Des 91   Y. Deswarte, L. Blain, J.C. Fabre, "Intrusion tolerance in distributed computing systems", *Proc. 1991 IEEE Symposium on Research in Security and Privacy*, Oakland (USA), 20-22 Mai 1991, pp.110-121

Dug 89   J.B. Dugan, K.S. Trivedi, "Coverage modeling for dependability analysis of fault-tolerant systems", *IEEE Trans. on Computers*, vol. 38, no. 6, June 1989, pp. 775-787.

Eck 91   D.E. Eckhardt, A.K. Caglayan, J.C. Knight, L.D. Lee, D.F. McAllister, M.A. Vouk, J.P.J. Kelly, "An experimental evaluation of software redundancy as a strategy for improving reliability", *IEEE Trans. on Software Engineering*, vol. 17, no. 7, July 1991, 692-702.

EEC 91   *Information Technology Security Evaluation Criteria, Provisional Harmonised criteria*, Office for Official Publications of the European Communities, June 1991.

Elm 72   W.R. Elmendorf, "Fault-tolerant programming", *Proc. 2nd IEEE Int. Symp. on Fault Tolerant Computing (FTCS-2)*, Newton, Massachusetts, June 1972, pp. 79-83.

Gas 88   M. Gasser, *Building a Secure Computer System*, Van Nostrand Reinhold, 1988.

Geb 88   J. Gebman, D. McIver, H. Shulman, "Faults with nonstationary observability are limiting avionics R&M", *Proc. 8th AIAA/IEEE Digital Systems Avionics Conf.*, San Jose, California, Oct. 1988, pp. 16-23.

Ghe 91   C. Ghezzi, M. Jazayeri, D. Mandrioli, *Fundamentals of Software Engineering*, Prentice-Hall, 1991

Gla 81   R.L. Glass, "Persistent software errors", *IEEE Transactions on Software Engineering*, vol. SE-7, no. 2, March 1981, pp. 162-168.

Gra 86   J.N. Gray, "Why do computers stop and what can be done about it?", *Proc. 5th Symp. on Reliability in Distributed Software and Database Systems*, Los Angeles, Jan. 1986, pp. 3-12.

Gra 90   J. Gray, "A census of Tandem system availability between 1985 and 1990", *IEEE Trans. on Reliability*, vol. 39, no. 4, Oct. 1990, pp. 409-418.

Gun 89   U. Gunneflo, J. Karlsson, J. Torin, "Evaluation of error detection schemes using fault injection by heavy-ion radiation", *Proc. 19th IEEE Int. Symp. on Fault Tolerant Computing (FTCS-19)*, Chicago, June 1989, pp. 340-347.

Hec 87   H. Hecht, H. Dussault, "Correlated failures in fault-tolerant computers", *IEEE Trans. on Reliability*, vol. R-36, no. 2, June 1987, pp. 171-175.

Hua 93   Y. Huang, C. Kintala, "Software implemented fault tolerance: technologies and experience", *Proc. 23rd IEEE Int. Symp. on Fault-Tolerant Computing (FTCS-23)*, Toulouse, June 1993, pp. 2-9.

Iye 82   R.K. Iyer, S.E. Butner, E.J. McCluskey, "A statistical failure/load relationship: results of a multi-computer study", *IEEE Trans. on Computers*, vol. C-31, July 1982, pp. 697-706.

Jac 91   J. Jacob, "The basic integrity theorem", *Proc. IEEE International Symposium on Security and Privacy*, Oakland, May 1991, pp. 89-97.

Jon 92   E. Jonsson, T. Olovsson, "On the Integration of Security and Dependability in Computer Systems", *Proc. IASTED Int. Conf. for Reliability, Quality Control and Risk Assessment*, 1992.

Jos 88   M.K. Joseph, A. Avizienis, "A fault tolerance approach to computer viruses", *Proc. 1988 Symp. on Security and Privacy*, Oakland, April 1988, pp. 52-58.

Kan 87   K. Kanoun, T. Sabourin, "Software dependability of a telephone switching system", *Proc. 17th IEEE Int. Symp. on Fault-Tolerant Computing (FTCS-17)*, Pittsburgh, Pennsylvania, USA, June 1987, pp. 236-241.

Kui 85   B. Kuipers, "Commonsense reasoning about causality: deriving behavior from structure", in *Qualitative Reasoning about Physical Systems*, D.G. Bobrow editor, MIT Press, 1985, pp. 169-203.

Lap 90   J.C. Laprie, J. Arlat, C. Beounes, K. Kanoun, "Definition and analysis of hardware- and software-fault-tolerant architectures", *IEEE Computer*, vol. 23, no. 7, July 1990, pp. 39- 51.

Lap 92a   J.C. Laprie, ed., *Dependability: Basic Concepts and Terminology*, Springer-Verlag, Vienna, 1992.

Lap 92b   J.C. Laprie, "Dependability: a unifying concept for reliable, safe, secure computing", *Proc. 12th IFIP World Computer Congress*, Madrid, Spain, Sept. 1992, vol. I, pp. 585-593

Lap 92c   J.C. Laprie, "For a product-in-a-process approach to software reliability evaluation", *Proc. 3rd Int. Symp. on Software Reliability Engineering*, Research Triangle Park, NC, Oct. 1992, pp. 134-139.

Max 86   R.A. Maxion, "Towards fault-tolerant user interfaces", *Proc. 5th IFAC Workshop on Safety of Computer Control Systems (SAFECOMP'86)*, Sarlat, France, Oct. 1986, pp. 117-122.

Nor 83   D.A. Norman, "Design rules based on analyses of human error", *Communications of the ACM*, vol. 26, no. 4, April 1983, pp. 254-258.

Par 90   D.L. Parnas, A.J. van Schouwen, S.P. Kwan, "Evaluation of safety-critical software", *Communications of the ACM*, vol. 33, no. 4, June 1990, pp. 636-648.

Pow 92   D. Powell, "Failure Mode Assumptions and Assumption Coverage", *Proc. 22nd IEEE Int. Symp. on Fault-Tolerant Computing (FTCS-22)*, Boston, July 1992, pp.386-395.

Rab 89   M.O. Rabin, "Efficient dispersal of information for security, load balancing and fault tolerance", *Jounal of the ACM*, vol. 36, no. 2, April 1989, pp. 335-348.

Ran 75   B. Randell, "System Structure for Software Fault Tolerance", *IEEE Trans. on Software Engineering*, vol. SE-1, no. 2, 1975, pp.220-232, .

Rou 87   W.B. Rouse, N.M. Morris, "Conceptual design of a human error tolerant interface for complex engineering systems", *Automatica*, vol. 23, no. 2, 1987, pp. 231-235.

Sie 92   D.P. Siewiorek, R.S. Swarz, *The Theory and Practice of Reliable System Design*, Digital Press, 1992.

The 91   P. Thévenod-Fosse, H. Waeselynck, "An investigation of statistical software testing", *Journal of Software Testing, Verification and Reliability*, vol. 1, no. 2, 1991, pp. 5-25.

Toy 78   W.N. Toy, "Fault-tolerant design of local ESS processors", *Proceedings of the IEEE*, vol. 66, no. 19, Oct. 1978, pp. 1126-1145.

Vog 88   U. Voges, ed., *Application of design diversity in computerized control systems*, Springer Verlag, Vienna, 1988.

Wil 92   C. Wild, S. Zeil, G. Feng, "Employing accumulated knowledge to refine test descriptions", *Software Testing, Verification and Reliability*, vol. 2, no. 2, July 1992, pp. 53-68.

VERIFICATION AND VALIDATION

Chair: B. Runge
STL Computer Automation, DK

The Rigorous Retrospective Static Analysis of the Sizewell 'B' Primary Protection System Software

N J Ward

TA Consultancy Services Ltd

Farnham, Surrey, England

1. Introduction

Sizewell 'B' is a Westinghouse designed Nuclear Pressurised Water Reactor (PWR) currently being built in Sizewell, Suffolk in the UK. It possesses two diverse protection systems whose role is to provide an automatic reactor trip when plant conditions reach safety limits and to actuate emergency safeguard features to limit consequences of a failure condition.

The Primary Protection System (PPS) is a microprocessor based system developed by Westinghouse in general agreement with IEC 880 [1]. The PPS is supported by a non-computer based Secondary Protection System (SPS), based on Laddic technology, developed in the UK by GEC. However, despite the presence of the SPS, the software within the PPS is considered to be 'safety critical' since the PPS on its own is required to meet an integrity requirement of 10^{-4} failures per demand.

The UK Nuclear Installations Inspectorate (NII), the regulatory body responsible for certificating the station, have two main criteria for software based safety systems, namely excellence of production and independent assessment. The rigour of each of these is required to be commensurate with the level of criticality and safety dependence of the software.

As a result of this requirement for independent assessment, Nuclear Electric, the utility responsible for Sizewell 'B', have had a range of assessments conducted on the PPS, independent of the equipment manufacturer. One of the main activities is the use of the static analysis tool MALPAS to rigorously verify the software against its specifications.

2. Scope of Independent Assessment Activities

The overall objectives of any rigorous independent software assessment are to demonstrate, to as high a level as is practicable, that the object code in the PROMs implements the user requirements, correctly, completely, safely and without side effects. However, with current software technology, no single activity is able to demonstrate this to the required level of confidence. Hence a range of activities are necessary, each providing assurance of individual steps in the software development life cycle such that, when taken together, the independent assessment activities provide the required level of assurance that the last step in the process meets the first.

On the Sizewell 'B' PPS five main activities are being conducted, as follows:

- **Engineering Confirmatory Analysis** conducted by NNC Ltd, involves the manual review of all specifications, from the system design requirements down to the low level code specifications, and the review of all source code and data, to ensure overall consistency and progressive implementation of the requirements.
- **Independent Design Assessment** conducted by Nuclear Electric's own team, ensures that all essential system functional requirements, including design principles derived from AGR reactor protection system experience, are correctly incorporated into the System Design Specification and down through the different stages of the software design.
- **MALPAS Analysis** conducted by TA Consultancy Services Ltd, involves the formal verification of the source code against its specifications and is discussed in detail throughout the rest of this paper.
- **Object/Source Code Comparison** conducted by Nuclear Electric's own Independent Design Assessment team and discussed in detail in reference [2], aims to eliminate the possibility of errors being introduced by the compiler and linker, by formally demonstrating (again using MALPAS) equivalence between object code in the PROMs and the source code from which it was generated.
- **Dynamic Testing** conducted by Rolls Royce and Associates Ltd, involves the conduct of an extensive series of tests (approximately 55000 randomly generated test cases) on one of the four identical channels of the PPS.

In total these activities are expected to have involved around 250 man years of effort, an amount equivalent to that spent by the software manufacturer in their own development and verification work, by the time that the software is certificated at the end of 1993. Although high, this level of effort is considered necessary, one reason being because the PPS software design pre-dates the practical application of the latest formal software design methods. Consequently the secondary aim of the static analysis of source code has been to impose formalism on the overall development process.

The main aim of the MALPAS analysis of the software is to verify, as formally as possible that the Sizewell 'B' software (source code) meets its specifications. This verification encompasses both the manual comparison of the analysis results against the design specifications and also the 'proof' of code against a mathematical representation of the detailed design specifications. However, it is important to appreciate from the above that, whilst the MALPAS analysis is the single largest activity, it is still one of a number of independent assessment activities which, taken together, are intended to provide comprehensive coverage and maximise confidence in the safety of the software.

3. Background to MALPAS Analysis

In 1988, following consideration of the activities necessary to demonstrate correctness of the PPS software, Nuclear Electric concluded that analysis methods were required which could demonstrate conformance with specifications and give high levels of confidence in freedom from errors. It was considered that dynamic testing on its own would not be able to give the required level of confidence because of known limitations of testing, such as the inability to achieve complete path coverage on any non-trivial piece of software. These same concerns had been addressed a few years previously by the UK Ministry of Defence and it was therefore decided by Nuclear Electric to follow the same approach adopted by the MoD, namely to use techniques known as Static Analysis.

The tool chosen by Nuclear Electric for the work was MALPAS [3] which itself originated from the UK MoD research in the 1970s and 1980s. MALPAS was chosen because it provided the level of formal analysis considered essential and was considered to be the tool most suited to retrospective analysis of code that has not been developed with formal verification in mind. The tool consists of three main sets of analysers, namely the Flow Analysers, the Semantic Analyser and the Compliance Analyser, each of which is able to provide a progressively more rigorous verification of the code against its specifications. It was Nuclear Electric's decision, right from the beginning, that all the MALPAS analysers should be used on the PPS software, with the emphasis being on the Compliance Analyser.

4. PPS Software Description

The Primary Protection System consists of four identical guardlines (channels), physically and electrically separated from each other, which perform coincidence (2 out of 4) voting to determine the need for action. Each guardline contains a number of sub-systems, providing facilities such as reactor trip, communications, engineering safety features, and auto test. Each subsystem comprises a general purpose 'host' processor and a number of slave processors. The host processor performs the unique functions required by the subsystem and is supported in this by the specialised slave processors which provide standard functions such as communications, analogue data acquisition and diagnostic monitoring.

The PPS software is primarily written in PL/M-86 with some ASM86 and small amounts of PL/M-51 and ASM51 variant code. In total there are approximately 100,000 lines of unique executable code with a typical host processor containing 40,000 lines of source code and a typical slave processor 10,000 lines. The software is highly modular and has been written mostly as 'general purpose' reusable software providing a range of common functions, configured through the use of configuration data. There is a relatively small amount of applications level code providing the particular functionality of the protection system.

The software is developed from two levels of specifications, namely a Software Design Requirements (SDR) which is a high level specification of the required functionality of each section of the code, and a Software Design Specification (SDS) which describes how the functional requirements are met by the design. Both of these sets of specifications are written in 'natural language' (American English). The SDS includes details of the precise functionality of each program section, and also contains data flow tables providing details of all program section variables and inputs and outputs.

5. Overview of MALPAS Analysis Process

The analysis of the PPS software is conducted on a procedure-by-procedure basis in a bottom-up manner. That is to say that the analysis commences with those procedures that call no others and then progresses up the call hierarchy until the top level applications code is reached. All the MALPAS analysers are run on each procedure and the results verified against the two levels of specification. The higher level specification (SDR) is taken to be the primary document against which the code is verified, with supporting information provided by the lower level SDS. All of the code that can be accessed during on-line operation is being subjected to MALPAS analysis, irrespective of the perceived criticality of individual sections of code within the system.

The analysis process has not changed in concept since the work began at the start of 1989. However, the detail of how the work is conducted has changed substantially both as experience has been gained over the years and with the progressive increase in the size of the analysis team. The following sections discuss various aspects of the analysis process.

6. Translation

It is necessary for source code to be translated into MALPAS's own input language, known as IL, before analysis can take place. IL is a strongly typed language with many features similar to those of Ada. One feature in particular, similar to Ada packages, is that all IL procedures comprise separate bodies and specifications (known as PROCSPECs). The advantage of this is that, following the analysis of the procedure body, all that is required for analysis of calls to that procedure is the procedure specification. This feature greatly facilitates bottom-up analysis.

As has been mentioned above, the majority of the code in the PPS is written in PL/M-86 and an automatic translator to convert PL/M-86 into IL was developed by TA Consultancy Services specifically for this project. Considerable care was taken in the derivation of the mappings between PL/M-86 and IL to ensure that they were a strictly correct representation of the semantics of the PL/M-86 language and that they were at an appropriate level of abstraction/precision to facilitate analysis. The translator was also designed to produce a number of checks, for example for exceeding array bounds or loop counter overflow, that could be formally checked during Semantic and Compliance Analysis.

The translator takes an input PL/M-86 source text, typically a module, along with other relevant inputs, such as PROCSPECs from previously analysed procedures, and automatically produces an IL translation. Error and warning messages are given where language features are encountered that the translator is unable to convert to IL automatically. The analyst then has to take appropriate action, checking the accuracy of the particular part of the translation or making suitable manual changes.

Pointers are one PL/M-86 language feature that can cause such problems, firstly, because there is no concept of pointers within IL and, secondly, because, they represent a source of aliasing which contradicts IL's philosophy of all variables being distinct and disjoint with all assignments having no side effects. Unfortunately PL/M-86 is a pointer-based language with the use of pointers being essential for a number of operations. The most significant of these concerns the passing of OUT or INOUT parameters to procedures which must be performed by reference (ie pointers must be used to pass any item to a procedure that may be modified by that procedure).

The solution adopted for the analysis is for all pointers to be 'dereferenced' (ie the pointer replaced by the item pointed at) during the translation process. Fortunately the use of pointers is fairly well constrained in the PPS software (through the application of Westinghouse's coding standards) and most pointers are used only at the procedure call stage and point to a single variable/memory location. The translator is therefore able to automatically dereference the majority of pointers used within the code. Those that the translator is unable to automatically dereference (for example pointers to templates) are brought to the attention of the analyst through error and warning messages and the analyst then has to implement an agreed (and subsequently reviewed) manual translation.

7. Analysis Process

Every program section/procedure in the source code is subjected to the three main stages of MALPAS analysis in sequence, with the following particular activities being conducted under each one.

7.1 Flow Analysis

This involves the analysis of the flow of control, data and information through each procedure, in particular:
- **Control Flow Analysis** The verification of safe control flow through each procedure (eg ensuring the absence of multi-entrant loops and black holes) and the confirmation that the code is well structured.
- **Data Use Analysis** The analysis of the use of all parameters and variables within each procedure to ensure that this use agrees with that detailed in the code specifications (SDS) and to ensure that the usage is safe and in accordance with general software engineering good practice rules.
- **Information Flow Analysis** The analysis of all dependencies between input and output parameters for each procedure to ensure that these agree with the code specifications (SDR/SDS), and the identification of redundant statements.

7.2 Semantic Analysis

This analysis involves the confirmation both that the functionality of the code conforms with the specifications (SDR and SDS) and is reasonable from an engineering knowledge of the system. The MALPAS Semantic Analyser converts the (potentially confusing) sequential logic of the code of each procedure into a clearer parallel form, giving a precise mathematical relationship between inputs and outputs for each path through that procedure. The analyst is then able to check the detailed semantics of every path through the program section and manually compare this against the specification to verify one against the other.

7.3 Compliance Analysis

The purpose of the Compliance Analysis is to obtain the highest level of confidence in the correctness of the software by formally verifying the code against a mathematical specification. Specific objectives are to demonstrate that the code of each procedure:

- conforms to the functional aspects of its specification (SDR, supported by SDS),
- respects any state invariants
- performs its specified functions without corrupting the computing environment and without corrupting data owned by any other module or procedure
- conforms to the static semantics of the language in which it is written

The MALPAS Compliance Analyser requires the specification to be represented as PRE and POST conditions for each procedure, along with any necessary ASSERT statements within the body of the code, and the analyser will then show whether the code meets the specification. The analyser identifies any differences between the code and specifications as a 'Threat', expressing this as a mathematical expression which details the domain over which the code and the specification disagree. When the code and specification agree the tool shows that the Threat is false.

The Compliance Analysis is by far the most important part of the MALPAS analysis of the PPS software and also involves the majority of the effort. This effort is expended partly in the derivation of the mathematical specification and partly in the provision of guidance to the tool to aid simplification of the Threat. The analyst is also required to derive invariants for each loop, expressing these as ASSERT statements, so that properties of each loop can be proven.

The first part of the Compliance Analysis work is the construction of the mathematical specification from the natural language SDR and SDS. This work represents a substantial challenge to ensure both correct interpretation of the existing specifications and that the important functionality and properties are modelled in the mathematical specifications.

Ideally the analyst should be able to define a high level abstract mathematical specification from the SDR and then derive refinement detail from the SDS. For example, a functional relationship could be defined between an output and appropriate inputs, using information from the SDR, and detailed semantics can then be defined (using an IL feature known as replacement rules - similar to OBJ rewrite rules) from information in the SDS. In practice, due to the SDRs and SDSs being at varying levels of detail and precision, the ideal is rarely possible and analyst skill and judgement is required to derive the mathematical specification from both specification documents.

The second, and most time consuming, part of the analysis is the use of techniques to assist the Compliance Analyser in its demonstration of whether the threat is false. MALPAS contains a powerful Algebraic Simplifier for simplifying expressions but, like all such tools, it has its limitations, particularly with the simplification of some forms of expressions. Analyst assistance may involve the re-writing of mathematical specifications into forms more amenable to simplification, the use of replacement rules to define additional theorems or the use of extra ASSERT statements. The derivation of a 'Threat False' statement is therefore an iterative process, involving running the tool, assessing the Threat condition and then making appropriate changes to reduce the expression on a subsequent run of the tool.

8. Ensuring Correctness, Consistency and Reproducibility

Despite using an automatic tool for the rigorous static analysis, it can be appreciated from the above description that substantial analyst skill and judgement is involved in interpreting the tool's output and in deriving the mathematical specifications. Because of this it is essential, given the size of the project, to ensure that all analysts are performing the work to the same standard. It is also considered important to ensure that all analysis is reproducible and could be re-run, for example if a query was ever raised by the regulatory authority on a particular piece of analysis.

Five main measures have been adopted by TA Consultancy Services to ensure that these requirements are met. The first, perhaps obvious one, is that all work is conducted within a defined quality system and to the requirements of BS5750 (ISO9001). The second, related, measure has been the derivation of a detailed 'standards and procedures' document, approximately 200 pages long, which defines in great detail precisely how the analysis is to be conducted. This document, analogous to a coding code of practice for software development, is followed by all analysts and ensures uniformity of approach.

The third measure is to ensure comprehensive recording of all aspects of analysis. In addition to the MALPAS analysis results themselves, analysts are required to fill out a series of forms to summarise the results and to record their interpretation of the results, the background to their derivation of the mathematical specification, their assessment of the conformance of the code and specifications, and details of all anomalies.

The fourth, and possibly most important measure, is the conduct of peer reviews of all analysis work. The main aims of these reviews are to ensure that the defined processes have been followed for the analysis and to ensure that Compliance proofs are well founded, that loop termination has been demonstrated and that replacement rules are correct. In addition they help to distribute knowledge of different parts of the PPS software and experience of the range of analysis techniques and to generally ensure consistency between analysts. The reviews are conducted on a controlled set of results, using a series of checklists and review deficiency forms.

The final measure is the enforcement of strict configuration control of all documents and analysis results, both in paper and magnetic media form. In terms of computer(VAX)-based configuration control a rigidly defined manual system is used involving controlled and frozen directories with a named librarian being the only person authorised to move files from one status to another.

9. Technique Refinement

The analysis techniques used on the project have been continuously refined and improved during the 4½ years of the project. This has covered everything from the forms on which the interpretation of the results are recorded (despite streamlining, there are still perpetual complaints about too much paperwork) to the format and depth of reviews. However, changes in the techniques used are necessarily slow to be implemented, firstly to ensure compatibility with the analysis of lower level procedures, and, secondly, because of the inertia and diverse views of such a large team.

The area in which there has undoubtably been the most improvement is Compliance Analysis. Substantial experience has been gained in the most effective ways to express mathematical specifications in order to be able to demonstrate conformance with the code. Numerous technical papers have been written internally to provide a series of hints and tips to all analysts on how to resolve particular problems that arise during Compliance Analysis, either in the expression of the mathematical specification, or in the resolution of the Threat expression to false.

It has also been of substantial benefit for TA Consultancy Services to themselves be the developers and suppliers of MALPAS, as well as the users of the tool on this project. Analysts have had access to the implementors of the tool and have been able to obtain expert advice on the best way to represent specifications in order to facilitate expression simplification. Furthermore, experience from the analysis project has been fed into the development of the algebraic simplifier to improve its performance with specific types of expressions, to the extent that speed improvements in excess of two orders of magnitude have been made to the Compliance Analyser in some areas.

One interesting aspect is that new techniques continue to be required even after 4½ years. This is primarily because different areas of code are encountered that present a new series of problems. One area that required considerable effort in the past was the analysis of ASM86 code involving double-length arithmetic. Another area where a whole range of new techniques has been necessary concerns the analysis of the shared memory communication system used on the PPS. It is possible that this could be the subject of a paper on its own at a later date.

10. Analysis Limitations

It is important to recognise that the technique being performed is Static Analysis which, by definition, is concerned with the non-dynamic aspects of the software. Although many dynamic and timing related aspects of the software are modelled and analysed in detail during the MALPAS work, others are not if it is considered more appropriate for them to be verified by other means. Where such cases arise, notification is given to Nuclear Electric that the specific aspects have not been verified statically and that other means are required.

The aspects not verified statically relate primarily to those concerned with real time operation and with interaction with hardware. For example the checking of a communications protocol with requirements of 'waits' of specified lengths of time may be more appropriate through dynamic testing on the target hardware. This is partly because testing of all paths through such procedures is likely to be possible but, more importantly, because dynamic testing will effectively also provide validation of the specification and will reveal whether the design actually works in practice.

11. Reporting and Resolving Results

The only deliverables to the customer (Nuclear Electric) resulting from the MALPAS analysis are comments detailing the anomalies (ie differences between code and specifications etc) found during the work. All comments are provisionally categorised according to their criticality by TA Consultancy Services prior to the comments being reported to Nuclear Electric. The following five categories are used for this reporting process:

Cat 1: Essential code change to address potential maloperation of the PPS

Cat 2: Specification or requirements change.

Cat 3: Code change to remove non-critical anomalies or to address necessary improvements.

Cat 4: Comments for which no action is required

Cat U: Interim categorisation to be resolved at sentencing

This last category is intended to cover comments where the analyst has insufficient information regarding the rest of the system's operation to be able to determine the severity of the anomaly.

Following the reporting of provisionally categorised comments a sentencing process is undergone between TA Consultancy Services, Nuclear Electric and Westinghouse (often also held with the NII in attendance). During the sentencing process each comment is discussed and resolved and given a mutually agreed final categorisation (using categories 1-4 above) which determines the corrective action, if any.

12. Project Status and Results to Date

The MALPAS analysis work started in January 1989 and is scheduled to complete at the end of October 1993. Because of the size of the analysis task and because the 'production' version of the software was not going to be ready much more than a year before the required certification date, the analysis commenced on 'pre-production' versions of the software.

Obviously, there are re-analysis cost implications of analysing early software versions and the analysis team has been growing to optimise the costs of the analysis and at the same time meet the required timescales. In particular the growth of the analysis team has been rapid since the start of 1992, when TA Consultancy Services had a team of 15 people working on the task, to the present time in May 1993 when the team has grown to in excess of 80 people working full time (including managers, analysts and dedicated support staff).

Considering the results, in terms of numbers of anomalies raised, the latest figures available relate to April 1993 at which time analysis had been completed and comments sentenced for 55% of the production version software. Just under

2000 comments had been raised from the analysis of this software, a rate of around one comment for every 30 lines of code. This rate compares well with other 'traditionally' developed and verified safety critical software on which TA Consultancy Services have conducted similar retrospective MALPAS analyses. After sentencing the majority of the comments (52%) were classified as category 4 (no action comment), 40% were category 2 (specification change) and 8% were category 3 (non-critical code change). There were no category 1 comments.

The no action comments cover a wide range of raised anomalies, some of which are suggestions for code or design improvements, for example to make code more defensive, but which are considered not to justify code changes. Others may relate to trivial specification deficiencies or points for clarification. Similarly the category 3 comments cover a wide range of instances where it has been considered necessary to make changes. Some have been because of straightforward deficiencies such as type mismatches at procedure calls and others relate to improvements in code defensiveness.

13. Conclusions

The MALPAS Analysis work conducted on the Sizewell 'B' Primary Protection system Software has shown the feasibility of conducting a rigorous retrospective analysis on a large software system that has not been developed with this form of analysis in mind. The tools, methods and techniques used for the work have been shown to be very suitable but the costs of their use in such a retrospective manner are high.

At a more detailed level the project has demonstrated the benefits of both Compliance Analysis and of in-depth reviews of all work. During Compliance Analysis, it has been found that the twin activities of deriving the mathematical specifications and directing the tool to show conformance, results in both the code and specifications being scrutinised in the minutest detail, thereby leading to the identification of subtle but potentially significant problems.

The benefits of the reviews in terms of the consistency and checking that they provide are considered to far outweigh their not insignificant cost (in terms of time taken for the reviews and any necessary rework). Similarly with the Compliance Analysis, although the costs of Compliance Analysis are high in absolute terms they are low in relation to the costs of any potential software malfunction. Consequently, the conduct of rigorous static analysis, including Compliance Analysis, is considered to be essential for software within systems of the criticality and potential failure consequences of the Sizewell 'B' PPS.

Whilst not claiming that the analysis provides a formal proof of the safety of the software, the analysis does provide a formal verification that the source code meets its low level and intermediate level specifications. The analysis has increased the integrity of the software through code and documentation modifications that have been made as a result of anomalies raised during the analysis. Furthermore, through its rigour, the analysis has greatly increased confidence in the correctness of the software and it is hoped that this, taken along with the four Independent Assessment activities will greatly contribute towards a successful certification of the software by the UK NII.

14. References

1. IEC 880. Software for Computers in the Safety Systems of Nuclear Power Stations. International Electrotechnical Commission. 1986.

2. Demonstrating Equivalence of Source code and PROM Contents. Paper by D J Pavey & L A Winsborrow of Nuclear Electric, presented at the Fourth European Workshop on Dependable Computing on April 8-10, 1992 in Prague, Czechoslovakia.

3. TACS/020/17/2. MALPAS Management Guide. TA Consultancy Services Ltd. Issue 3, February 1992.

© TA Consultancy Services Ltd, 1993

A Safety Critical Computer System in a Railway Application

Bernhard Stamm, René Baumann, Martin Kündig-Herzog
Siemens Integra Verkehrstechnik AG
8304 Wallisellen, Switzerland

Abstract

This paper describes a safety critical computer system used for automatic train control. It has been developed during the last three years and is currently in the phase of final testing and validation. After a short system overview, the paper will concentrate on safety aspects in system design and on the description of the verification and validation process that was chosen. This specifically includes the problems and aspects of the selection of applicable norms, the definition of a validation and verification plan and the upper levels of verification.

Keywords

automatic train control, verification & validation plan, fault tree analyses, risk assessment, design rules

1   Introduction

In the past decade suburban public transportation has experienced an enormous rise in popularity. In Switzerland, many existing railway companies are confronted with the need of increased train capacity and train frequency.

As many of these networks consist of mostly single track lines and are built in densely built areas, increasing the capacity of the existing lines is often the only solution to suiting these demands.

Railway safety has reached a very high standard. One of the biggest remaining safety problems today is the supervision of train drivers. The above mentioned increased train speeds and traffic density of today's railroading leads to an increased pressure on the train drivers and therefore to an increased chance of human error. Also due to the mentioned factors the probability and consequences of accidents in case of human errors have drastically increased.

Based on these circumstances four railway companies under co-ordination of the Swiss Federal Bureau of Transportation defined a specification for a new, advanced automatic train control system.

2   System Overview

2.1   General System Requirements

A study of incidents in the past, and of the present safety situation led to a system requirement specification. It is helpful for the general understanding of the proposed solution to summarise the major requirements.

- Train speed has to be supervised continuously, based on signal aspects, track conditions and permitted train speed. This supervision has to include the brake curves in order to maintain a speed permitted by these sometimes overlapping constraints at any time.

- A speed of 10 km/h in excess of the permitted value has to be prevented at any time.

- Trains must be prevented from unauthorised departures in both directions as well as from overrunning of signals at danger (e.g. stop indication).

- Signalled stops must be adhered to within a limit of 10 m.

- Disturbance of train operation has to be kept to a minimum. Apart from entering train data prior to the first departure train crews must not manipulate the system during train operations.

- A change to a less restrictive signal aspect has to be recognised immediately by the system. The capacity of existing lines has to be utilised fully. This includes the speed profile given by track geometry.

- The proposed system has to be fail safe, e.g. a safety certification similar to that for solid state interlocking systems is required.

As none of the currently produced ATC-systems fulfilled the given specification nor had the capacity of getting modified up to an appropriate level it was decided to develop a new system. This also allowed us to take benefit of the enormous advances in technology during last few years.

2.2 System Functionality

The following paragraphs will give a short description of the functionality of the ATC-system .

The proposed new ATC-system is used to supervise train movements according to signal aspects and track conditions and to prevent any unauthorised exceeding of the given speed limits at any time.

It is based on a data base which contains a route map that describes the network of a supervised railway. This data base contains the track geometry (distances, gradient, permitted track speed, points, track arrangement etc.), the location of signals, level crossings etc. and general data like the braking characteristics of vehicles etc..

This data base is stored in a data module on board of each vehicle. This module is easily changeable in case of changes in the network data.

Signal indications and point positions of a certain area, usually a station, are transmitted cyclically to any vehicle within this area. Individual addressing of specific vehicles as well as communication from train to track is not necessary. A pilot line was chosen as a transmission system but data radio or similar technology could also be used.

Computers on board of the vehicles are permanently updating their position within the network data with the aid of wheel rotation encoders. Wheel slip gets suppressed by an algorithm similar to the ones used in wheel slip monitoring systems. The position measurement is readjusted at certain locations with the aid of passive synchronisation points. Their exact location is recorded in the data base.

The on-board computer of each vehicle permanently evaluates its route based on the current position of the vehicle, the direction of movement, the data base and the received point positions. It searches the route for speed restricting elements like signals, track speeds, speeds over points etc. and calculates the maximum permitted speed at any moment (Figure 1). If the actual speed reaches the permitted value, a warning is issued to the driver requiring him to reduce the train speed. When the speed exceeds the permitted speed by a defined value (e.g. 5 km/h) the computer actuates the regenerative/rheostatic brake. As soon as the speed is below the limit of the speed curve the braking is discontinued. In the event of a speed excess greater than what is considered tolerable (e.g. 10 km/h) the computer actuates an emergency brake application. The latter should only occur if a driver doesn't react properly to an issued warning and if the vehicle doesn't react to the application of the regenerative/rheostatic brake, whether due to insufficient adhesion or with older vehicles because they are not equipped with a regenerative/rheostatic brake.

Temporary restrictions (e.g. construction sites etc.) are programmed into the stationary computers and transmitted to the vehicles together with the signal indications and point positions.

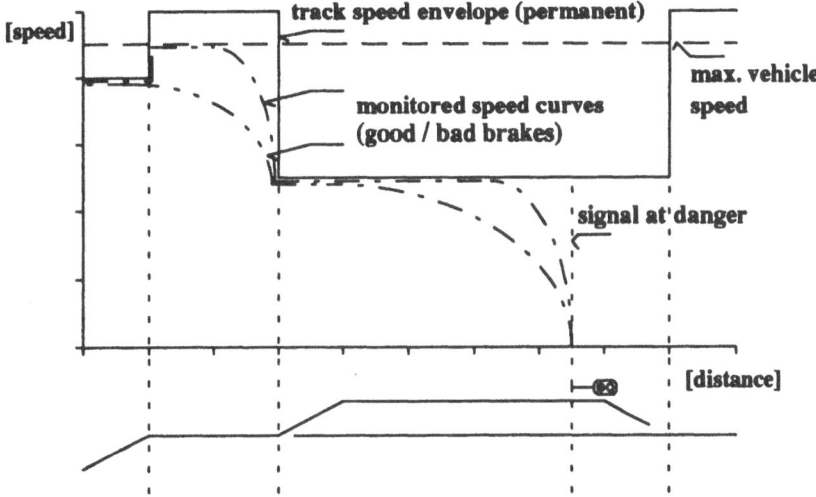

Figure 1: Examples of monitored speed curves with overlapping restrictions

The actual system structure is shown in Figure 2. The stationary computer in the station interlocking as well as the computer on board of each vehicle are designed with a 2 out of 3 structure.

Each sub-computer of the 2 out of 3 systems contains the full system functionality including all the necessary hardware and software. While performing the functionality of the system the three sub-computers interchange input data and final results of their calculations. In case of differences between the results of the three, the one with the differing data gets switched off by the two others. The remaining sub-computers continue as a 2 out of 2 system. In case of an additional failure, the system shuts down entirely. In the case of the stationary computer this means that no telegrams are transmitted. In the case of the on-board computer this leads to an application of the emergency brake.

There is of course a large number of additional functions like emulation of the existing ATP-system, special modes for depots or industrial spurs, shunting mode, overriding of signals at danger in case of an interlocking failure, exception handling etc. but their description is neither needed for the general understanding of the system design nor for the description of the verification and validation process.

186

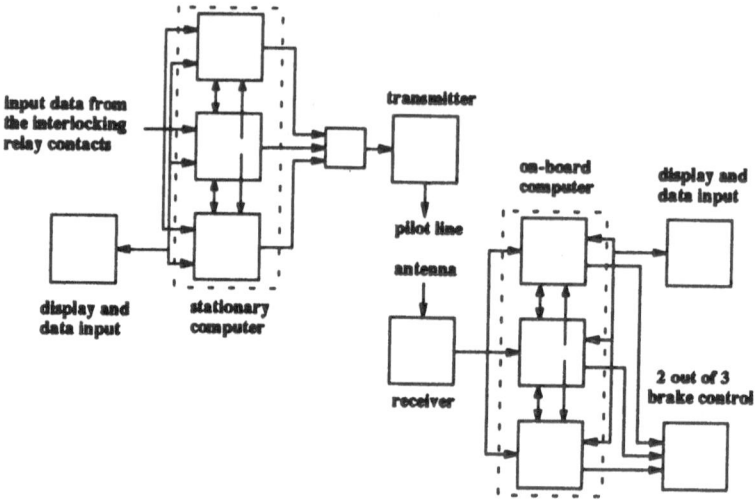

Figure 2: General System Architecture

3   Safety Aspects of the System Design

The described ATC-system includes cab signalling of the permitted speed. At a later stage of the installation and usage of the system it is intended to remove the existing conventional signalling. This leads to the requirement for a fail safe system design, as the correct observation of lineside signals by the driver can no longer be assumed. Together with the required functionality, reliability and redundancy, the following safety concept was chosen:

- A 2 out of 3 system structure is used to detect hardware failures and to give the required level of redundancy. A data exchange between the sub-computers allows evaluation of the correct functioning of the hardware including all the data input circuitry. This method is supported by additional self tests. Output circuitry for safety relevant outputs is designed based on conventional rules for fail safe hardware (relays etc.). Otherwise, the use of hardware components designed after the classical rules for fail safe hardware is wherever possible avoided. This allows the usage of industry standard hardware in most components and eases later hardware upgrades.

- Data transmission between track and train is designed taking advantage of the 2 out of 3 system structure. Each of the three sub computers in the interlocking generates data telegrams including data encoding. These telegrams are transmitted by each of the three sub-computers in a cyclic manner. The on-board computers require the reception of at least two identical telegrams from two different sources to accept the received data. This procedure tolerates the transmission of incorrect telegrams by one sub-computer. It is also immune against the falsification of single telegrams into different but correctly encoded telegrams.

- The necessary correctness of the system software is ensured by the strict application of the chosen norms and by rigid testing. This includes modern testing methods and procedures like for example static software analysis and path coverage analyses.

- Data entry of safety relevant data like train characteristics is secured by the chosen data entry procedure. This procedure includes the echoing of the entered data after validation by the three sub-computers. This echo gets displayed via an independent hardware channel for verification and confirmation by the train driver. This procedure additionally allows the detection of data distortion or hardware failures in the data entry channel.

- The correctness of the route map is ensured with several independent procedures. First, data collection is performed with the aid of a computer based data collection software. This software ensures systematic working procedures by forcing the user to generate and to update the data base following well defined paths. Second, the data collection software performs a whole set of data verification procedures. This includes boundary checking of entered data elements and data consistency testing after data entry. The software automatically generates indices for new data releases to ensure the exclusive use of valid data. Once a new version of the data base has been generated a back transformation software regenerates a route map for an additional visual data verification.

4 Norms and Standards used for System Verification and Validation

Any new development has to consider the requirements set by the opening of the European Market to international competition. This includes of course norms and standards required for verification and validation. One of the major problems in this field is the current lack of harmonisation of standards.

In several European countries, national railway companies and industry have in the past developed local norms, standards or rulebooks. Some of these are based on long term experience with conventional relay based technology or with electronic hardware.

Some norms for electronic systems including microprocessors and software are currently under development, but they follow different approaches and philosophies. This depends on the country from which they originate and on the understanding and the definition of the term safety on which they are based.

A study was carried out to compare the currently used norms in Great Britain, France and Germany with the ones used in Switzerland. The study also had a look at the proposals in work in the different committees of the CEN/CENELEC. This comparison showed that bigger differences between the philosophies of the proposed standards still have to be overcome. As a result of the study, the following norms were chosen as a base of the verification and validation plan:

- the British norm RIA 23 [1], which was based on the Draft IEC SC65A WG9, for the implementation of the safety analyses and the risk assessment,

- the proposed DIN norm DIN V 19250 [2] for the evaluation of the required safety level

- and the proposed DIN norm DIN V VDE 0801 [3] for the definition of the further implementation.

In the mean time the work groups WGA1 and WGA2 of the CENELEC committee SC9XA, which is responsible for the proposal of a new norm, chose two IEC papers [4], [5] as a base of their future standardisation work. These papers take reference to the chosen DIN norms.

5   The Verification and Validation Plan

Based on the chosen norms a verification and validation plan was defined. This was done in close co-operation with the Swiss Federal Bureau of Transportation, which will be responsible for the final system approval. The verification and validation plan covers the following topics:

- project management

- development

- commissioning

- maintenance

- documentation

- configuration management.

For each of the above mentioned topics individual chapters cover the following topics:

- Norms and standards to be applied for each of the defined steps.

- Definition of a detailed plan for the verification and validation of each step to reach system approval.

- Definition of the persons or offices responsible for the implementation, verification and validation of each of the defined steps.

- Definition of the responsibilities of each of the participating persons or offices to perform the above mentioned steps.

Figure 3 shows as an example the proposed plan for the system development:

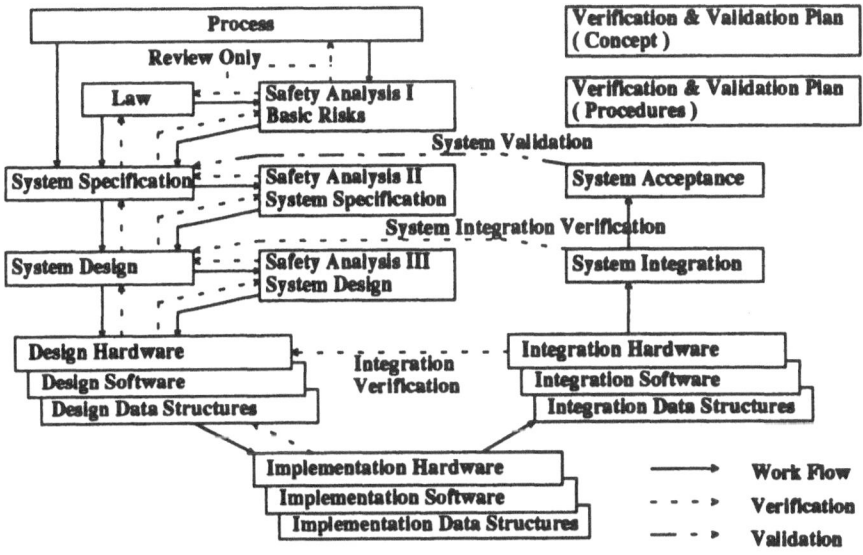

Figure 3: Verification and Validation Plan

Each step (work flow) of the development as shown in the plan above is verified with the defined rules and according to the chosen norms of the verification and validation plan and documented in an individual paper. Each of these papers is presented individually to the supervisory authority for approval. This procedure allows an early detection of disagreement about the contents of the safety certification and also fast certification after final testing, even in the case of such a fairly complex ATC-system.

6 Examples of Verification Procedures

This chapter gives a more detailed look into two of the large number of different areas of system verification. These examples got chosen to show the spectrum of measures necessary to fully implement a verification and validation plan.

6.1 Example 1: Verification of the System Specification

The first example shows the safety analyses that were carried out to verify the system specification. It is based on a fault tree analyses of the current operations at the participating railway companies. The fault trees were verified with the available accident statistics and with a review with railway experts from different branches.

Figure 4 shows as an example a fraction of one of the fault trees.

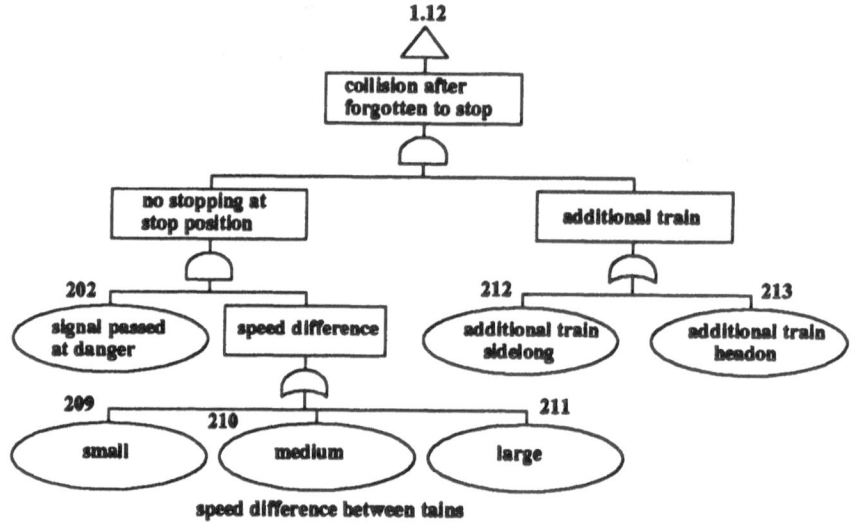

Figure 4: Example of Fault Tree Analyses

The different possible accidents deducted from this fault tree analyses got weighted, based on the probability of occurrence of the different causes leading to these accidents and the scale of the damage that could likely result. This procedure is based on the model of RIA 23 [1], which again is based on the Draft IEC SC65A WG9 [4]. It led to tables like the one shown in Figure 5.

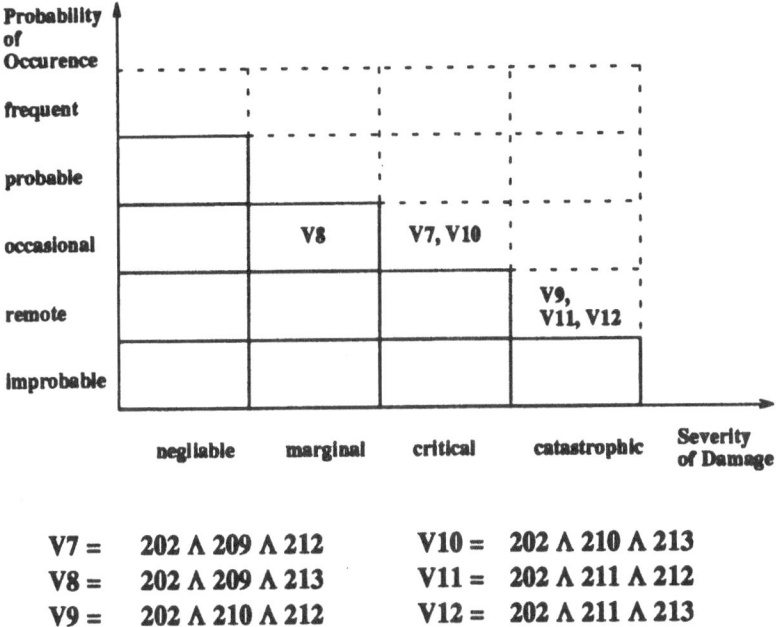

$$V7 = 202 \wedge 209 \wedge 212 \qquad V10 = 202 \wedge 210 \wedge 213$$
$$V8 = 202 \wedge 209 \wedge 213 \qquad V11 = 202 \wedge 211 \wedge 212$$
$$V9 = 202 \wedge 210 \wedge 212 \qquad V12 = 202 \wedge 211 \wedge 213$$

Figure 5: Example of Failure Classification

The norm DIN V VDE 0801 [3] contains more detailed rules for the implementation of the verification and validation plan compared to the RIA 23 [1] and got therefore chosen as a base for further steps. To allow this, a transformation of the required safety level via DIN V 19250 [2] got undertaken. Some assumptions based on this norm had to be made. As the possible severity of accidents with trains in suburban traffic is limited, catastropic events were defined to correspond to S3. The probability remote was assigned to W2. These assumption led with the definition of a frequent to permanent collective stay within the systems sphere to a safety requirement level 6.

With assumptions on the possible effect of an additional safety system on the probability and the resulting damage of accidents, the system specification was verified. This verification also led to the definition of the required integrity level for the system according to the chosen norm [2].

6.2   Example 2: Software Implementation

The second example gives an overview of the procedures defined for the implementation and the testing of the software.

As already mentioned each step (work flow) of the development as shown in the corresponding verification and validation plan is performed and verified according to a reviewed and approved guideline. The guideline for software implementation covers the following topics:

- Definition of the chosen programming language (in this case Modula 2). This includes rules, restrictions and recommendations for the use of language elements which could lead to difficulties in the verification of the code, like pointers, variant records etc..

- Definition of the structure and the look of source files to allow several programmers to share code without the risk of misunderstanding or miss-interpretations due to personal style.

- Definitions for the naming and the marking of language elements like variables, constants, subroutines etc. for the same reasons.

- Definitions for the required documentation of program code including style, depth, volume etc..

- Definition of the procedures and tools used to verify the correctness of the code including check lists. The tools used include for example: syntax check of the source code including the above mentioned rules and restrictions, static code analyses including software metrics and path coverage analyses including the definition of the used tools and the registration of test cases and test results.

7   Conclusions and Acknowledgements

The experience gained in the development of the described ATC-system in the last three years is summarised in the following list:

- As long as there is no harmonisation and standardisation for the verification and validation of computer based safety systems in railway applications in Europe, a selection of the norms currently under development has to be made individually for new projects. Once a harmonisation has been reached the transition time for the application of the new norms will be very short compared to the development cycle of a larger project. It is therefore advisable to already today use norms that will probably closely conform to the new norms. Existing norms are only applicable in some countries and usually represent a country specific approach to safety. They are often not applicable in other countries.

- A verification and validation plan has to be defined, approved and followed from the very beginning of a new development to avoid later problems, delays and cost for system approval.

- The experience shows that with the application of the above mentioned plan a development can get carried out in a shorter time due to a permanent co-ordination between developer, customer and supervisory authority.

- The experience also shows that the application of the above mentioned plan allows system validation at a significantly lower cost due to an early involvement of the supervisory authority. Wishes and requirements for changes can get respected at the appropriate level of the development cycle.

As a final acknowledgement the authors would like to express special thanks to the Swiss Federal Office of Transport and to the participating railway companies, which with their knowledge, experience and openness to new proposals made significant contributions to the project.

References

1. R.I.A 23, Draft, Standard for Safety Related Software for Railways (Signalling), Version 1.0, Issued 23 January 1991
2. DIN V 19250, Grundlegende Sicherheitsbetrachtungen für MSR Schutz-einrichtungen, Vornorm, Januar 1989
3. DIN V VDE 0801, Grundsätze für Rechner in Systemen mit Sicherheitsaufgaben, Vornorm, Januar 1990
4. Draft IEC SC65A WG9, Software for Computers in the Application of Industrial Safety-Related Systems, 1989
5. Draft IEC SC65A WG10, Functional Safety of Electrical/Electronic/Programmable Electronic Systems: Generic Aspects, Part 1: General Requirements, 1989

Session 6

TESTING

Chair: B. Cronhjort
Royal Institute of Technology, S

Confidently Assessing a Zero Probability of Software Failure

Jeffrey M. Voas, Christoph C. Michael*   Keith W. Miller†

ABSTRACT

Randomly generated software tests are an established method of estimating software reliability [5, 7]. But as software applications require higher and higher reliabilities, practical difficulties with random testing have become increasingly problematic. These practical problems are particularly acute in life-critical applications, where requirements of 10^{-7} failures per hour of system reliability translate into a probability of failure (pof) of perhaps 10^{-9} or less for each individual execution of the software [4]. We refer to software with reliability requirements of this magnitude as ultra-reliable software.

This paper presents a method for assessing the confidence that the software does not contain any faults given that software testing and software testability analysis have been performed. In this method, it is assumed that software testing of the current version has not resulted in any failures, and that software testing has not been exhaustive. In previous publications, we have termed this method of combining testability and testing to assess a confidence in correctness as the "Squeeze Play" and "Reliability Amplification," [15, 13] however, we have not formally developed the mathematical foundation for quantifying a confidence that the software is correct. We do so in this paper.

1   Introduction

The probability of failure of a program is conditioned on an input distribution. (Another term for an input distribution is "operational profile" [14].) An input distribution is a probability density function that describes for each legal input the probability that the input will occur during the use of the software. Given an input distribution, the *probability of failure (pof)* is the probability that a random input drawn from that distribution will cause the program to output an incorrect response to that input.

Software reliability is defined as the probability of failure-free operation of the software in a fixed environment for a fixed period of time. Note the differ-

---

*Reliable Software Technologies Corporation, Penthouse Suite, 1001 North Highland Street, Arlington, VA 22201 USA, (703) 276-1219, e-mail: jmvoas@mothra.isse.gmu.edu.

†Correspondence address is: Department of Computer Science, College of William & Mary, Williamsburg, VA 23187 USA, (804) 221-3464, e-mail: miller@cs.wm.edu.

ence between the definition for reliability and that for software probability of failure: probability of failure is time independent. However, both reliability and probability of failure are tied to a specific environment.

Even if ultra-reliable software can be in theory *achieved*, we cannot comfortably depend on this achievement unless we can assess its reliability in a convincing, systematic, and scientific manner. As pointed out in [2], black-box testing is impractical for establishing these very high reliabilities. In general, by executing T random tests, we can estimate a probability of failure in the neighborhood of $1/T$ when none of the tests reveal a failure [3]. If the required reliability is in the ultra-reliable range, random testing would require decades of testing before it could establish a reasonable confidence in this reliability, even with the most sophisticated hardware. Based on these impracticalities, some researchers contend that very high reliabilities can not be quantified using statistical techniques [1]. In dismissing all possible statistical techniques because of the practical problems with random testing, we think Butler and Finelli are being premature, and this paper describes how a statistical technique *in addition to* random testing may be brought to bear on the problem of assessing ultra-reliability.

Two purposes of software testing are establishing a reliability estimate and finding software faults. When software does not fail during non-exhaustive testing, there is good news and bad news. The good news is that we suspect that the software no longer has gross faults. The bad news is that testing is no longer as effective at estimating reliability or at uncovering the remaining faults (if they exist).

It is disheartening to realize that it is more difficult to assess the reliability of a program that has not failed than it is to assess the reliability of a program undergoing some proportion of failures. Previous work has tackled the problem of assessing the probability of failure of software that has not failed [3]. In this paper, we consider a closely related problem using an approach that is distinct from traditional testing. We are interested in the confidence that the software is correct given that:

1. the program has not failed in T tests, and

2. we have a prediction of the minimum non-zero failure probability (from testability analysis that has been performed on the program).

We agree that testing, on its own, cannot be used to establish ultra-reliabilities. However, testing is not the only statistical technique possible for analyzing software reliability. We believe that software testability, as a quantifiable measure of software quality, can be used in conjunction with testing to assess reliability.

2 Software Testability Analysis

We now discuss "sensitivity analysis," a statistical technique *complementary* to testing. Sensitivity analysis is one algorithm for performing testability analysis. Used in conjunction with testing, sensitivity analysis may allow us to estimate reliability to a much higher precision than was possible with testing alone. A preliminary model for doing this reliability assessment was previously presented at [13].

Sensitivity analysis uses program mutation, data state mutation, and repeated executions to predict a minimum fault size [6]. The minimum fault size is the smallest probability of failure likely to be induced by a programming error (with respect to the program, testing distribution, and simulated faults that are injected). Sensitivity analysis does not use an oracle, and can therefore be completely automated as implemented by the *PiSCES* tool [18].

Testing establishes an upper limit on the software's probability of failure; sensitivity analysis establishes a lower limit on the probability of failure that is likely to occur. Together, the estimate and the prediction can be used to establish confidence that software does not contain any faults.

Sensitivity analysis is based on separating software failure into three phases: execution of a software fault, creation of an incorrect data state, and propagation of this incorrect data state to a discernible output. This three part model of software failure [8] will be referred to as PIE, for Propagation, Infection, and Execution. In this paper we examine how to apply PIE to the task of finding a realistic minimum non-zero probability of failure prediction, α, when random testing has discovered no errors.

If a location contains a fault, and if the location is executed, the data state of the execution may or may not be changed adversely by the fault. If the fault does change the data state into a data state that is incorrect for this input, we say the data state is *infected*. To predict the probability of infection, the second phase of sensitivity analysis performs a series of syntactic mutations on each location [9]. After each mutation, the program is re-executed with random inputs; each time the monitored location is executed, the data state is immediately compared with the data state of the original (unmutated) program at that same point in the execution. If the state differs, infection has taken place [12].

The third phase of the analysis estimates propagation. Again the location in question is monitored during random tests. After the location is executed, the resulting data state is changed by assigning a random value to one data item using a predetermined distribution. (Research is ongoing as to the best distribution to use for this random selection. Current experiments use an equally likely distribution over the range of values for this variable during random testing.) After the data state is changed, the program continues executing until an output results. The output that results from the changed data state is compared to the output that would have resulted without the change. If the outputs differ, error propagation has occurred.

Each phase produces a probability estimate based on the number of trials divided by the number of events (either execution, infection, or propagation). Execution, infection, and propagation must all occur to result in a failure at this location. Thus the product of these estimates yields an estimate of the probability of failure that would result if this location had a fault.

Sensitivity analysis is a new empirical technique. Since sensitivity analysis does not require an oracle, it can be completely automated. Preliminary results of the accuracy of the lower bounds produced by sensitivity analysis have been encouraging [9, 11, 10].

Sensitivity analysis produces 3 probability estimates for each location examined. In order to assess testability for a program or module, we must be able to give *one* probability prediction: the "latent failure rate." The *latent failure rate* is a prediction of the minimum non-zero failure probability that can occur from any possible fault in the program given the fault classes that were simulated by sensitivity analysis. It is possible that an actual fault will induce a lower failure probability than the latent failure rate. This occurs when the fault classes simulated by SA do not include a particular fault, and that fault has a lower failure rate associated with it than any of the faults that are simulated. This is an unavoidable possibility that is incurred by all fault-based techniques; fault-based techniques are only as powerful as the range of the fault classes that they simulate. However, in the experiments cited above, such "surprisingly small faults" were rare.

3   Hoeffding's Inequality

When we test a program randomly, success is defined by whether or not it performs correctly on all tests. However, since test cases are chosen randomly, there is a certain probability of drawing a test sequence that fools us into believing that a "bad" program is "good." An extreme example is the case in which the same test case is drawn N times, an unlikely but possible outcome of N draws with replacement. Although we can honestly say in this case that we have performed N random tests, in truth our testing has told us little about the quality of the program.

Our analysis assumes that sampling is with replacement. However, it is normally not likely that we will draw the same test N times, and we might hope that there is only a small probability of drawing any sort of grossly unrepresentative set of tests. In fact, we can bound this probability as follows: we first note that if the probability of a program failure is ϵ, then the probability that it will fail in exactly k out of N tests is given by a binomial distribution:

$$\binom{N}{k} \epsilon^k (1 - \epsilon)^{N-k}.$$

Therefore, the probability that the program will not fail on any test is obtained by setting $k = 0$; (1) then becomes $(1 - \epsilon)^N$, the confidence bound that we have already used. To state this result in mathematical terms, let ϕ be an empirical estimate for the probability of an event (in this case the event is a program failure), which is constructed by dividing the number of times the event occurred by the number of tests that were made. Also let ν be the true probability of the event. Then we have just shown that

$$\Pr(|\phi - \nu| \geq \epsilon) \leq (1 - \epsilon)^N$$

given that ϕ is 0. In other words, we have shown that the difference between the true probability of failure and the estimated probability of failure is greater than or equal to ϵ only with probability $(1 - \epsilon)^N$.

In testability analysis we are faced with a problem that is similar to the one we face in testing. We obtain an empirical estimate of the probability of some event (in this case the event is no longer a program failure but a propagation, an infection, or an execution) and we wish to know how likely it is that we have drawn an unrepresentative sequence of tests, and thus obtained an estimate that is actually far off the mark. We require an upper bound on

$$\Pr(|\,\phi - \nu\,| \geq \epsilon)$$

that applies in the general case, and not just when $\phi = 0$. One such bound was given in [17] and is commonly known as Hoeffding's inequality. It states that

$$\Pr(|\,\phi - \nu\,| \geq \epsilon) \leq 2e^{-2\epsilon^2 N}.$$

(Note that this is only an upper bound. Even though $\Pr(|\,\phi - \nu\,| \geq \epsilon)$ is a probability and cannot be greater than 1, the right side of (2) can be as large as 2. But if $2e^{-2\epsilon^2 N}$ were, say, 1.6, then (2) would not tell us anything new, because we already knew that $\Pr(|\,\phi - \nu\,| \geq \epsilon)$ is less than or equal to 1.0, and hence less than 1.6. In that case, however, (2) would not give us any confidence that $|\,\phi - \nu\,|$ was less than ϵ, because, as far as we could ascertain by looking at (2), the probability that $|\,\phi - \nu\,| \geq \epsilon$ could be as large as 1.0.)

For the purposes of quantifying our confidence in a latent failure rate, we apply Hoeffding's inequality and get that

$$\Pr{}_{XD}[|\,\phi - \nu\,| \geq \epsilon] \leq 2e^{-2\epsilon^2 N} \tag{1}$$

where

1. X is the space of mutants and perturbation functions used, i.e., X is the space of programs based on P for which D was used during testing.

2. ν is the true latent failure rate (unknown).

3. ϕ is the predicted (empirical) latent failure rate found via sensitivity analysis.

4. ϵ is the "fudge factor" that we set for how much we believe we have miscalculated ϕ.

For ϵ, our intuition suggests the use of:

$$\epsilon = \phi - (10^{-1} \cdot \phi) \tag{2}$$

Equation 2 provides a fudge factor of one order-of-magnitude; although ϵ is independent of ϕ, it makes sense that ϵ is near ϕ's order-of-magnitude. As $-2\epsilon^2 N$ decreases, $2e^{-2\epsilon^2 N}$ increases; this undesirable situation will be fully explained later. For now, it is enough to know that ϵ is the dominant argument in $2e^{-2\epsilon^2 N}$, and for a small ϵ, N must be enormous for $2e^{-2\epsilon^2 N}$ to be small. (Tables 1 and 2 show the relationship between the parameters of Equation 4.)

4   The "Squeeze Play"

As we have explained in Section 2, sensitivity analysis provides an empirical prediction of the minimum non-zero failure probability. And from testing we have an upper bound, θ, on the probability of failure. If the upper and lower bounds hold, we have bracketed the true probability of failure. Note that both the prediction and estimate must be based on the input distribution D.

To understand what this prediction and this estimate tell us, it is often easier to visualize these probabilities as *fault sizes*, meaning the upper bound can be viewed as the largest fault that *is likely to* still be remaining after testing T times, and the latent failure rate can be viewed as the smallest fault that is likely to still be in the code. Again it is important to remember that we do not know whether any faults exist in the code.

If we have done enough testing to be very confident that the true probability of failure is less than ϕ, the smallest fault we expect to see, then we believe that there are no remaining faults with respect to the fault classes that were simulated during sensitivity analysis using D. Section 5 presents the mathematical confidence that we have in this belief. This idea of increased testing to increase our confidence that the true pof is less than ϕ is termed the "Squeeze Play." [15, 13]. The same effect can be realized by increasing ϕ via either software *design-for-testability* or removing those PIE estimates from the latent failure rate that are driving ϕ down. We can confidently ignore PIE probability estimates *only* when are sure that the location that they are associated with is correct. After all, if we know there is no fault at a location, then we do not care what the predicted ability of that location is to hide a fault from us, since none is there. Such certainty for very small portions of code might be possible in isolated cases using formal methods such as proof of correctness. Since such proofs are more easily carried out for small code segments, the use of sensitivity analysis may encourage future applications of formal techniques often dismissed as impractical today.

5   Quantifying Absolute Correctness

Testing down to the level of $\phi - \epsilon$ as well as performing testability analysis for ϕ are processes that are statistically subject to error. As we have shown, Hoeffding's inequality provides a mechanism for quantifying this error.

Hamlet's probable correctness model [16] provides a mechanism for quantifying how likely testing is to give us a confidence that the true probability of failure is greater than θ:

$$\text{Prob}[\Theta_D > (\theta)] = (1 - (\theta))^T \tag{3}$$

where the T successful tests are selected according to distribution D. This allows us to determine how much confidence we place in the results of testing the program T times successfully.

The following equation gives us a lower bound on our confidence that the software is correct. If the lower bound is less than zero, this means that we have zero confidence.

$$\text{Confidence}[\Theta_D = 0]_{XD} = 1-$$

$$[(1 - (\theta))^T + 2e^{-2\epsilon^2 N}] \tag{4}$$

There are several aspects of Equation 4 that are noteworthy. First, in the case where $\theta \leq \phi$, the size of ϕ impacts our confidence in correctness given fixed T and N. Equation 4, then, suggests that just having a cross-over of θ and ϕ is not enough; we also need a sufficiently large ϕ. Second, as ϕ decreases, the required N to overcome a small ϵ in order for $2e^{-2\epsilon^2 N}$ to be near zero becomes virtually intractable. Thus if we are to have any confidence that our software is correct we will need $2e^{-2\epsilon^2 N}$ to be near zero and $(1 - (\theta))^T$ to be near zero.

6 Conclusions

We contend that the preliminary results of experiments in software sensitivity are sufficient to motivate research into quantifying sensitivity analysis. Although the technique will likely require revision, the ideas that motivate sensitivity analysis dispute the contention that random testing is the only method of experimentally quantifying software reliability. We cannot guarantee that this new technique will make it possible to assess reliability to the precisions required for life-critical software. However, we do think it is premature to declare such an assessment impossible. In the preceding sections we have argued that if testability predictions can be quantified accurately, then it is plausible to combine random testing results with testability results to assess reliability more precisely than is possible with testing alone.

Both random testing and testability analysis gather information about possible probability of failure values for a program. However, the two techniques generate information in distinct ways: random testing treats the program as a single monolithic black box while sensitivity analysis examines the source code location by location; random testing requires an oracle to determine correctness but sensitivity analysis requires no oracle because it does not judge correctness; testing that reveals no failures focuses on the possibility of no faults existing while sensitivity analysis focuses on a "what if a fault exists in this location" analysis. The two techniques provide independent data about how frequently the program should fail if any faults exist.

This paper has primarily focused on the case where $\theta \leq \phi$, because this is the easiest situation in which to gain confidence in the software's correctness. And as Tables 1 and 2 have shown, it is preferable that $\theta < \phi$. Unfortunately, this situation may be infrequent, and more frequently $\theta > \phi$. This situation makes it almost impossible to provide a confidence in the absolute correctness of the code. Two main strategies can push θ and ϕ closer: decreasing θ by increasing T, the number of tests; or increasing ϕ by rewriting code locations that have low sensitivity. Software design-for-testability is one avenue of research that we are exploring that we hope will generally cause systems to have higher ϕs.

One interesting side-effect of Equation 4 has revealed that not all cross-over cases are equivalent in the derived confidence. This is because as ϵ decreases

Confidence$[\Theta_D = 0]_{XD}$	ϵ	T	N	θ	ϕ
0	0.00009	10	1000000	0.001	0.0001
0	0.00009	100	1000000	0.001	0.0001
0	0.00009	1000	1000000	0.001	0.0001
0	0.00009	10000	1000000	0.001	0.0001
0	0.00009	100000	1000000	0.001	0.0001
0	0.00009	1000000	1000000	0.001	0.0001
0.009955	0.009	10	1000000	0.001	0.01
0.095	0.009	100	1000000	0.001	0.01
0.632	0.009	1000	1000000	0.001	0.01
0.999	0.009	10000	1000000	0.001	0.01
1.0	0.009	100000	1000000	0.001	0.01
1.0	0.009	1000000	1000000	0.001	0.01
0.00995	0.09	100	1000000	0.0001	0.1
0.0951	0.09	1000	1000000	0.0001	0.1
0.632	0.09	10000	1000000	0.0001	0.1
0.999	0.09	100000	1000000	0.0001	0.1
1.0	0.09	1000000	1000000	0.0001	0.1
1.0	0.09	10000000	1000000	0.0001	0.1
0.999955	0.0009	10000	1000000000	0.001	0.001
0.604157	0.0009	10000	1000000	0.001	0.001
0.0	0.0009	10000	100000	0.001	0.001
0.236	0.0009	100000	1000000	0.00001	0.001
0.468	0.0009	200000	1000000	0.00001	0.001
0.5544	0.0009	300000	1000000	0.00001	0.001
0.585	0.0009	400000	1000000	0.00001	0.001
0.597	0.0009	500000	1000000	0.00001	0.001
0.604	0.0009	1500000	1000000	0.00001	0.001
0.604	0.0009	5000000	1000000	0.00001	0.001
0.9216	0.0009	5000000	2000000	0.00001	0.001
0.965	0.0009	5000000	2500000	0.00001	0.001

Table 1: Various Parameters for Equation 4

Confidence$[\Theta_D = 0]_{XD}$	ϵ	T	N	θ	ϕ
0.604	0.0009	100000	1000000	0.001	0.001
0.604	0.0009	200000	1000000	0.001	0.001
0.604	0.0009	300000	1000000	0.001	0.001
0.604	0.0009	400000	1000000	0.001	0.001
0.604	0.0009	500000	1000000	0.001	0.001
0.604	0.0009	1500000	1000000	0.001	0.001
0.604	0.0009	5000000	1000000	0.001	0.001
0.921	0.0009	5000000	2000000	0.001	0.001
0.965	0.0009	5000000	2500000	0.001	0.001
0.9844	0.0009	5000000	3000000	0.001	0.001
0.993	0.0009	5000000	3500000	0.001	0.001
0.997	0.0009	5000000	4000000	0.001	0.001
0.001	0.05	100	10000	0.00001	0.1
0.095	0.05	10000	10000	0.00001	0.1
0.393	0.05	50000	10000	0.00001	0.1
0.632	0.05	100000	10000	0.00001	0.1
0.632	0.05	100000	20000	0.00001	0.1
0.864	0.05	200000	10000	0.00001	0.1
0.950	0.05	300000	10000	0.00001	0.1

Table 2: Additional Parameters for Equation 4

(when ϕ increases), the required N to overcome this deficiency becomes intractable. Thus we have discovered a new argument for attaining a higher ϕ: *not only does a higher ϕ require fewer T tests but a higher ϕ means a higher ϵ which a greater confidence in ϕ can be achieved with a smaller N.* This is crucial to the success of Equation 4, since a tiny N will almost certainly destroy any chance of getting a greater than zero confidence in the software's correctness.

Acknowledgement

This work has been partially supported by NASA Grant NAG-1-884. The authors would like to thank Larry Leemis of the mathematics department of the College of William & Mary for reviewing a copy of this paper. Also, we thank Jeffery Payne of RST for producing the tables.

References

[1] R. Butler and G. Finelli. The infeasibility of experimental quantification of life-critical software reliability. *Proceedings of SIGSOFT '91: Software for Critical Systems* (December 4-6, 1991), New Orleans, LA., 66-76.

[2] D. R. Miller. Making statistical inferences about software reliability. NASA Contractor Report 4197 (December 1988).

206

[3] K. Miller, L. Morell, R. Noonan, S. Park, D. Nicol, B. Murrill, and J. Voas. Estimating the probability of failure when testing reveals no errors. *IEEE Trans. on Software Engineering* 18(1):33-44, January, 1992.

[4] I. Peterson. Software failure: counting up the risks. *Science News*, Vol. 140, No. 24 (December 14, 1991), 140-141.

[5] T. A. Thayer, M. Lipow, and E. C. Nelson. *Software Reliability* (TRW Series of Software Technology, Vol. 2). New York: North-Holland, 1978.

[6] J. Voas, L. Morell, and K. Miller. Predicting where faults can hide from testing. *IEEE Software* (March 1991), 41-48.

[7] S. N. Weiss and E. J. Weyuker. An extended domain-based model of software reliability. *IEEE Trans. on Software Engineering*, Vol 14, No. 10 (October 1988), 1512-1524.

[8] L. J. Morell. Theoretical Insights into Fault-Based Testing. *Proc. of the Second Workshop on Software Testing, Validation, and Analysis*, July, 1988, 45-62.

[9] J. Voas and K. Miller. The Revealing Power of a Test Case. *Journal of Software Testing, Verification, and Reliability* 2(1), 1992.

[10] J. Voas and K. Miller. *PA*: A Dynamic Method for Debugging Certain Classes of Software Faults. To appear in *Software Quality Journal*, 1993.

[11] J. Voas. *PIE*: A Dynamic Failure-Based Technique. *IEEE Transactions on Software Engineering* 18(8):717-727, August, 1992.

[12] Richard A. DeMillo, Richard J. Lipton, and Frederick G. Sayward. Hints on Test Data Selection: Help for the Practicing Programmer. *IEEE Computer*, April, 1978, 11(4):34-41.

[13] J. Voas and K. Miller. Improving the Software Development Process Using Testability Research. *Proc. of the 3rd International Symposium on Software Reliability Engineering*, October, 1992, Research Triangle Park, NC.

[14] John Musa. Operational Profiles in Software-Reliability Engineering. *IEEE Software*, March, 1993, 10(2):14-32.

[15] R. Hamlet and J. Voas. Faults on Its Sleeve: Amplifying Software Reliability Testing. *Proc. of the International Symposium on Software Testing and Analysis*, June 28-30, 1993.

[16] R. Hamlet. Probable Correctness Theory. *Information Processing Letters*, 25(1):17-25, April, 1987.

[17] W. Hoeffding. Probability Inequalities for Sums of Bounded Random Variables. *American Statistical Association Journal*, March, 1963, p.13-30.

[18] J. Voas, K. Miller, and J. Payne. PISCES: A Tool for Predicting Software Testability. *Proc. of the 2nd Symposium on Assessment of Quality Software Development Tools*, May, 1992. IEEE Computer Society.

A Knowledge-Based Approach to Program Testing and Analysis

Igor M. Galkin
Computer Center, Academy of Sciences
Minsk, Republic of Belarus

Abstract

An approach to computer support organization of program testing and analysis is considered. The approach is based on a semantic net representation and usage of knowledge about a program. The possibilities and benefits of this approach application in different kinds of program analysis and usage of the Prolog language as the tool of such analysis implementation are described. Also the possibility of the approach spreading over different program representations analysis and other problem areas, connected with the program engineering, is indicated.

1 Introduction

It is known that important factors of software reliability and safety improvement are program testing and analysis. As objects of studying and analysis may serve such program features as interrelations of program objects, properties of ones, control flow, data flow and structure, quality characteristics, results of execution. Models traditionally used for such analysis are control flow graph (c-graph), data flow graph, call graph. The offered approach envisages various program features modeling by semantic nets and frame ones. Using of these formalisms of knowledge representation allows to uniformly represent information about program features which is necessary for different kinds of program analysis. This information is represented by a set of facts — instances of relations (e.g. relations of calls, nesting, declaration, usage, location, precedence, characterization, edge coverage), existing or arising under program execution between entities of program (e.g. procedures, variables, constants, c-graph nodes) or between ones and their characteristics (as examples of the latter may serve "coordinates" of location in the program for procedures, "coordinates" of declaration or usage for variables and arrays, quality metric values for procedures and programs). Facts of these relations simulate a state (features) of program and form its model; program analysis actions are described by a set of appropriate rules and are based on construction and analysis of such models. The main tool of this approach implementation is Prolog.

Different papers have influenced the approach formation. First of all these are [1–5], which are based on selection and usage of relations between program objects, [6, 7], in which some aspects of the Prolog usage for program analysis were considered, and also [8], in which the possibility of graphs and graph grammars usage for program representation modeling was given.

2 Representation of Knowledge about Program Features

Discussing the modeling of program features it is useful to distinguish between two levels of knowledge representation: "user" level, associated with description, and "system" level, connected with implementation. On the former level, which corresponds to the requirements of convenience and simplicity of knowledge representation, the using of a simple semantic net formalism is preferable, on the latter level, which takes into consideration unambiguity of representation of program objects in the model, model size and other realization aspects, frame net formalism is more adequate [9].

Semantic net formalism is based on the idea of knowledge representation in the form of the oriented graph with named nodes and edges, where nodes correspond to the objects of the problem area studied, and edges to relations between them. There is a sufficiently large number of kinds of semantic nets. Let us limit our consideration by non-homogeneous semantic nets, which contain different, not only one kind relations, and simple (non-hierarchical) ones, whose nodes don't have its own structure. From the logic view the main functional element of semantic net (relation and two nodes connected by it) is equivalent to predicate with two arguments. For this reason, semantic net may be represented both in a graphic form and in a predicate form.

So, knowledge about some features of an imaginary Pascal program, containing procedure sample, on the "user" level may be represented by the next facts of semantic net relations (in infix predicate form):

    sample uses-the-variable index
    index is-declared-in-line 52
    index has-type integer
    index is-referred-to-in-line 56
    sample has-length-in-lines 8
    sample has-commentedness 0
    sample has-complexity 2

Frame formalism, which also implies a choice of objects and relations of problem area and in a certain sense may be considered as a particular case of semantic nets, is oriented towards the representation of stereotype situations. Frame is the composition of situation components (slots) having its own name and value and united by a frame name. As a slot value may be a name (individual label) of a frame, then a frame set turns into a frame net. Like the semantic nets frame ones may be expressed in graphic and symbolic forms.

So, on the "system" level semantic net presented above may be transformed into the frame set (in a symbolic form)

(TYPE   object index   procedure sample   declaration-line 52   type integer)
(REFERENCE   object index   line 56)
(CHARACTERISTICS   procedure sample   length-in-lines 8
                                 commentedness 0   complexity 2).

Prolog program consists of facts, rules and queries. Facts fix the existing of some relations between objects; rules set common dependencies for using relations and allow to get new facts from ones taking place; queries require to confirm the existence of the relation between concrete objects or point out the objects connected by a certain relation with other objects. Objects may be represented by ones' lists. For example, the frame set given above is compactly represented by Prolog facts (here and below syntax and terminology [10] are used)

type   (index sample 52 integer)
reference   (index 56)
characteristics   (sample 8 0 2),

relationships between frames are set by Prolog rules.

Program testing and analysis actions are described by Prolog rules. Program state models are analyzed by execution of Prolog program, which includes an appropriate set of rules and queries.

3 Contents and Ways of Forming of the Model

Apparently, treating program features modeling it is necessary to answer at least three important questions: WHAT, HOW and WHY must be represented in the model? The model reflected features of some really existing program is, nevertheless, quite independent entity with own internal properties, physically separated from modeling program, and itself is a possible object for investigations. Given above, the explanations of the common idea of the approach mainly answer a question HOW (how are program features represented in the model?). Now let's try to answer a question WHAT (what program features must be represented in the model?).

Purpose of the model is to reflect program features adequately to analysis requirements (having as experience shows tendency to permanent growing). For this reason, it has to contain maximum possible (but at the same time practically acceptable) volume of useful information about various program features. Since Prolog allows to infer new knowledge from existing one, it is worthwhile to reveal some "base" (necessary) volume of knowledge about a program. In a certain sense contents of the model is some compromise between requirements of analysis execution effectiveness and memory economy. Intuitively it is clear that from program analysis and testing view the model must contain information about all program modules and their relationships (such as calls and parameter transmissions), about all places and contents of program data objects' declarations and usages, about control structure in a program, about values of some metrics (e.g. of a complexity), about coverage of program paths by tests, etc.

Information about program module relationships, data objects' declarations and usages and control structure may be extracted from program source code under its syntactic parsing; values of necessary metrics are computed on the base of this parsing data; test run result data are obtained by instrumented program execution.

4 Dependencies between Elements of the Model

To answer (naturally, not exhaustively) a question WHY (why just such kind of information must be represented in the model?) let's profit by mathematical apparatus of relations, based on the set theory.

Let A and B be two sets, $A \times B$ is their cartesian product, then any its subset $R \subseteq A \times B$ defines some binary relation between elements of A and B. Objects (elements) x and y are in a binary relation R that is denoted as xRy, if $(x, y) \in R$.

Let's as a base relation set take the next collection:

$R_1("\text{begins-at}") \subseteq M \times P,$

$R_2("\text{terminates-at}") \subseteq M \times P,$

$R_3("\text{is-directly-nested-in}") \subseteq M \times M,$

$R_4("\text{is-called-at}") \subseteq M \times P,$

$R_5("\text{is-declared-at}") \subseteq V \times P,$

$R_6("\text{is-defined-at}") \subseteq V \times P,$

$R_7("\text{is-referred-to-at}") \subseteq V \times P,$

$R_8("\text{is-undefined-at}") \subseteq V \times P,$

$R_9("\text{starts-at}") \subseteq G \times P,$

$R_{10}("\text{ends-at}") \subseteq G \times P,$

$R_{11}("\text{is-directly-preceded-to}") \subseteq G \times G,$

where M, V, P, G are non-intersected sets of the model elements, and besides M-elements represent program modules (relatively independent program entities, such as head modules, subprograms, procedures, functions, etc.), V-elements represent data objects (e.g. variables and arrays) of a program, P-elements represent points (placements) in a program, G-elements represent c-graph nodes; sense of the relations is clear from its linguistic meaning. Not concerning concrete details of representation of program objects in the model, we shall confine ourselves to the requirement of unambiguity of this representation and regularity of P-elements for the possibility of its comparison (before-after, less-greater), that may be ensured by their number nature.

On the base of relations $R_1 - R_{11}$ and also common order relations "not-greater-than" (R_{12}) and "not-less-than" (R_{13}), defined, in particular, on the set $P \times P$, with the help of appropriate operations under relations it is possible to define some auxiliary relations, promoting to the inference of existing dependencies between model elements (and accordingly, between program objects). So, with the help of relations

$R_{14}($"is-used-at"$) \subseteq V \times P,$
$R_{15}($"is-nested-in"$) \subseteq M \times M,$
$R_{16}($"belongs-to-the-module"$) \subseteq P \times M,$
$R_{17}($"is-declared-in-the-module"$) \subseteq V \times M,$
$R_{18}($"is-used-in-the-module"$) \subseteq V \times M,$
$R_{19}($"is-known-in-the-module"$) \subseteq V \times M,$

defining as

$R_{14} = R_6 \cup R_7 \cup R_8,$
$R_{15} = R_3 \cup R_3^2 \cup R_3^3 \cup \ldots = R_3^+$
$R_{16} = ((R_{13} \circ R_1^{-1}) \cap (R_{12} \circ R_2^{-1})) \setminus (((R_{13} \circ R_1^{-1}) \cap (R_{12} \circ R_2^{-1})) \circ R_{15}^{-1}),$
$R_{17} = R_5 \circ R_{16},$
$R_{18} = R_{14} \circ R_{16},$
$R_{19} = R_{17} \cup (R_{17} \circ R_{15}^{-1}),$

where $\cup$, $\cap$, $\setminus$ and $\circ$ mean union, intersection, difference and composition (product) of relations, respectively, R^k — kth power of a relation R, R^{-1} — the inverse relation to relation R, R^+ — the transitive closure of a relation R, the rule of the obligatory declaring of data objects used within a program may be written as the restriction $R_{18} \subseteq R_{19},$ and the rule of the obligatory usage of data objects declared in a program — as the restriction

$$R_{17} \subseteq R_{18} \cup ((R_{18} \circ R_{15}) \setminus (R_{17} \circ R_{15})).$$

By similar manner through the use of the relations from the base collection some useful auxiliary relations may be defined, in particular,

$R_{20}($"belongs-to-the-node"$) \subseteq P \times G,$
$R_{21}($"precedes"$) \subseteq G \times G,$
$R_{22}($"dominates-over"$) \subseteq G \times G),$
$R_{23}($"belongs-to-a-node-preceding-to-a-node-containing-the-point"$) \subseteq P \times P,$

and with their help it is possible to formulate some dependencies peculiar to a program control and data flow. So, the rule of the obligatory initialization of referring data objects may be expressed by the restrictions

$(R_{20}^{-1} \circ R_6^{-1} \circ R_7 \circ R_{20}) \cap R_{21} \subseteq R_{22},$
$(R_8^{-1} \circ R_7) \cap R_{23} \subseteq ((R_8^{-1} \circ R_6) \cap R_{23}) \circ ((R_6^{-1} \circ R_7) \cap R_{23}),$

(it is assumed that undefinitions of data objects form separate c-graph nodes), and the sufficient condition of assignments' nonredundancy — by the restriction

$$(R_6^{-1} \circ (R_6 \cup R_8)) \cap R_{23} \subseteq ((R_6^{-1} \circ R_7) \cap R_{23}) \circ ((R_7^{-1} \circ (R_6 \cup R_8)) \cap R_{23})$$

(in present case it is assumed that c-graph of a program consists of nodes-statements, but not of nodes-blocks).

In the same way, by introducing new sets of model elements reflecting types and the size of program data objects, and by determining relations on them reflecting characteristics of program data objects and their use as formal and actual parameters one can formulate the restrictions expressing requirements of correspondence of some program elements' characteristics in certain situations, in particular, of correspondence of a type and length of formal and actual parameters, of a type of left and right assignment parts, etc. As the such sets may be suggested the sets, say, C, F, E, T and S representing, respectively, constants, functions, expressions, and also possible types and sizes of program elements, and as the relations on them supplementing the base collection — the relations

$R_{24}("has-type") \subseteq (V \cup C \cup F \cup E) \times T$,

$R_{25}("has-size") \subseteq (V \cup C \cup F \cup E) \times S$,

$R_{26_i}("is-the-ith-actual-parameter-in-the-call-at") \subseteq (V \cup C \cup F \cup E) \times P$,

$R_{27_i}("has-the-ith-formal-parameter") \subseteq M \times V$.

In the context of notions we introduced, many kinds of a static program analysis are reduced to checking of formulated above correlations between elements of respective models.

Let us illustrate the possibility of Prolog-implementation of checking model restrictions we introduced through the following example. Let relations of a base set, specifically, $R_1 - R_{11}$ be determined in a Prolog-program (via setting a respective number of facts). Then, for instance, the relations $R_{14} - R_{19}$ are determined as:

x is-used-at y if (either x is-defined-at y
                            or (either x is-referred-to-at y
                                    or x is-undefined-at y))

x is-nested-in y if (either x is-directly-nested-in y
                          or x is-directly-nested-in z and
                                 z is-nested-in y)

x belongs-to-the-module y if y begins-at z and
                                      x not-less-than z and
                                        y terminates-at $z1$ and
                                        x not-greater-than $z1$ and
                                      not ($y1$ is-nested-in y and
                                              x belongs-to-the-module $y1$)

x is-declared-in-the-module y if x is-declared-at z and
                                          z belongs-to-the-module y

x is-used-in-the-module y if x is-used-at z and
                                   z belongs-to-the-module y

x is-known-in-the-module y if either x is-declared-in-the-module y
                              or x is-declared-in-the-module z and
                                      y is-nested-in z)

(relations "not-greater-than" and "not-less-than" in Prolog program is determined by means of rules using built-in arithmetic relation LESS, for example:

x not-greater-than x

x not-greater-than y if x LESS y

x not-less-than x

x not-less-than y if y LESS x),

and checking of restriction $R_{18} \subseteq R_{19}$, i.e. revealing of model elements (and consequently, of program ones), not satisfying this restriction (in this case — used, but not declared program data objects), is implemented by the query

which ($x : x$ is-used-in-the-module y and not x is-known-in-the-module y).

5 SAIL System

The considered approach is the basis of a static and dynamic program analysis in the SAIL system [11], which analyses programs written in Fortran-77 and consists of three principle parts:

1) Fortran-program analyzer, making restricted parsing, construction of c-graph, instrumentation, extraction of information from specially organized introductory comments, cyclomatic complexity [12] and commentedness measurement and on its basis — construction of program state net model analyzed (in fact, it is a translator of syntax correct Fortran-programs into Prolog-models);

2) program state net models analyzer, providing the solution of different tasks in program analysis;

3) monitor, which is a mediator between user and above analyzers and providing user-friendliness of the system.

Micro-Prolog [10] is the language of Fortran-program state models representation and analysis and the one of monitor implementation, and Turbo-Pascal is the language used for Fortran-program analyzer implementation.

The implemented version of SAIL system defines infeasible parts of code and latent cycles, reveals the usage of non-initiated variables and redundant assignments, checks up the usage of the variables declared and declaration of ones being used, reveals untested program parts and proposes available plans of testing, and also provides a user with various common information about program and its characteristics. So, it allows the users to get data about size, number of entries and exits, complexity and commentedness by means of requests including complex conjunctive ones setting different combinatorial variants of program characteristics. An example of conjunctive request, formed by user with the selection and refining of corresponding menu lines is "Point the program module, which has length more than 100, complexity more than 20 and commentedness less than 10".

Besides, a user of SAIL has the possibility of asking the system different questions connected with a program (in Russian), having simple or complex (conjunctive) conditions. The question form is fixed, though it avails some "liberties" (e.g. commas, prepositions, inflexional endings), which make it closer to natural one. This possibility, which may make easier a task of program maintenance, is based on user "understanding" of program representation at his level and ensured by simplicity of this representation, by its likeness to natural language clauses and easiness of such clauses parsing implementation with Prolog.

Examples of questions with simple conditions are: What is r1? What type does every variable have? Which module uses the variable index? What modules use arrays? Which variables are declared implicitly? In which line is every array declared?

Examples of questions with complex conditions are: What type and length does the variable index have? What type, length and dimension does every array have? What module calls module $s1$ and is called by module $s2$? What variable has type real, length 4 and belongs to common block?

Form and contents of answers also are oriented to the support of natural language dialogue. So, the answer to the question similar to the first of ones given above will contain complete information about the program object mentioned, for example:

```
r1 is the variable of the module def, which
    is a formal parameter,
    is declared implicitly,
    has the type real,
    has length (in bytes) 4,
    is defined in lines 4 14 20,
    is referred to in lines 21 22,
r1 is the variable of the module quad, which
    is declared in line 40,
    has the type integer,
    . . .
```

and the answer to the second question may be the following:

```
in the module def:
        yes — character,
        i — integer,
        r1 — real,
        r2 — real,
        . . .
```

While typing a user question the system offers him a number of convenient prompts and while answering all the requests it makes the research area more precise, allowing the user to choose the names of modules in which he takes interest from submenu, or to denote a search mode in all program modules.

6 Possible Fields of the Application

Fortran, due to its specific features, is the language traditionally selected for illustration of possibilities of program analysis systems [13], but most of the analysis kinds implemented in SAIL system is common for a sufficiently large number of programming languages, in the first turn, for ones of a procedural type [14]. For this reason, many judgments given above are true for analysis of programs written in other languages.

Program life cycle usually includes its specification, design, coding, testing and maintenance. The principles of the approach, given above in application to program code analysis, naturally spread over other representations (specification, project) analysis, used during the program life cycle. Besides, with such representations models it is possible also to analyze program representations interrelations in different stages of program life cycle for these interrelations co-ordination support in the case, for example, of modifying some of them.

The main principles of this approach, which are based on semantic net model construction in knowledge domain and on making available the intelligent interface with the given model, can be used for organization of support of different processes attendant to program making and development [15]. Such processes are program development control, program configuration control, etc.

7 Conclusions and Future Work

The approach to computer support organization of program testing and analysis has been considered.

Main advantages of the offered approach are:

1) reflection of various program features in a uniform model and diversity of supporting analysis kinds (including common ones for a sufficiently numerous set of programming languages);

2) possibility of spreading the main approach ideas over the other programming languages (by appropriate program analyzers' development) and over the other (specific for every language) analysis kinds, and also over the support of analysis of program representations differed from program code (i.e. specification, design);

3) ensured by Prolog use, the possibility of inference of new knowledge about a program from the basic one represented in the model, that allows, on the one hand, to decrease (to some necessary minimum) a model size and, on the other hand, to add new analysis kinds without model extending, in particular, to get new quality appraisals when additional criteria are adopted;

4) based on the simplicity of knowledge representation at the "user" level, similarity of semantic net facts to natural language sentences and well known Prolog ability for such sentences parsing, possibility of organization question-answering dialogue of user and system with the help of some subset of natural language.

Works on the SAIL project are still in progress. Main directions of the further developing of SAIL are:

1) increasing of the number of supporting programming languages (by appropriate program source code analyzers development and making more precise an analysis kinds totality for each concrete language);

2) increasing of the number of deciding tasks (by extending a set of the used analysis kinds, testing criteria and coding style metrics);

3) increasing of the number of languages for communication with the system;

4) differentiation of the system for two directions, namely, its developing, on the one hand, as analysis, testing and maintenance tool for programmers, and, on the other hand, as means of teaching students basic technological aspects of high quality program development.

References

1. Ince DC. The provision of procedural and functional interfaces for the maintenance of program design and program language notations. SIG-PLAN Not 1984; 19, 2: 68-74.

2. Woodman M. A program design language for software engineering. SIG-PLAN Not 1984; 19, 8: 109-118.

3. Ince DC. A program design language based software maintenance tool. Softw Pract Exper 1985; 15, 6: 583-594.

4. Ince DC, Woodman M. The rapid generation of a class of software tools. Comput J 1986; 29, 2: 151-160.

5. Meier B. The software knowledge base. In: 8th Int. Conf. Softw. Eng., London, August, 28-30, 1985. Proc. Washington, 1985, pp 158-165.

6. Ince DC. Module interconnection language and Prolog. SIGPLAN Not 1984; 19, 8: 89-93.

7. Leung CHC, Choo QH. A knowledge-base for effective software specification and maintenance. In: 3-rd Int. Workshop Softw. Specif. and Des., London, Aug., 16-17, 1985, pp 139-142.

8. Yau SS, Nicholl RA, Tsai JJ, Liu SS. An integrated life-cycle model for software maintenance. IEEE Trans Softw Eng 1988; 14, 8: 1128-1144.

9. Galkin IM. Net modeling, static and dynamic program analysis. Prepr. No.5(455), Minsk, The Inst. of Mathematics of Byelorussian Academy of Sciences, 1991; in Russian.

10. Clark KL, McCabe FG. Micro-Prolog: programming in logic. Prentice-Hall, 1984.

11. Galkin IM. Semantic nets in program analysis. In: Mixed computations and transformation. Novosibirsk, 1991, pp 112-120; in Russian.

12. McCabe TJ. A complexity measure. IEEE Trans Softw Eng 1976; SE-2, 4: 308-320.

13. DeMillo RA, McCracken WM, Martin RJ, Passafiume JF. Software testing and evaluation. Menlo Park, 1987.

14. Galkin IM. Usage of semantic nets in a process of program making and maintenance. Prepr. No 32(432), Minsk, The Inst. of Mathematics of Byelorussian Academy of Sciences, 1990; in Russian.

15. Galkin IM. Usage of semantic nets for program modeling and analysis. USiM (Control Systems and Machines) 1991; 5: 55-61; in Russian.

DEPENDABLE SOFTWARE

Chair: W. Ehrenberger
GRS Forschungsgelande, Garching, D

Robust Requirements Specifications for Safety−Critical Systems

Amer Saeed, Rogério de Lemos and Tom Anderson
Department of Computing Science, University of Newcastle
Newcastle upon Tyne, NE1 7RU, UK

Abstract

Experience in safety−critical systems has shown that deviations from assumed behaviour can and do cause accidents. This suggests that the development of requirements specifications for such systems should be supported with a risk analysis. In this paper we present an approach to the development of robust requirements specifications (i.e. specifications that are adequate for the risks involved), based on qualitative and quantitative analyses.

1 Introduction

During software development, the phase of requirements analysis provides the system context in which the software requirements must be considered. This is a fundamental issue for safety−critical systems because "safety" is essentially an attribute of the system rather than just software. The work in this paper enhances a methodology for the requirements analysis of safety−critical process control systems [1] by incorporating techniques for the production of *robust requirements specifications*, and by providing means to evaluate these specifications against the *system risks*. A robust requirements specification is constructed by modifying a specification to take into account violations in the assumptions upon which the specification is based, and the possibility of specifications being violated due to faults that might be introduced during later stages of software development. System risk is related to the likelihood of a system entering into a hazard state, the likelihood that the hazard will lead to an accident, and the expected potential loss associated with such an accident [2].

Robust requirements specifications are obtained by conducting qualitative and quantitative analysis of the requirements. Analysis aims to provide confidence that the level of risk is acceptable. The qualitative analysis seeks to identify those circumstances that can lead to violations of a specification, and subsequently take the system into a hazard state. The quantitative analysis attaches probabilities to the occurrence of the identified circumstances, in order to estimate the risk associated with a specification. The risk estimates provide the basis for conducting risk assessments, that compare alternative specifications and judge if the risk is acceptable. For the process of requirements analysis we adopt the approach of analysing the system from different perspectives and using different techniques [3]. This approach enables the extraction of different (and complementary) information concerning the robustness of the requirements specifications.

In summary, the enhancements proposed, in this paper, for the basic methodology are as follows. For each level of abstraction at which the analysis is performed, the assumptions are identified and recorded, and the fault analysis of the specifications is conducted, with the aim of analysing the circumstances in which the specifications are unable to maintain safe behaviour from the system. In other words, apart from checking how good the specifications are, the aim is to identify their weakness, and modify the specifications, in to make them more robust.

The methodology and its enhancements will be presented as follows. The next section describes a methodology for requirements analysis. Section 3 describes how quality can be attained, in terms of risk, by performing qualitative and quantitative analysis using viewpoints. Finally, section 4 contributes some concluding remarks.

2 A Methodology for Requirements Analysis

In this section we overview a methodology for requirements analysis; a more detailed discussion is given elsewhere [1]. The methodology consists of a framework with distinct phases of analysis, a graph that depicts the relationship between the specifications produced during the analysis, and a set of formal techniques appropriate for the issues to be analysed at each phase.

2.1 Framework for Requirements Analysis

The framework adopts the approach of separating the mission from the safety requirements during an initial phase, and then partitioning the analysis of the safety requirements into distinct phases. Each phase of analysis is focused onto a specific domain, where the identification of the relevant domains follows directly from the components (i.e. operator, plant and controller) of a general structure for safety−critical systems, and the relationship between the phases is dictated by the interactions between these components. The analysis of the phases will take into account non−standard behaviours of the entities of a domain; a basis for the analysis is provided by establishing the standard, exceptional and failure behaviours of the entities [4].

- **Conceptual Analysis.** The objective of this phase is to produce an initial statement of the aim and purpose of the system and determine those failure behaviours of the system which constitute accidents. As a product of this phase we obtain the *Safety Requirements*, enumerating the *accidents*. The accidents are the basis for separating mission from safety issues. Another activity to be performed during this phase is the identification of the modes of operation of the system; these are classes of states that group together related operational functions.

- **Safety Plant Analysis.** During this phase the plant properties relevant to the Safety Requirements, such as the physical laws and rules of operation that govern plant behaviour and potential *hazards*, are identified. The outcome is the *Safety Plant Specification* which contains *safety constraints* (conditions over the physical process that are the negations of hazards modified to incorporate safety margins) and *safety strategies* (schemes to maintain safety constraints defined as a set of conditions, in terms of controllable factors, over the physical process).

- **Safety Interface Analysis.** The objective of this phase is to delineate the plant interface, and specify the behaviour that must be exhibited at that interface. This phase leads to the production of the *Safety Interface Specification*, containing the *interface safety strategies* (refinements of safety strategies, incorporating properties of sensors and actuators).

- **Safety Control System Analysis.** During this phase we establish a top level organization for the control system in terms of the properties of its components, and their interactions. This phase leads to the production of the *Safety Control System Specification*, containing the *control system safety strategies* (refinements of interface safety strategies incorporating the components of the control system).

2.2 Safety Specification Graph

The specifications produced at the different phases of the requirements analysis, are organized into a *Safety Specification Graph* (SSG). The structure embodied in modes of operations can be reflected in the organization of the requirements specifications by constructing a separate SSG for each mode. An SSG is a directed acyclic graph, in which the *vertices* represent the safety specifications (requirements specifications for safety) and the *edges* denotes relationships between the specifications. For a system with p accidents, the SSG consists of p component graphs. Each component graph is an evolutionary graph [5]; the evolution is related to the phases of the framework. At each phase a set of new specifications is added to the graph of the previous phases, by connecting the specifications to the terminal vertices (representing the specifications of the previous phase) of the graph.

On completion (see figure 1) the top element of each component graph is an accident (denoted by AC_i) and is related to a set of hazards ($HZ_{i,j}$) that can lead to it. Each hazard is related to the safety constraint ($SC_{i,j}$) that negates the hazard, and each safety constraint related to the safety strategies ($SS_{i,j,k}$) that maintain the constraint. Then the safety strategies are related to their refinements into interface safety strategies ($ISS_{i,j,k,l}$), and a similar relation is depicted for control system strategies ($CSS_{i,j,k,l,m}$). When more than one strategy is related to a specification of a previous level either the strategies are exclusive and a choice has to be made in later stages of development to implement a single strategy, or the strategies complement each other and all are needed to attain the confidence required for the risk involved.

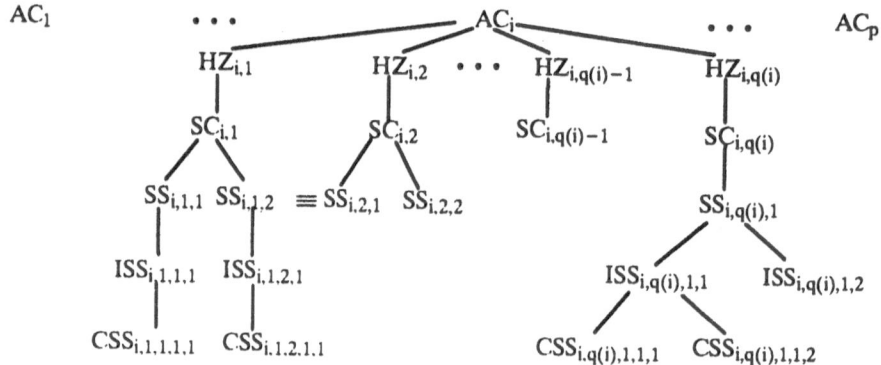

Figure 1. Example safety specification graph

The SSG of a system provides support in conducting a *qualitative analysis* and provides the basis for a systematic approach to the *modification* of the specifications. For the qualitative analysis, support is provided by establishing the conditions (which follow from the edges of the component graphs) that must be confirmed to ensure that the specifications maintain safe behaviour. A key concern in modification is *traceability*, that is the ability to trace back from a strategy to its origins and to trace forward to the strategies which are derived from the strategy. Support for traceability is provided by constructing reachability and adjacency matrices for the SSG. These matrices enable the localization of the side−effects of a modification and identification of the relationships that must be reconfirmed, thereby increasing the assurance that when changes are necessary they will be complete and consistent.

2.3 Techniques for the Framework

For the application of formal notations and techniques, the approach adopted is to employ notations in accordance with the characteristics of the system to be analysed during the different domains of analysis. Within the context of the framework, the relevant formalisms are grouped into two classes: *descriptive* and *operational*. A descriptive formalism specifies the behaviour of a domain in terms of axioms (representing system properties) over a model of the domain, whereas an operational formalism is used to model the activities and interactions between the entities of a domain. Real Time Temporal Logic and Timed History Logic are examples of descriptive formalisms, and Statecharts and Predicate−Transition Nets are examples of operational formalisms. The extent to which each class of formalism is applied in a specific domain depends on the level of abstraction at which the domain resides. At higher levels, descriptive formalisms should be more prominent, however at the lower levels operational formalisms become increasingly relevant.

In order to describe the behaviour of systems at different levels of abstraction, we adopt an event/action model (E/A model) [1]. The main features of the E/A model are that its primitive concepts (events, actions, states and the concept of a time line) can be expressed in both descriptive and operational formalisms, and it supports both discrete and dense time structures. When employed in the framework, the models of system behaviour, constructed at the different phases, are built on top of a common foundation providing support for verification between the different levels of abstraction

3 Quality Analysis of Safety Specifications

One important factor in determining the quality of the specifications for safety−critical systems is the risk analysis of the safety specifications; this aims to determine if the contribution of the software to the overall system risk is acceptable. In order to achieve this aim, a bridge has to be established between the risk analysis of the system and the software. Within the context of the methodology, this bridge is established through the SSG by relating the system requirements to the software requirements. To perform the risk analysis, those circumstances which can violate a specification, and cause the system to enter into a hazard state, have to be identified and their probability of occurring calculated. Once the risk is quantified we are able to judge whether the risk associated with a specification is acceptable or not (risk assessment). If not, the specification has to be modified or combined with other specifications in order to reduce the risk. As a result, we obtain a robust safety specification which is a specification that can be violated only within an acceptable risk. It should be noted that the risk analysis presented in this section does not take into account the consequences of an accident.

During the operation of the system, the occurrence of an initiating event (an event which can lead the system into a hazard state) of an accident sequence [6] distinguishes two kinds of system state: safe and unsafe state. An *unsafe* state is a state which could lead the system into a hazard state in the absence of corrective action and in the absence of subsequent initiating events. If a state is not an unsafe state then it is said to be *safe*. These definitions ensure that a hazard cannot occur subsequent to a safe state if no initiating event occurs. In terms of the requirements specifications, the concept of initiating event refers to those circumstances which can lead to the violation of a safety specification.

The quality analysis of the requirements, in each domain of analysis, is performed from two different perspectives: qualitative and quantitative. The purpose of the qualitative analysis is to identify those circumstances which can violate a specification, and analyse the impact of these violations upon the safety of the system. These circumstances are related to the violation of assumptions upon which a specification is based and to the violation of certain conditions of a specification. The quantitative analysis complements the qualitative analysis by attaching occurrence probabilities to these circumstances. In order to ensure that essential system behaviour is not precluded, the restrictions that a safety specification will impose on the mission must also be considered.

3.1. Qualitative Risk Analysis

At each level of abstraction, analysis is conducted over safety specifications (descriptions of safe behaviour at the level) and assumptions (properties assumed at the level). In the proposed approach the qualitative analysis is conducted in two stages; firstly we perform the preliminary analysis and secondly the vulnerability analysis of the safety specifications.

3.1.1 Preliminary Analysis

In this paper, we consider the preliminary analysis to be the analysis that must be conducted prior to the risk analysis. This analysis will involve confirming that the specifications at a particular layer of the SSG comply with those of the layer that precede it, and that the specifications in a layer are consistent. The relationships that must be confirmed, to ensure compliance between the layers, follow from the edges of the SSG. Demonstrating compliance between the layers involves employing both *verification* (formal analysis) and *validation* (informal analysis) techniques. The hazards are *validated* against the accidents and the safety constraints are *verified* against the negation of the hazards. Subsequently the strategies are *verified* against the specifications of the previous layer. At each layer, any assumptions required to confirm the relationships, depicted by the edges of the SSG, are recorded. As an example of the relations that must be verified, we examine the edge (from the SSG in figure 1) that connects the safety constraint $SC_{i,1}$ to the safety strategy $SS_{i,1,1}$. Let us suppose that the strategy is based upon assumption A (which represents a property of the physical process); the relationship to be confirmed is then:

$$A \wedge SS_{i,1,1} \Rightarrow SC_{i,1} \hspace{4cm} f1$$

A result of the preliminary analysis is that the circumstances under which safe behaviour is maintained, are clearly scoped and organised in accordance with their contribution to each phase of the analysis. This activity ensures that the knowledge gained during the development and validation/verification of the safety specifications can be applied effectively during the risk analysis.

3.1.2 Vulnerability Analysis

After performing the preliminary analysis of the safety specifications, the qualitative risk analysis consists of performing the vulnerability analysis of the specifications which probes the safety specification, and associated assumptions, to identify the circumstances under which the specification is unable to maintain safe behaviour, i.e. the violation of a specification. Once these circumstances are identified, the safety specifications can be modified to become more robust against possible violations. An initial step in the vulnerability analysis is to negate the relationships obtained during the preliminary analysis and to identify the system states which can lead to the violation of the specification when the above circumstances occur.

For the relationship $f1$, the logical assertion is negated and those plant states (PS) which can lead to the violation of $SC_{i,1}$ are identified:

$$\neg SC_{i,1} \Rightarrow (\neg SS_{i,1,1} \wedge PS) \vee (\neg A \wedge PS) \qquad f2$$

For this relationship, which is associated with the plant level, the subsequent vulnerability analysis of the safety strategy $SS_{i,1,1}$ will identify those conditions that can lead to the violation of $SC_{i,1}$.

Although logical formulae are useful in obtaining a high−level view of the relationship between the specifications and assumptions, such formulae provide limited support for a failure analysis. A suitable representation, for such analysis, is one which supports the identification of possible failure behaviours that can lead to the identified hazardous states. In this paper, to perform the vulnerability analysis of the safety specifications, we employ fault tree analysis (FTA) [7] which has been used extensively in the analysis of system safety and more recently in the analysis of software safety [8]. A key feature of fault tree analysis that makes it suitable for the analysis to be conducted here, is that the analysis is restricted to the identification of system components and conditions that lead to one particular undesired system state.

To construct a fault tree for the relationship $f2$, the initial step is to identify the undesired state, in this case the negation of the safety constraint $SC_{i,1}$, and then to determine the set of possible causes which can lead to the undesired state (refer to figure 2). For the logical formula $f2$, we identify the violation of the assumption and the violation of the safety strategy $SS_{i,1,1}$. The latter has to be further refined in order to identify its primary events.

Qualitative risk analysis provides a basis for obtaining more robust safety specifications which will lead to a risk abatement of the overall system. In the approach adopted, the analysis is performed by employing both formal analysis and fault tree analysis in order to determine the weaknesses of the safety specifications. Once these weaknesses are identified the safety specifications can be modified to incorporate mechanisms which aim to reduce their vulnerability.

3.2 Quantitative Risk Analysis

In this section we discuss how a quantitative analysis complements the qualitative analysis by introducing a measurement of confidence in the quality of the safety

226

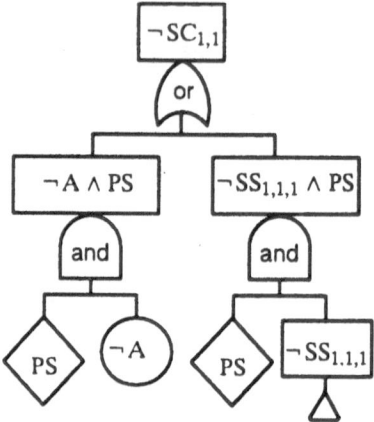

Figure 2. Fault tree for formula f2

specifications. While the latter identifies circumstances which can lead to the violation of the specifications, the former associates probabilities to these circumstances.

Although the qualitative approach strives to achieve total assurance for the safety specifications, there are three basic limitations which indicate that this aim may not be realised. A first limitation stems from the process of capturing user requirements: some faults introduced during the requirements stage may not be removed during the verification process, nor can it be guaranteed that all such faults will be removed during validation. A second limitation arises from observing past experience in the utilization of formal techniques, which shows that a formal verification may itself contain faults [9]. The third limitation is related to the confidence that can be placed on the assumptions upon which a specification is based. From these limitations we infer that even after performing the qualitative analysis we are still faced with uncertainties concerning the quality of the safety specifications, hence the necessity to quantify the uncertainties in order to obtain a level of confidence in the quality of the safety specifications. In other words, the aim is to obtain an early prediction of the contribution of the software to the risk of the system.

To associate occurrence probabilities to those circumstances which can lead to the violation of the specification, such as plant states and violation of assumptions, might not be a difficult task. On the other hand, associating occurrence probabilities to the violation of certain conditions that depend on a software implementation is more problematic because during the requirements phase of software development sufficient design and implementation information is not yet available. Instead of estimating the probability of a condition to be violated, target probabilities demanded from the higher level safety specifications such as hazards, can be used. However, once a specification is sufficiently detailed, currently available techniques which attempt to make early predictions of the software reliability can be used, such as: metrics [10], product−in−a−process [11], and execution of the specifications [12].

After conducting the quantitative risk analysis, the last stage of the quality (risk) analysis is to perform the risk assessment of the safety specifications. This is a judgement based on the estimated risk which provides guidance for high level decisions, usually associated with the process of requirements analysis. The results obtained from the quantitative analysis should be considered as a relative measurement of how effective a given strategy is in reducing the risk of a hazard, compared to the results obtained for alternative strategies. Hence it is most useful in determining which strategy or combination of strategies is most suitable for the risks involved (the choice of a strategy might also be influenced by constraints imposed by the implementation, e.g. availability of sensors and actuators with the required properties). Also, if the utilization of more than one strategy is required, this preliminary risk analysis facilitates the search for a suitable combination of the available strategies in order to avoid common mode failures.

3.3 Mission and Safety Analysis

The primary aim of the quality analysis presented in this paper is to reduce the system risk. However, it is usually impossible to maintain a complete dichotomy between the mission and safety aspects, and it would be futile to impose safety requirements which were so stringent that the system could not satisfy its mission. To complement the risk analysis, the impact of the safety specifications on the mission of the system must also be considered. Such an analysis involves relating the different safety specifications to the mission requirements that can be affected by them. If analysis of the mission requirements follows the framework described in section 2.1, leading to the construction of a *Mission Specification Graph* (MSG), a comparison between the safety specifications and mission specifications (requirements specifications for the mission) is made possible. During the development of a robust safety specification it would be possible to identify the mission specifications that may be influenced by inspecting the variables that are restricted by the safety specifications and relating these to the mission specifications at the same level of abstraction. Once the relevant mission specifications have been identified an informal analysis of the restrictions that the safety specification imposes on the mission can be conducted. An example of such an analysis is presented elsewhere [13].

4 Conclusions

This paper describes a systematic approach for the quality analysis of the requirements specifications, in the context of a methodology for the requirements analysis of safety−critical systems. The approach is based on an analysis, at each level of abstraction, of the risks introduced by the various decisions (that are based on assumptions) made in establishing the requirements. The quality analysis follows the structure of a traditional safety study and incorporates both qualitative and quantitative techniques.

The results of the risk analysis provide estimates of the risk associated with a specification and predictions of the software's contribution to the system risk. These results are used to guide the construction of robust requirements specifications, increase the confidence (assurance) that the level of risk is acceptable and provide the basis for a feasibility study. The approach to risk analysis brings the safety studies of the system and software closer together and delineates the contribution of the software to the overall system risk. Some aspects of the approach have been applied to a train set example [14].

Acknowledgements

The authors would like to acknowledge the financial support of BAe (DCSC) and the SERC (UK) SCHEMA project.

References

1. Anderson T, de Lemos R, Fitzgerald J S, Saeed A. On Formal Support for Industrial—Scale Requirements Analysis. In: Ravn A P, Rischel H (eds) Proceedings of the Workshop on Theory of Hybrid Systems. Lyngby, Denmark. Springer—Verlag, 1993 (Lecture notes in computer science — to appear)

2. Leveson N G. Software Safety: Why, What and How. ACM Computing Surveys 1986; 18: 125—163

3. Finkelstein A, Kramer J, Nuseibeh B, Finkelstein L, Goedicke, M. Viewpoints: A Framework for Integrating Multiple Perspectives in System Development. International Journal of Software Engineering and Knowledge Engineering 1992; 1: 31—57

4. de Lemos R, Saeed A, Waterworth A. Exception Handling in Real—Time Software from Specification to Design. Proceedings of the 2nd International Workshop on Responsive Computer Systems. Saitama, Japan. October, 1992. pp 108—121

5. Marshall C W. Applied Graph Theory. Wiley—Interscience, 1971

6. Draft Interim Defence Standard 00—56. Hazards Analysis and Safety Classification of the Computer and Programmable Electronic System Elements of Defence Equipment. UK Ministry of Defence. London, UK, 1991

7. Vesely W E, Goldberg F F, Roberts N H, Haasl, D F. Fault Tree Handbook. US Nuclear Regulatory Commission NUREG—0492. Washington, DC, 1981

8. Leveson N G, Cha S S, Shimeall T J. Safety Verification of Ada Programs using Software Fault Trees. IEEE Software 1991; 4:48—59

9. Miller D G. The Role of Statistical Modeling and Inference in Software Quality Assurance. In: de Neumann B (ed) Software Certification. Elsevier Applied Science, 1990, pp 135−152

10. Ramamoorthy C V, Tsai N−T, Yamura T, Bhide A. Metrics Guided Methodology. Proceedings 9th International Computer Software and Applications Conference − COMPSAC'85. Chicago, IL. October, 1985. pp 111−120

11. Laprie, J−C. For a Product−in−a−Process Approach to Software Reliability Evaluation. Proceedings of the 3rd International Symposium on Software Reliability Engineering. Research Park Triangle, NC. October, 1992. pp 134−139

12. Wohlin C, Runeson, P. A Method for Early Software Reliability Estimation. Proceedings of the 3rd International Symposium on Software Reliability Engineering. Research Park Triangle, NC. October, 1992. pp 156−165

13. de Lemos R, Saeed A, Anderson T. A Train set as a Case Study for the Requirements Analysis of Safety−Critical Systems. The Computer Journal 1992; 35: 30−40

14. Saeed A, de Lemos R, Anderson T. An Approach to the Assessment of Requirements Specifications for Safety−Critical Systems. Computing Laboratory TR 381. University of Newcastle upon Tyne, UK, 1992

Software Failure Data Analysis of two Successive Generations of a Switching System

Mohamed Kaâniche and Karama Kanoun
LAAS-CNRS, 7 Avenue du Colonel Roche
31077 Toulouse Cedex — France

Abstract

Experimental studies dealing with the analysis of data collected on families of products are seldom reported. In this paper, we analyse the failure data of two successive products of a software switching system during validation and operation. A comparative analysis is done with respect to: i) the modifications performed on system components, ii) the distribution of failures and corrected faults in the components and the functions fulfilled by the system, and iii) the evolution of the failure intensity functions.

1 Introduction

Most current approaches to software reliability evaluations are based on data collected on a single generation of products. However, many applications, not to say the great majority, result from evolutions of existing software: there are families of products, the various generations resulting from evolutions for implementing new functionalities. A new approach that is aimed at the incorporation of past experience in predicting the reliability of a new, but similar, software has recently been proposed in [1]. This approach requires the identification of parameters which characterize past experience to be incorporated in the evaluation of the software reliability. Clearly, the identification of these parameters will be based on the analysis of data collected over the whole family of products. Experimental studies dealing with the analysis of families of products are seldom reported [2, 3]. The data considered in this paper were collected on the software of two successive generations of the Brazilian Electronic Switching System (ESS)—TROPICO. Throughout this paper, the two products will be identified as PRA and PRB. PRA was first developed and allows connection of 1500 subscribers. The processing capacity of the TROPICO system was subsequently increased with the release of PRB which allows the processing of up to 4096 calls. Many PRA software components have been reused for the development of PRB and additional components were developed.

The failure data collected on each one of these products have been considered respectively in [4] for PRA and [5] for PRB. While our previous work was mainly devoted to reliability analysis and evaluation, this paper is concerned with the qualitative as well as quantitative analysis of the failure data. Our objective is to do a comparative analysis of the two successive products based on the data collected during the end of validation and the beginning of operation. Emphasis will be put on the evolution of the software and the corresponding failures and corrected faults.

This paper is composed of five sections. Section 2 gives a general overview of the TROPICO switching system. It describes the main functions performed by the system and presents some statistics about the evolution of PRB with respect to PRA. Section 3 describes the test environment and the failure data collected. Section 4 presents some of the results derived from the collected data. Finally, Section 5 outlines the main results obtained from the analysis of both products.

2 Software Description

2.1 General description

The TROPICO ESS software features a modular and distributed structure monitored by microprocessors. The software can be decomposed into two main parts, that is, the **applicative software** and the **executive software**.

Two categories of components can be distinguished in the TROPICO ESS software: i) **elementary implementation blocks** (EIB), which fulfil elementary functions and ii) **groups** of elementary implementation blocks according to the main four functions of the system. These groups are:

- Telephony (TEL): call processing, charge-metering, etc.
- Defense (DEF): on-line testing, traffic measurement, error detection, etc.
- Interface (INT): communication with local devices (memories, terminals),...
- Management (MAN): trunk and subscriber signalling tasks, communication with external devices,...

2.2 Evolution of PRB with respect to PRA

The development of PRB started while PRA was under validation. Many PRA components have been reused for the development of PRB and additional ones were developed. Three types of EIBs can be distinguished:

- *new*: specifically developed for PRB;
- *modified*: developed for PRA and modified to meet the requirements of PRB;
- *unchanged*: corresponding to PRA EIBs included in PRB without modification.

Figure 1 gives the number of EIBs and the size of the software for PRA and PRB. The software of PRA and PRB was coded in Assembly language. A 10 percent increase of the PRB size can be noticed relative to PRA. Only 4 new EIBs were developed for PRB. All the EIBs of PRA have been reused with or without modifications for PRB.

	#EIB	size (kbytes)
PRA	29	319.416
PRB	32	350.800

Figure 1: Number of EIBs and size of PRA and PRB

Figure 2 shows the amount of modification performed on PRB with respect to the number of EIBs and to the size of the software. About 67% of PRB code results from the modification of the PRA code. About 75% of the modified EIB's belong to the applicative software and 84% of unchanged EIB's to the executive. Thus, the increase of the TROPICO capacity mainly led to major modifications of the applicative software with only minor modifications of the executive.

When considering the four functions and the distribution of the three types of EIB of PRB, we notice that most of the unchanged modules belong to INT (about 60%).

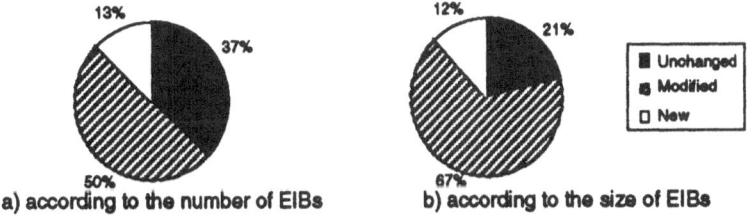

a) according to the number of EIBs   b) according to the size of EIBs

Figure 2: Distribution of unchanged, modified and new EIBs in PRB

3 Test Environment and Failure Data

3.1 Test Program

The software test program for TROPICO consists of four steps: 1) unit test, 2) integration test, 3) validation test, and 4) field trial test. The first three steps correspond to the test phases usually defined for the software life cycle. Field trial consists of testing a prototype in a real environment, which is similar to the operational environment. It uses a system configuration (hardware and software) that has reached an acceptable level of quality after completing the laboratory tests.

The description of the whole quality control program for TROPICO is given in [6, 7]. The test program carried out during validation and field trial test is decomposed into four kinds of test (functional, quality, performance and overload tests). PRA and PRB validation were carried out according to this program. Figure 3 shows, for the period of data collection on PRA and PRB, the length of validation in months, field trial and operation phases. As can be seen, no field trial tests were performed for PRB. This is because many PRA components were reused for the development of PRB, and PRB was put in operation while PRA had already been operating for several months.

	validation	field trial	operation
PRA	10	4	13
PRB	8	0	24

Figure 3: Validation, field trial and operation length for the period of data collection (months)

During the operational phase, the number of PRAs and PRBs installed on operational sites was progressively increased (see Figure 4). At the end of the data collection period, up to 15 PRAs and 42 PRBs had been installed.

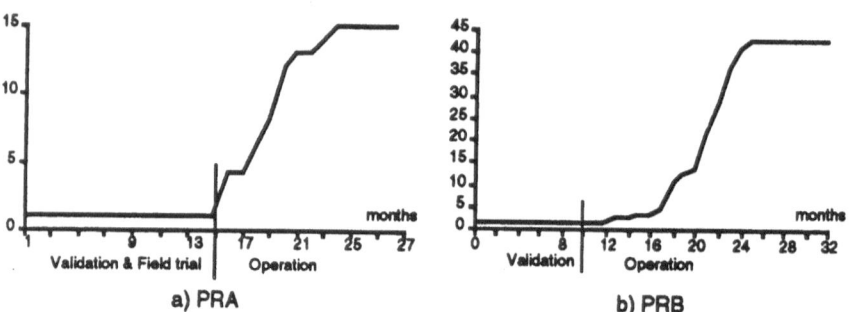

a) PRA   b) PRB

Figure 4: Number of installed sites versus time

3.2 Data Collection

Handling of failure data affecting the TROPICO ESS is through use of an appropriate failure report (FR) sheet containing the following:
- date of failure occurrence;
- origin of failure: description of system configuration in which the failure was observed and of the conditions of failure occurrence;
- type of FR: hardware, software, documentation with indication of affected elementary implementation blocks;
- analysis: identification and classification of the fault(s) which led to failure (coding, specification, interface,...);

- solutions: the proposed solutions and those retained;
- modification control: control of the corrected elementary implementation blocks;
- regression testing: results of the tests applied to the corrected elementary implementation block(s).

Only one FR is kept per observed failure: rediscoveries are not recorded. In other words, if several FR's cover the same failure, only one (the first) is entered into the database. In fact, an FR is both a failure report and a correction report since it also contains information on the fault(s) that resulted in failure.

The results presented in the following sections are based on the analysis of the data collected on the observed failures and on the corrections performed.

4 Relationships derived from the data

This section presents and discusses some of the results obtained from the data.

4.1 Statistics on failures and corrected faults in PRA and PRB

Figure 5 gives the number of failures and corrected faults in PRA and PRB. It can be seen that less failures occurred in PRB even though: i) the period of data collection for PRB is longer than that of PRA (see Figure 3) and ii) a greater number of systems have been in use during the operation phase (see Figure 4). Furthermore, the number of corrected faults exceed the number of failures. This is due to fact that some failures led to the modification of more than one EIB.

	# FR	# corrected faults
PRA	465	637
PRB	210	289

Figure 5: number of failures and corrected faults in PRA and PRB

Figure 6 shows the statistics concerning the number of EIBs that have been corrected because of a software failure. As can be seen, the results for PRA and PRB are similar. For both products about 80% of the failures led to the correction of only one EIB. This is really in favor of software modularity and equally shows that there is little failure interdependence among EIBs.

The analysis of the data corresponding to failures involving more than one component allowed us to identify two pairs of EIBs that are strongly dependent with respect to failure occurrence. For these two pairs, we noticed that the probability of simultaneous modification of both EIBs given that a failure was due to a fault located in one of them, exceeds 0.5. This result was obtained for both PRA and PRB. This type of analysis can be of a great help for software maintenance. It allows software debuggers to identify the stochastically dependent components and to take into account this information when looking for the origin of failures.

# corrected EIBs	# FR in PRA	# FR in PRB
1	362 (77.8%)	165 (78.6%)
2	72 (15.5%)	33 (15.7%)
≥3	31 (6.5%)	12 (5.7%)

Figure 6: Statistics on the number of EIBs affected by a failure

4.2 Distribution of failures and corrected faults per functions

Figure 7 gives the number of failures and corrected faults attributed to the four functions: TEL, DEF, INT and MAN (as defined in Section 2.1). The sum of failure

234

reports attributed to the functions is higher than the total number of failure reports indicated in Figure 5: this is because when a failure is due to the activation of faults in different functions, an FR is attributed to each of them.

	# FR	# corrected faults
TEL	146	190
DEF	138	164
INT	170	191
MAN	78	92
Sum	532	637

a) PRA

	# FR	# corrected faults
TEL	74	102
DEF	67	71
INT	61	68
MAN	31	41
Sum	233	282

b) PRB

Figure 7: Failure reports and corrected faults in TEL, DEF, INT and MAN

When looking at the distribution of corrected faults per functions (Figure 8), we obtain similar figures for both products, in particular DEF and INT. It can be seen that most of the corrections were performed in TEL and INT. This can be explained by the fact that these functions are more activated than DEF and MAN.

Figure 8: Fault distribution in TAP, DEF, INT and MAN

Furthermore, most of the failures reported led to the modification of only one function (90 %). Among the 465 FRs (resp. 210 FRs) recorded for PRA (resp. PRB), only 54 FRs: 31 during validation, 10 during field tests and 13 during operation (resp. 21 FRs: 10 during validation and 11 during operation) led to the modification of more than one function. This shows that the functions are not totally independent with respect to failure occurrence, although, only a weak dependence was observed. Note that this result, compared to those reported in Section 4.1, shows that less dependence is observed between functions than between EIBs.

4.3 Distribution of PRB faults per EIB type

Figure 9 shows the distribution of corrected faults in PRB when considering the unchanged, modified and new EIBs. Thus more than 80 percent of corrected faults were attributed to modified EIBs. It is noteworthy that almost the same distribution was obtained when considering data from validation or from operation only.

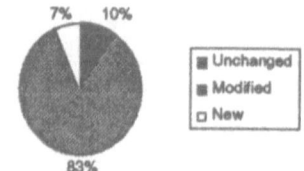

Figure 9: Distribution of FR per type

When reviewing the mean values of fault density (number of faults per Kbyte)[*] in the three types of EIBs, we obtain the following figures: 0.95 for modified EIBs, 0.75 for new ones and 0.49 for unchanged ones. One may think that the modified EIBs are more error prone than the new and unchanged ones, and would conclude that it is better to create new components than to modify already existing ones. However, we should be careful when analysing this type of result. In fact, as only 4 new EIBs were developed for PRB, no significant conclusions with respect to this particular point could be derived from this analysis.

An analysis of the average values of fault density presented in Figure 10 shows a significant decrease of the fault density of PRB EIBs when compared to PRA. Also, it can be seen that the fault density of all unchanged and modified PRB EIBs significantly decreased when compared to the values computed for PRA. This indicates an enhancement of the quality of the software. The experience cumulated during the validation and operational use of PRA leaded to a better understanding of the system and contributed to the improvement of the quality of PRB code.

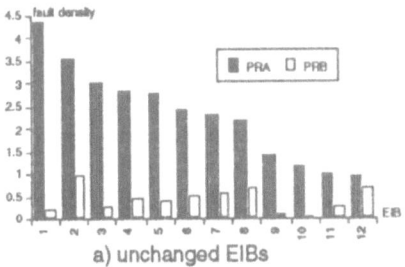
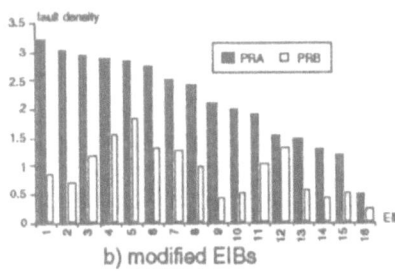

a) unchanged EIBs   b) modified EIBs

Figure 10: Fault densities of PRB EIBs compared to PRA

4.4 EIB size and fault density in PRA and PRB

Scatter plots of fault density per EIB (number of faults per Kbyte) versus the size of the EIB were plotted for PRA and PRB. It was difficult to ascertain any trend within these plots. Our objective was to analyse a possible significant dependence between the EIB fault density and their size.

Figure 11 gives for PRA and PRB the fault density average values for three categories of EIB size. The fault density is almost constant, it is around 2 faults per Kbyte for PRA and 1 fault per Kbyte for PRB. This illustrates the improvement of the quality of PRB code with respect to PRA and thus confirms the results reported in Section 4.3. As the size of PRA and PRB EIBs is measured in Kbytes and not in kilo lines of code, it is difficult to compare these values to other fault density values obtained which are reported for instance in [8, 9].

Size	PRA	PRB
EIB size > 15 Kb	1.80	1.08
10 Kb <EIB size< 15 Kb	2.02	0.68
5 Kb <EIB size< 10 Kb	2.31	0.60
EIB size< 5 Kb	2.56	0.71

Figure 11: Average values of PRA & PRB fault density versus EIB size

Figure 12 shows that the PRB modified EIBs exhibit higher fault densities on

---

[*] Note that the fault density as defined here is different from the commonly used one (i.e., number of faults per kilo lines of code); the latter is not available for this application.

average than unchanged EIBs. However, it should be noticed that the number of EIBs in each category of size is small. Also, it can be seen that most of unchanged EIBs have a small size (less than 5 kbytes) compared to modified EIBs.

Size	modified	unchanged
EIB size > 15 Kb	1.25 (6 EIBs)	0.06 (1 EIB)
10 Kb <EIB size< 15 Kb	0.86 (5 EIBs)	0.75 (1 EIB)
5 Kb <EIB size< 10 Kb	0.8 (4 EIBs)	0.4 (4 EIBs)
EIB size< 5 Kb	0.26 (1 EIB)	0.55 (6 EIBs)

Figure 12: Fault density average values of modified and unchanged PRB EIBs versus size

4.5 Evolution of failure occurrences with respect to time

Figure 13 shows the evolution of the failure intensities of PRA and PRB during the period of data collection: for both products—even though the failure intensity is globally decreasing during the operational phase—the trend is not monotone. The local variations observed are due to the progressive installation of new systems (see Figure 4). It is noteworthy that the impact of the number of operational systems on the evolution of the failure intensity has been reported in several papers, see for instance [10, 11].

In order to evaluate the reliability of PRA and PRB as usually perceived by the users, we need to consider the failure intensities corresponding to an average system (i.e., the failure intensity divided by the number of systems in use). Figure 14 shows the evolution of the failure intensities of PRA and PRB for an average system. It can be seen that the failure intensities of both products decreased globally during operation thus exhibiting reliability growth.

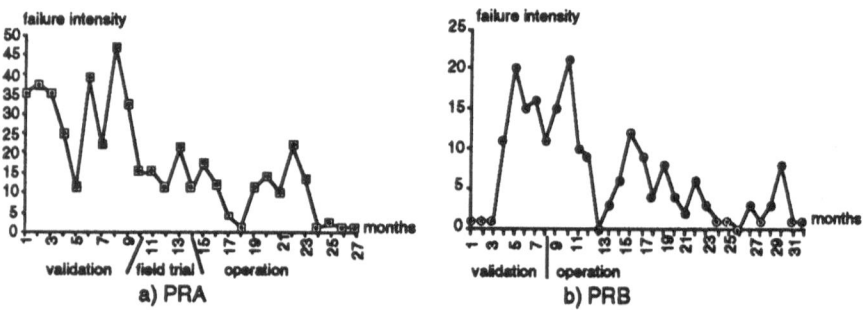

Figure 13: PRA and PRB failure intensities

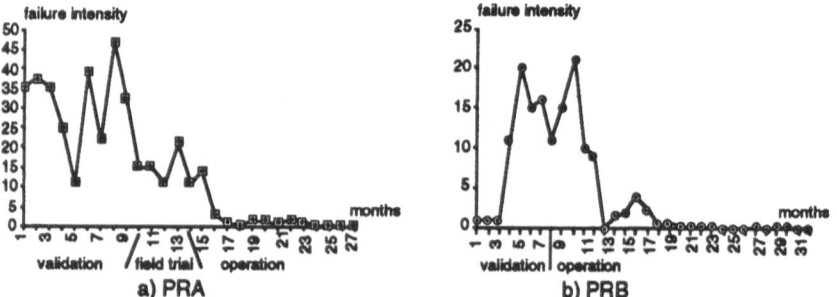

Figure 14: PRA and PRB failure intensities for an average system

In order to compare the reliability of PRA and PRB, we plot in the same figure (Figure 15) the failure intensities observed for an average system during operation. Unexpectedly, the reliability of PRB is worse than that of PRA. The same holds for the groups of functions TEL, DEF and INT (Figure 16). This is surprising because, as PRB has been developed from PRA which has been validated and extensively used one would anticipate that its reliability would be better than that of PRA. This may be explained by the fact that major modifications had been performed on PRA in order to adapt the system to the new specifications and no field trial test had been performed before the introduction of the system in the field. Note that about 80 % of PRB failures recorded during operation occurred during the first year of operation.

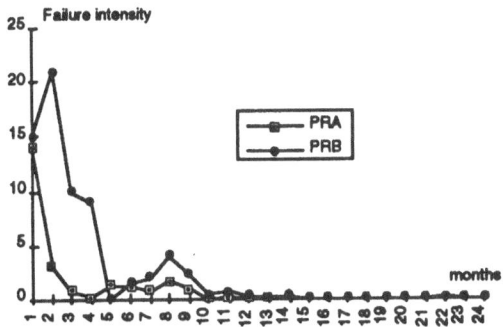

Figure 15: PRA and PRB failure intensities for an average system during operation*

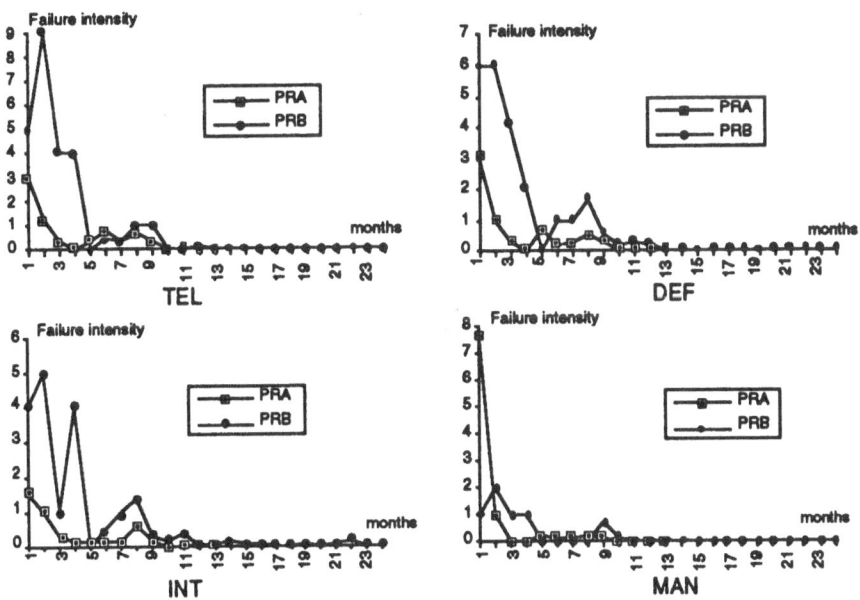

Figure 16: Component failure intensities for an average system during operation*

Typically, a new system experiences a maturing period during which its reliability is relatively low but afterwards, reliability keeps improving and becomes better than

---

*   Note that for Figures 15 and 16, the X-axis indicates the number of months since the system was put in operation.

that of its predecessors. In fact, if we look at the long term evolution of the failure intensity functions of PRA and PRB (see Figure 17) it can be seen that the residual failure rate of PRB evaluated by the hyperexponential model [12] is less than the residual failure rate evaluated for PRA. Similar results are also obtained for TEL, DEF, and INT (Figure 18). For both products, the evaluations are based on the data collected during the last year of operation.

It is noteworthy that the same was noticed in [10] for successive releases of a wide-distribution software product and in [2] for three successive products of a family of ultra-available computers designed by AT&T Bell Laboratories.

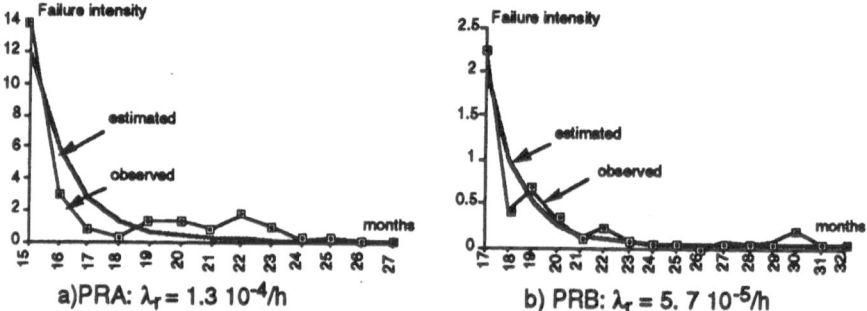

a) PRA: $\lambda_r = 1.3 \ 10^{-4}$/h   b) PRB: $\lambda_r = 5.7 \ 10^{-5}$/h

Figure 17: Estimation of PRA & PRB failure intensities

	PRA	PRB
TEL	$2.6 \ 10^{-5}$ /h	$1.2 \ 10^{-6}$ /h
DEF	$4.3 \ 10^{-5}$ /h	$1.4 \ 10^{-5}$ /h
INT	$4.2 \ 10^{-5}$ /h	$2.9 \ 10^{-5}$ /h
MAN	$1.4 \ 10^{-6}$ /h	$8.5 \ 10^{-6}$ /h

Figure 18: Residual failure rates evaluated by the hyperexponential model

5 Concluding Remarks

The data considered in this paper allowed us to analyse the evolution of the software and the failures of two consecutive products of the TROPICO ESS. The main results derived are as follows:

- A high percentage of failures was attributed to modified EIBs.
- For both products, about 80 % (resp. 90 %) of the failures led to the correction of only one EIB (resp. function). Therefore, only a weak dependence with respect to failure occurrence was observed between components.
- The fault density of PRA and PRB is almost constant with respect to size. It is about 2 faults per Kbyte for PRA and 1 fault per Kbyte for PRB.
- The fault density values of all modified and unchanged PRB EIBs are lower than those of PRA EIBs. This shows an improvement of the quality of PRB code with respect to PRA.
- Comparison of the PRA and PRB failure intensities during operation shows that PRB experienced a maturing period during which its reliability was relatively low but afterwards, its reliability improved and became better than that of PRA.

The comparative analysis provides insight into the evolution of the software and the reliability of two successive products of the TROPICO ESS. However, the results obtained did not allow us to identify the various factors that influence the evolution of the reliability of a family of products. In order to reach this objective, additional information is needed concerning for instance: i) the development process and ii) more than two successive generations of products. Furthermore, the collection and analysis of several failure data sets relative to different families of products will be of great help in this learning phase.

6 Acknowledgements

This work was partially supported by the ESPRIT Basic Research Action on Predictably Dependable Computing Systems (Action no. 6362). The authors are indebted to Marta Bastos Martini and Jorge Moreira de Souza from CPqD for their cooperation in supplying the information needed for this study and their assistance in analysing the data. We also would like to thank Sylvain Metge for his help in processing the PRB data within the framework of a student's project, and Jean-Claude Laprie and Jean Paul Blanquart for their constructive comments.

7 References

[1] Laprie J. C., "For a Product-in-a Process Approach to Software Reliability Evaluation", in Proc. *3rd Int. Symp. on Software Reliability Engineering*, pp.134-139, Raleigh, NC, USA, October 1992.

[2] Wallace J. J. and Barnes W. W., "Designing for Ultrahigh Availability: The Unix RTR Operating System", *Computer*, pp.31-39, August 1984.

[3] Sullivan M. and Chillarege R., "A Comparison of Software Defects in Database Management Systems and Operating Systems", in Proc. *22nd IEEE Int. Symp. on Fault-Tolerant Computing (FTCS-22)*, pp.475-484, Boston, MA, USA, July 1992.

[4] Kanoun K., Bastos Martini M. and Moreira de Souza J., "A Method for Software Reliability Analysis and Prediction—Application to The TROPICO-R Switching System", *IEEE Trans. Software Engineering*, SE-17 (4) April 1991.

[5] Kanoun K., Kaâniche M. and Laprie J.-C., "Experience in Software Reliability: From Data Collection to Quantitative Evaluation", in Proc. *4th Int. Symp. on Software Reliability Engineering (ISSRE'93)*, Denver, CO, USA, November 1993.

[6] Vianna B., "R&D at TELEBRAS-CPqD: The TROPICO System", in Proc. *Int. Conf. Communications (ICC 88)*, pp.622-626, Philadelphia, PA, USA, June 1988.

[7] Vianna B., Cunha E. C. and Boin F. F., "Hardware Quality Control in the TROPICO System", in Proc. *Int. Conf. Communications (ICC 88)*, pp.632-636, Philadelphia, PA, USA, June 1988.

[8] Basili V. R. and Perricone B. T., "Software Errors and Complexity: An Empirical Investigation", *Communications of the ACM*, 27 (1), pp.42-52, January 1984.

[9] Grady R. B. and Caswell D. L., *Software Metrics: Establishing a Company-Wide Program*, 288p., Prentice-Hall, Inc., Englewood Cliffs, New Jersey, USA, 1987.

[10] Kenney G. Q. and Vouk M. A., "Measuring the Field Quality of Wide-Distribution Commercial Software", in Proc. *3rd Int. Symp. on Software Reliability Engineering (ISSRE'92)*, pp.351-357, Raleigh, NC, USA, October 1992.

[11] Musa J., Iannino A. and Okumoto K., *Software Reliability: Measurement, Prediction, Application*, Computer Science Series, McGraw-Hill, New-York, 1987.

[12] Laprie J. C., Kanoun K., Béounes C. and Kaâniche M., "The KAT (Knowledge-Action-Transformation) Approach to the Modeling and Evaluation of Reliability and Availability Growth", *IEEE Trans. Software Engineering*, SE-17 (4), pp.370-382, April 1991.

Software validation with CASE-TOOLS

Dipl.-Ing. Johannes RAINER
Austrian Federal Test & Research Centre Arsenal (BVFA)
Dept. System Reliability & Traffic Electronics
A-1030 Vienna, Faradaygasse 3, Austria

Abstract

For the software validation with CASE-tools this paper gives an overview about the requirements for the software and the different validation methods for safety critical software. The requirements and validation methods are discussed on a case study with CASE-tools. Also a critical assessment on the use of CASE-tools is given.

1 Introduction

The department of the author works since a lot of years on the field of system validation of electronic equipment used by the surface traffics (e.g. the railway). The complexity of control applications for many systems has nowadays grown up so that computer systems are required and software is used in the system elements. In safety critical system elements the implemented software has to be a "safe software". The efforts on achieving safety in software are going in different directions. Measures can be employed to avoid errors in the design process of software or to neutralize safety critical effects of possible errors. To discover errors in the software, validation methods are required. This paper will discuss on a case study of software validation with CASE-TOOLS the safety requirements for the software and software validation methods.

2 Safety requirements of the software

Before speaking about the validation methods it should be clear what are the requirements for the safety relevant software. These requirements will be shortly discussed in this chapter. The requirements concern a big group of software attributes. They could be split under different points of view, e.g. in the following way:

- Requirements for the software as a consequence of requirements for the complete computer system.

- Requirements for the structure of the software.
- Requirements for the coding.
- Requirements for the protective measures against errors.
- Requirements for the documentation.
- Other requirements.

Short examples for the requirements will be given in the following chapter. The enumerated aspects for each subgroup of requirements are only exemplary.

2.1 Requirements for the software as a consequence of requirements for the complete computer system.

The requirements for the complete computer system have a great influence on the safety concept of the software. The following requirements result from the structure of the whole system:[1][2]

- It should be clear if the computer system has a redundant structure, or diversity, if the system concept is a fault tolerance or fault avoidance concept.
- The interfaces to other systems must be specified in a clear and simple way.
- The safety relevant part of the software shall be separated from the non critical software.
- The number of interrupts should be minimized to simplify the validation.

2.2 Requirements for the structure of the software.

The structure of the software has a great influence on the complexity of the validation process and on the possibility of errors. Therefore the following requirements should be considered:[1][2]

- The software shall consist of small modules for simplifying the validation.
- The software elements shall only be sequences of statements, loops, condition clauses.

2.3 Requirements for coding.

The coding action is the transfer from a logical structured program (structure diagram Nassi Shneiderman diagram) into an executable program language. At this stage the errors that can occur would be systematic errors.

To avoid errors in this stage the following aspects shall be observed:[2]

- The variables must be clearly separated in output, input, output/input variable, global and local variables.
- The addressing of variables and jumps must be clear. Complex calculating of the conditions and addresses for branches and jumps shall be avoided.
- Jumps (branches) shall only go to the beginning of a loop.
- The next executed statement after the end of a loop or subroutine must be the next statement after the call of the loop or subroutine.
- Dynamic modifications of instructions in the operational programs shall not be allowed.
- It must be assured that used compilers do not generate new errors into the code.

2.4 Requirements for the protective measures against errors.

This measures could help to reduce the error rate and can be used to detect errors in an early process stage:

- Use of diverse software
- Using redundant bits for coding,
- Mutual comparison of checksums from parallel channels.
- The software module shall run automatic tests in specified time intervals.

2.5 Requirements for the documentation.

The fulfillment of the requirements for the documentation is important for the understanding and it supports the testability of the program. The documentation will be also important for the reusability of software units. The most important requirements for the documentation are the following ones:[3]

- The minimum documentation for the code is the text of the equivalent step in the structure diagram
- The documentation of each statement must be clear, redundant information shall be avoided.
- The use of each variable shall be described exactly, which procedures use these variables, and to which physical address do the variables correspond.
- The structure diagrams for each software module shall be clearly arranged, and the connection between the modules shall be obvious

- The documentation shall correspond to the newest version of the code.

2.6 Other requirements

The requirements which don't fit into the above enumerated classification are for example: [4]
- The requirements demanded by the shut down procedure of the system and the restarting procedure.
- The program segments shall have the same size if overlays are used. If program segments require less memory, the unused memory shall be filled with a defined bit pattern.
- Critical bit patterns (e.g. all bit 0s or 1s) shall be avoided. The output from defect pieces of hardware has these patterns.

3 Validation methods

In this chapter a short overview is given about the validation methods and a classification of the different methods is carried out. However, the fulfillment of the above mentioned requirements does not give enough guarantee, that there are no errors in the software. Therefore each software product has to be validated. The validation methods are widely independent from the field of application. The methods can be subdivided in two main subgroups: "black box" validation methods (also called functional validation) and "white box" validation methods [5], which consist of qualitative and quantitative ones. Both subgroups include static and dynamic validation methods. (The division of each subgroup into static and dynamic methods has been discussed in [6])

3.1 Black box testing

The black box testing focuses on the reaction of the test object, which depends on the different external parameters of the test object.[7][8]

Functions tests
    Boundary value tests
        At this method the boundaries and extremes of the input domains are tested if there is a coincidence with the specifications. The use of the value zero (direct as well as in indirect translation) shall be included in these tests.

Probabilistic tests

> With this method the distribution of the input data shall be simulated. At this test also data out of the specified domains shall be included.

Input output requirements tests

> The fulfillment of the input output requirements is proved by comparing the output data with the specified data.

Interface tests

> With this test method errors in subprograms and errors that can lead to failures in particular applications shall be found out. This is realized by boundary value tests and probabilistic tests.

Performance tests

> At this test the boundary of the system efficiency is tested.

3.2 White box testing

At the white box methods the tests are focused on the structure and the internal parameters of the test object.[7][8][9]

Analytical methods

Semantic analysis

> There a relationship between the output and input variables is delivered.

Compliance analysis

> It helps to find out differences in use of functions, variables, procedures against the specifications in the program.

Structural analysis

> The structural analysis is used to find out jumps into a loop that are not allowed, or unreachable statements

Control flow analysis

> This method is used to find out inaccessible code segments (unconditional jumps that leaves statements unreachable)

Data flow analysis

> The data flow analysis helps to find variables that are read before written, or to find variables that are written more then once without reading, or variables that are written but never read.

Listing inspection

> At the listing inspection the program is reviewed concerning inconsistency, incompleteness of development directions.

Walkthrough method

> Hence the test of the program focuses on finding out contradictions by carrying out the functions mentally in a group.

Syntax check

> This method is used to find out if the declarations of the variables, types, functions, procedures, are correct and

> to find out if the sequence of variables (input/output) is correct.

Time testing

> With this validation method the worst case in adjustment of running times will be tested to find out collisions in running times.

Analysis of memory access

> This method shall find out if some software modules write to a memory area that is already reserved for a variable used by other procedures. It can also be usable for probabilistic analysis of the internal variables.

Specification test

> This test proves the fulfillment of the specification.

Structural test

> This test shall find out if the structure of the software is appropriate to the structure of the specification.

4 Case study with CASE-tools

4.1 Preparation phase

The used CASE-tools cannot interpret an assembler language. They use a special language. The source code has to be translated into the CASE-tool specific language. The translation process can be simplified by realizing a model of the processor (written in the CASE-tool specific language) that simulates the used instructions.

The input file needs some additional informations e.g. procedure specifications, mainprogram specifications, function specifications, derive relationships, assert statements. The derive relationships simplifies the analysis at a procedure call. The assert statements can be used for refining the analysis. These informations have great influence to the results of the analysis.

4.2 Analysis phase

The output of the CASE-tool depends on the specification in the command line, (which keywords were used). The CASE-tools that have been used cover (see also 3.1 White box testing) the control flow analysis, data use analysis, information flow analysis, semantic analysis and compliance analysis. The information flow analyser delivers as result the dependency of the variables from the used variables and constants. For each variable the dependency from conditional nodes is stated. Also a list of possible errors and redundant statements is given.

At the compliance analysis the relations between input and output variables are calculated and compared with the specifications at the begin and the end of the program block. These specifications have to be inserted by the operator of the CASE-tool. The quality of the result from this analysis depends on these specification statements. By implementing more conditions in the code the result of the analysis can be simplified.

The control flow analysis is used to find out the structure of the code and to find out unreachable statements, multiple entries into loops. The control analyser simplifies the graph. The stage of simplification (if only sequences of nodes are removed, or also self loops) can be controlled by the used keywords.

The data use analysis shows how often a variable is read before written. Hence variables which are written more then once without reading could indicate omitted code. Also the data use analysis shows if variables have been written and never read. This could indicate redundant code. As result also possible errors are stated, which have to be confirmed by the user of the CASE-tool.

The semantic analysis generates the relation between input and output variables of each executable path. The user of the CASE-tool has to compare the results of the semantic analysis with the requirements specification.

4.3 Evaluation phase

This is the most difficult section of the validation with CASE tools. There the results of the different analysis methods have to be compared and conclusions must be made.

Some problems about evaluating are given in this chapter. For example the communication between the individual subroutines of the tested software is realized for many times by using the accumulator and flag register. In this case the CASE-tools can deliver an error statement that the register is not defined. This handover procedure was not done randomly it was used systematically. There a violation of the software requirements occurs. The main question here is now is this violation acceptable or shall it be treated as a safety critical violation. The decision whether the use of a register as variable is acceptable or not, has to be made by the proofing person and the orderer of the validation (contractor).

At a procedure the data use analyser indicated that a variable was written for sometimes with no intervening read. A review of the procedure showed that these writing actions to the variable were correct. The used processor model simulates the flag register by using boolean variables for each flag. According to the operation of the processor the flags have to be set. To avoid such results by the CASE-tool the variables which should be proved can be selected. This option simplifies the analysis results. It should be taken in account, that the selection of the variables is a critical decision done by the user of the CASE-tool.

The software, that has been validated by the department of the author, has as documentation of the code only a structure diagram (Nassi Shneiderman Diagram) and an incomplete variable list. It is clear that the requirements concerning documentation of the software were not fulfilled. The question is, shall the software be treated as a safe software or as unsafe software. The incomplete documentation will increase the necessary time for the validation. The presentation of the validation report has been discussed by G.List[10].

4.4 Advantages for the use of CASE-tools

The use of CASE-tools for software validation supports a formalizing of the analysis results. This formalizing make the analysis of the code easier. To utilize the simplification of the analysis the CASE-tool user has to investigate some time into the preparation of the code before using the CASE-tool (see also 4.1 preparation phase). The time profit T_{pr} by using CASE-tools can be described mathematically as time profit margin

$$T_{pr}(t, t_{case}) = t - t_{case}.$$

Where t is the required time for validation without CASE-tools and t_{case} the required time for validation with CASE-tools. The time profit depends mainly on the efficiency use of the CASE-tools.

The use of CASE-tools has not only an influence on the validation time, it also influence the rest error rate of the validation. The rest error ratio of the validation will be reduced by using CASE tools. This quality improvement depends mainly on the person who carries out the preparation of the validation object and the assessment of the analysis results. The rest error ratio r_r can be quantitatively described as

$$r_r = \frac{N_u}{N}$$

where N is the number of all items and N_u is the number of all undetected errors. The quantitative view of the error ratio has been discussed in more detail by A. Sethy [11][12]. The quality improvement can be described as the ratio of the rest error ratio without CASE-tools and the rest error ratio with CASE-tools

$$V(r_r, r_{r\text{-case}}) = \frac{r_r}{r_{r\text{-case}}}$$

called also improvement factor V [13]. The time profit and the quality improvement can also be seen in the economical view. The CASE-tools represent a bigger investment for firms. Therefore a economical justification for such investment is required. The CASE-tools can be used for example 5 years. In this time interval the user has to do M numbers of validation.

The costs per validation C_{val} are then

$$C_{val} = \frac{INV}{M}$$

where INV is the investment for the CASE-tool (including costs for training and price for CASE-tool). This costs per validation can be transfered in a time equivalent T_{val} as followed

$$T_{val} = \frac{C_{val}}{C_{man}}$$

where C_{man} is used for the cost manpower per hour.

The investment of the CASE-tool will be justified if following condition is fulfilled:

$$T_{pr}(t, t_{case}) > T_{val}$$

in words: the time profit has to be greater then the time equivalnt T_{val}. The quality improvement of the rest error ratio can be quantified economically by using the mean costs of error consequences. The quantification of these costs depends on the user of the validation object and they can vary in a wide range. Discussing this proplem would go far byond this paper and cannot be done here therefore.

For software metrics some usable results can be easily obtained by the use of CASE-tools. The property of the analysis results are well specified, so that for example the complexity of the software can be determined reproducibly.

5 Conclusion

The use of CASE-tools makes the analysis of the code easier. As it has been shown in this article, the use of CASE-tools for software validation delivers an improvement of the time- and the quality aspect. It should be considered that the time for the analysis is reduced and the time for assessing of the results increase. The quality improvement of the validation results depends also on the person, who validates the software (see also chapter 4.4).

As it has been discussed above, the CASE-tools cover methods of the white box testing group. For a complete validation of a safety critical software some methods of the black box testing group also must be carried out. The CASE-tools will point at possible errors in the code. These errors have to be confirmed by other validation methods. The use of CASE-tools may replace some parts of the conventional test methods. However, it must be clear that the understanding of the code is still necessary.

Hopefully it was possible to show in this paper that the use of CASE-tools can simplify the life of the validating person.

6 References

1 ORE A155.2/RP3 Software for safety systems - an overview. 1985

2 ORE A155.2/RP9 Software design for computer based safety systems. 1987

3 DB Mü 8004 Allgemeine Richtlinien für signaltechnisch sichere Schaltungen und Einrichtungen der Elektronik (General guidelines for signalling safe circuits and equipments of electronics).

4 Keene S. "Assuring Software Safety" Reliability and Maintability Symposium 1992 IEEE proceedings pp. 274-279 , Las Vegas.

5 IEC 56 (sec) 307 Software test methods 1990

6 Sethy A. "Methodische Frage bei der Prüfung der Softwaresicherheit." e&i Verlag Springer New York Wien 1991/3 pp 80 - 82.

7 IEC 65 (sec) 122 Software for computer in application of industrial safety related system. 1991

8 IEC 65 (sec) 123 Functional safety of programmable electronic systems: generic aspects. May 1992

9 ORE A155.2/RP11 On the proof of safety of computer based software systems. 1987

10 List G. "Methodological aspects of critics during safety validation."Safecomp'90 Gatwick Pergamon Press, Oxford, New York pp 99-103.

11 Sethy A. " The actual change in questions of proof of safety and availability in the railway technics." IFAC Conference 1983/4 Baden-Baden, Preprintes VDE-Verlag, Düsseldorf pp 275-281.

12 Sethy A. "Connection between Reliability and Signalling-Safety in Railway Technology" Reliability and Maintability Symposium 1992 IEEE proceedings pp. 75-79, Las Vegas.

13 Sethy A. "Fragen zur Messung der Qualität von Datenübertragungssystemen." Nachrichtentechnische Zeitschrift, 1962/2, VDE-Verlag, Berlin, pp 85 - 87.

Dependability of Scaleable, Distributed Systems: Communication Strategies for Redundant Processes

W. Kuhn, E. Schoitsch
Austrian Research Center Seibersdorf
Department of Information Technology
A-2444 Seibersdorf, Austria

Abstract

The key element of dependable distributed systems is the communication strategy. Communication between distributed and/or redundant system components (processes) may use standard network tools and protocols. The absence of a multicast-support in the ISO network model above the Network Layer requires special provisions for software that must distribute data over a network to an unknown number of network partners. A method is presented which combines the benefits of different network layers to cover the needs of such a distributed, dependable, and redundant system.

1 Introduction

The Austrian Research Center Seibersdorf (ARCS) has specified, designed and implemented the distributed security, alarm and control system called CSS (Scaleable Security System) for Philips Industry [1]. The task of this system is to protect an area, a plant or a building complex from threats from the environment (therefore is is sometimes called a "Risk Management System"). The properties of such a system depend on its ability to get information about the environment and its inner status (by peripheral subsystems, sensors, etc.) and the thrustworthiness of the CSS itself. The system has as primary goals high availability and scaleability (i.e. configurable freely within any topology, and network). One of the key ideas of the concept is, that processes and processors may be distributed freely according to the principles enumerated above, and the configuration is scaleable from a single workstation to a redundant network of n processors. Dependability and fault tolerance [2, 3] are implemented via distributed, redundant (software-) processes using a multilayer software structure for communication and message exchange and functional software interfaces between the various external subsystems. One topic is the communication strategy chosen to support any redundant hardware and software structure, fail-over strategies, and dynamic reconfiguration. The experiences with the implementation of the communication mechanisms on different levels of the underlying network protocol and the problems encountered during the implementation and installation phase under hard real-time and load restrictions will be discussed.

2 System Overview

The design and development of the CSS system started in 1988, when Philips Industry decided to plan a new "Alarm and Control System" which should take the place of the hitherto existing PDP11-based system XLSS (Extended Local Supervisor Station). The primary goal of such a system is to protect an area, a plant or a building complex reliably from break-in, fire and other undesired events. In addition, the system should be able to control parts of the building, plant, or area, and, of course, reflect (and show) the current state of the "outer world" as well as the "inner status" at any time.

The overall system is composed of the central CSS for processing, managing, visualization and operating, and the peripheral subsystems, which are partially autonomous sources of information (and control), provided by different vendors and following different communications and control strategies. The overall system dependability is limited by the dependability characteristics of the peripheral subsystems; the goal of the design was, that peripheral subsystems as well as human operators, guards etc. can justifiably rely on the CSS services.

Several constraints concerning the environment and the target hardware components were given by Philips, so it was not possible to choose special reliable computing elements. The most important restrictions were the following:

R1 The usage of standard *off-the-shelf* Digital Equipment Corporation hardware and software, especially VAX computers running the VAX/VMS operating system.

R2 The usage of a standard Ethernet Local Area Network (LAN), including standard network controllers and protocols (CSMA/CD).

R3 The usage of customer-defined and/or pre-installed subsystems (redundant or non-redundant communication lines), which represent the interface to the real world.

R4 The usage of standard software components wherever possible (e.g. a standard database system and the Graphical Kernel System GKS [4]).

R5 A restricted development budget and a target-date for completion.

Restriction R2 prevents the use of special (hardware) network attachment controllers as described in [5], and it also implies the discussion of Ethernet being adequate for real-time (e.g. [6]), and dependability [7].

Restriction R3 also means that some subsystems are not available for "off-line" (lab) software tests; the software for these subsystems must be carefully checked out in the "living" system.

All these limitations lead to the challenging task of designing and building a cheap, re-usable, and dependable system consisting only of standard components [8].

2.1 Building Blocks

The main strategy for the system was to split up the software and the hardware into building blocks.

2.1.1 Software Building Blocks

The basic decision while designing the software was (i) to separate the CSS-software into an arbitrary number of processes (about 40 at the moment), and (ii) to provide a single logical communication path for inter-process communication. To avoid an uncontrolled information exchange between processes, primary communication paths have been introduced. These primary communication paths define groups of processes that may exchange information. The processes are grouped into two classes, namely central processes and peripheral processes. Central processes are typically the Central Coordinator or the Database Access Module, peripheral processes are Human Interface processes or the Subsystem Access processes [1]. Communication can only take place (i) between a peripheral and a central process, and (ii) between central processes. The main difference of central and peripheral processes is, that central processes *must* be available in the system, and peripheral processes *may* be available in the system. So the peripheral processes can be seen as one of the "scaleable parts" of a CSS. Fig. 1 shows a simple structure of CSS processes in a single node and the primary communication paths.

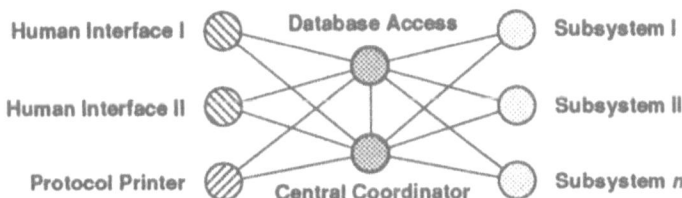

Figure 1   CSS Processes in a Single Node

This structure has the following advantages:

A1 The system may grow as completely new processes (e.g. new subsystems or new human interfaces) can be added very easily.

A2 The processes may reside all on one node, or may be distributed over several network nodes as the process communication can be seen as remote procedure calls.

A3 Single (or all) processes may be replicated according to the needs of a specific CSS installation.

2.1.2 Hardware Building Blocks

As mentioned above, one of the main characteristics of the system is its scaleability. There are several ways in which this takes place. For the hardware, two requirements had to be fulfilled, first the requirement to build a *n*-fold redundant system, and second to provide interfaces (typically RS232) to a conceptually unlimited number of subsystems. The first characteristic can simply be realized by connecting n nodes to a network, where each node runs a CSS. The second characteristic is of more interest, because it is not possible to put an arbitrary number of interfaces to a computer, so, Terminal Servers were used (Fig. 2).

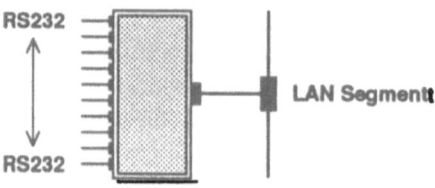

Figure 2   Terminal Server

In addition to the benefit of having an arbitrary number of communication lines, Terminal Servers offer the advantage of being able to control a (server) communication port from several nodes. This feature makes a special line-switching hardware obsolete. The switching of lines to different nodes can be controlled by the software. Disadvantages are discussed in Sec. 5.1.

3 Process Communication

Process communication takes place by means of mailboxes. The principal idea for mailbox communication between CSS processes is that each CSS process has exactly *one* mailbox where it receives information (messages). The basic view of two communication CSS processes is show in Fig. 3.

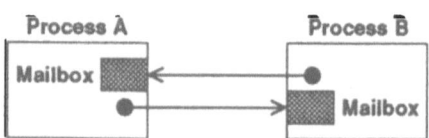

Figure 3   Basic Communication of Two CSS Processes

This kind of communication strategy ensures that each process has precisely one input channel. Handling messages in this way would cause some situations where a process may be blocked. For instance, when process *A* is busy (i.e. not ready to empty its own mailbox) and another process *B* would fill up the mailbox of process *A*. In this case, process *B* would be blocked because it cannot get rid of its messages.

3.1 Receive Queues

To avoid these situations, it must be guaranteed that a process is always able to empty its mailbox. This is done by splitting a CSS process into two (or more) parallel-working threads [9]. Thread 1 reads the mailbox asynchronously and puts the message into into an arbitrary large FIFO queue, which is read in thread 0. Thread 0 (the application program) does not read from the mailbox, but from the receive queue (Fig. 4).

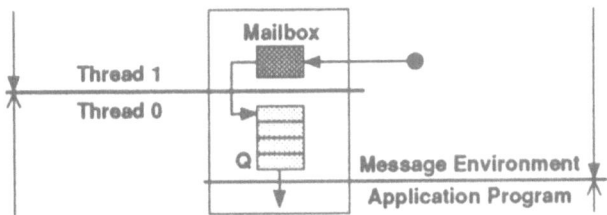

Figure 4   Receive Queue of One CSS Process

Even if thread 0 is blocked (e.g. if the application program is doing some calculation), thread 1 is still working.

A problem arises when the system must handle (soft) real-time events with this kind of communication structure. Assuming that a burst of non-real-time events followed by a real-time event would cause the real-time event to be delayed. A solution for this problem is to introduce several receive queues as shown in Fig. 5.

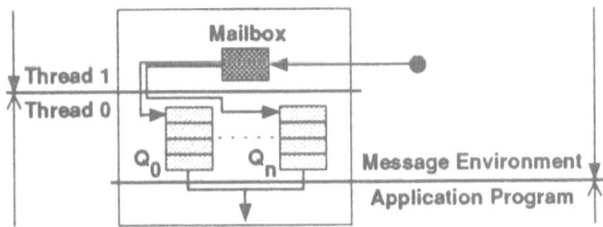

Figure 5   Multiple Receive Queues of One CSS Process

The events (messages) in the CSS have assigned several priorities, and each priority has a separate receive queue. The receive part of the CSS message handling environment first scans the the first receive queue (with priority 0, which means "real-time" priority) and delivers the message to the application program. Then all other queues are scanned. This method still requires the use of bounded loops (e.g. [10]).

258

3.2 Real-World Interface

The previously described mechanisms are also used to implement a non-polling (event-driven) form of real-world communication. The communication layers are extended by an additional thread. This thread may handle different protocols, e.g. protocols for external devices, or the X-Protocol (Fig. 6). After completion of protocol handling in thread 2, the final "packet" is processed by the system as any other message.

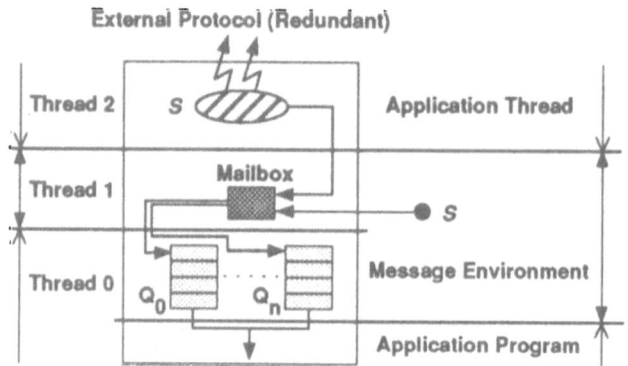

Figure 6   External Protocols

As mentioned above, this stack is used to build an interface to foreign protocols. These protocols are widely used in the CSS as a link to different subsystems, about 30 at the moment.

4 Network

The principal communication layers of a CSS process are shown in Fig. 6. The sending object S can be any other process or even the receiving process itself. So it is (or appears to be) very easy to expand the system for network usage by simply adding a new sending object which performs a network operation. Network capability may be introduced to such a system for the following two reasons:

N1   to build a client-server system

N2   to distribute data to process replicas

of which only N2 is of interest here.

The main difference between these two items is that N2 requires (parallel) communication with an usually unknown number of (network) partners. Several techniques have been introduced to perform these multicast communications [11, 12]. But due to some restrictions given in Sec. 2, it was not possible to implement a complete reliable group, or multicast, communication protocol.

4.1 ISO Network Layers

The method described now tries to combine the benefits of different ISO network layers to cover the needs of N2 and to minimize the expenditure of implementation.

The following prerequisites were given:

(i) An Ethernet Local Area Network, that provides multicast and broadcast functionality on the Data Link Layer.

(ii) An ISO network, that provides reliable communication above the Network Layer (with the Network Service Protocol NSP, *"A protocol that provides reliable message transmission over virtual circuits. Its functions include establishing and destroying logical links, error control, flow control, and segmentation and re-assembly of messages"* [13]).

4.2 Simulated Multicasts

The idea now was to use these two network features, namely

- *real* multicasts on the Data Link Layer,
- and an ordered set of reliable unicasts on the Network Layer

together as *simulated* multicasts in the following manner. Each process transmits a unique *"hello"* multicast packet (unique means (i) a group-unique Ethernet Protocol ID, and (ii) a group-unique multicast address; both administrated by the CSS) on the Data Link Layer periodically, which is responded to by the instances of that processes on other nodes. The response to this real multicast packet (a unicast packet) is used to build a list of network partners. Both activities are performed in different threads. Further network calls use this list to transmit data with the NSP as simulated multicasts, i.e. a sequence of unicasts, which are

(i) strictly *sequential*,

(ii) *synchronous*, and

(iii) do not require any protocol handling.

If a process does not respond to subsequent *"hello"* packets, or if a call on the Network Layer fails, the process that caused that failure is assumed to be down and will be removed from the list.

Since communication on the Data Link Layer is not reliable and packets may get lost during heavy network load, the receiving partner has to wait a certain time before he is allowed to assume a non-responding process being down. The CSS application process that uses the multicast mechanism has to call one *MULTICAST* routine only which then performs all necessary processing steps.

An overview of the simulated multicast mechanisms is given in Fig. 7. Note that this figure does not show the relation to the CSS process communication as described in Sec. 3, Fig. 6.

Voting is also performed on this level within segments (a) and (b). As a result of this voting, an appropriate message is delivered to the calling process.

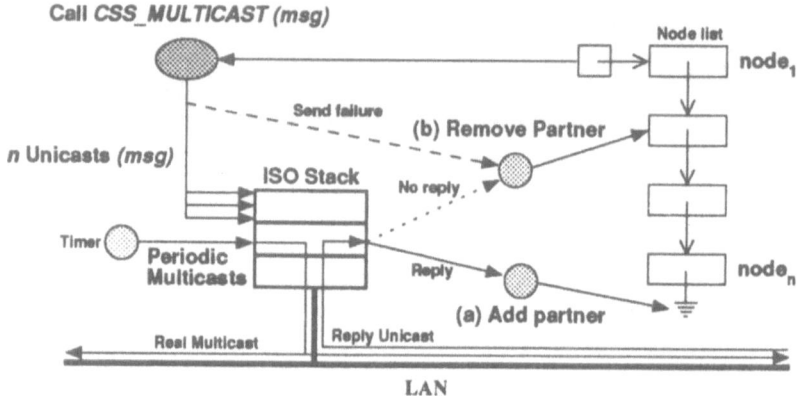

Figure 7   Simulated Multicasts

It is clear that the overall transmission time T for a simulated multicast packet increases with the number of nodes n involved, with

$$T = nt_0 + \sum_{i=1}^{n}(t_i + \tau_i)$$

where t_0 is the time consumed by the sender node, t_i is the individual error-free transmission (and receive) time to node i, and τ_i is an additional delay caused by retries due to network errors. t_0 and the individual transmission time t_i depend on the CPU power only, τ_i depends on the network load *and* the CPU power, with $\tau_i = 0\ \forall i (i \in \{1...n\})$ for an error-free transmission. Table 1 gives an overview of individual transmission rates $(t_0 + t + \tau, n = 1)$.

CPU Type	CPU Power			R (Mbps)	x	σ
	CSS	VUP	SPECmark			
VAXstation 2000	1.0	0.9	-	0.06	144	2.608
MicroVAX 3600	2.5	3.2	-	0.34	144	0.110
VAXstation 3100/76	10.0	7.6	-	0.67	144	0.252
VAXstation 4000/90	25.0	-	32.8	1.93	144	0.032

Table 1   Network Transmission Rates of Application Data

Data are based on the communication between two computers of the same type over an Ethernet Local Area Network (10 Mbps). The transmission time T of user-level packets with a constant size of 128 octets of application data has been converted into the transmission rate R given in Mbps. (Note we have to distinguish between a user-level information packet and a LAN-level information packet!) The

CPU power is given in different terms (CSS = CSS-specific computing power relative to a VAXstation 2000; VUP = VAX Units of Processing ~MIP), x is the number of measurements, σ is the standard deviation.

5 Experience and Field Data

The system is now installed at 6 sites with a sum of 10 system-years of operation. Experience and field data refer to these systems.

5.1 Terminal Servers

Terminal Servers as mentioned in Sec. 2.1.2 turned out to be the most problematical components in the CSS. The problems encountered were

(i) questionable real-time characteristics

(ii) unmotivated "port stops"

(iii) different behavior of different types of Servers

of which we will discuss (i).

5.1.1 Timing Problems

Fig. 8 shows the typical I/O timing behavior of a Terminal Server with the following setups: DECserver 200/MC (V3.1 BL37, LAT V5.1, ROM BL20), RS232, 1200 baud transmit/receive speed, one start-bit, one stop-bit, even parity, 23 characters message length, VAXstation 3100/76, VAX/VMS V5.5-2.

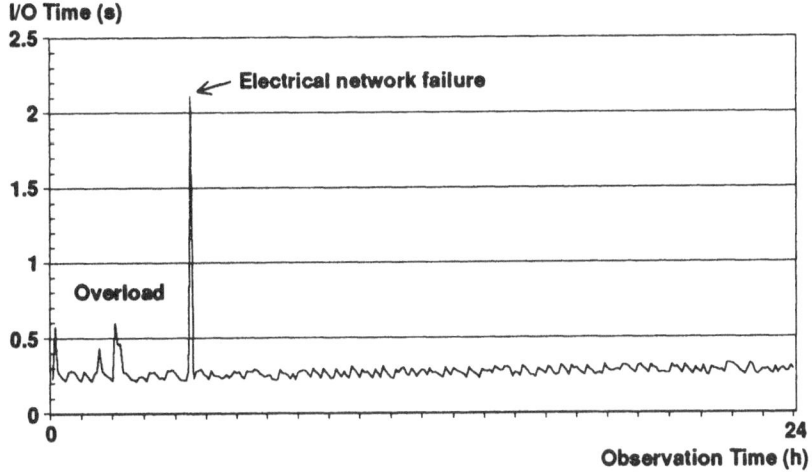

Figure 8 I/O Time on a Terminal Server

A "Schauer PDU/DCF77" clock was used as a data-generator. Every five minutes this clock transmits a 23 character packet containing the current date and time over a serial line. This time information was compared with the current system time, as the line clock was assumed to be accurate, and the system-clock was assumed to be a "good" clock [14]. The result of this measurement was that the transmission and processing time of a packet was between 0.2 seconds and 0.3 seconds during normal network load (2%) and low CPU load (<1%). Heavy network load (50%) increased the transmission time up to 0.6 seconds, and electrical network failures caused delays of over 2 seconds.

Using a direct communication link (UART) gives a constant transmission time of 0.2 seconds (not shown in Fig. 8).

As the CSS was not designed to be a hard real-time system, is was possible to solve these timing problems within the CSS software.

5.2 System Field Data

Table 2 gives an overview of the systems, nodes, disks, and Terminal Servers currently installed and operating in the field.

	Systems	Nodes	Disks	Servers
Number of elements	6	12	16	8
Operational years	10	22	30	14
Damages	n/a	0	2	1

Table 2 CSS Elements

To date, three accidents have occurred; two head crashes on disks, and one Terminal Server breakdown. The head crashes have been tolerated by the CSS since the VAX/VMS operating system automatically shut down and the CSS processes running on those nodes have been recognized as being unavailable. The Terminal Server breakdown has been tolerated for those subsystems with redundant communication lines to different Terminal Servers.

Down reasons	System		Node downs
	downs	UA (hours)	
1. Power fails (test, service)	0	0	7
2. CSS software failures	1	24	8
3. VMS software failures	0	0	5
4. Hardware failures	0	0	2
5. Maintenance	21	3.3	61
Σ	22	27.3	83
Availability	99.95%		

Table 3 CSS Downs

Table 3 shows the down-times of single nodes and the whole system. The unavailability (UA) of the whole system had two reasons; first a fatal software failure in a central CSS process that caused all nodes to be inoperable. The failure was repaired within 24 hours. The second reason was (and is) the down time due to system maintenance and software upgrades. All other node failures have been tolerated by the CSS.

6 Conclusion

A system overview of a dependable, scaleable distributed system has been given (for a more broader description see [1]). The design goals of scaleability, flexibility of configuration, ergonomy of human interfaces, flexibility to integrate new peripheral subsystems and the application of standards as far as possible have been reached by "modularization through distribution", which includes the concept of hardware and software building blocks, process replication and standard ISO network layers. System maintainability and the possibility of easy implementation of a variety of fault tolerant architectures are further results of that concept. These goals cannot be reached by a single processor/single layer approach, although there are some tradeoffs with respect to some of the dependability attributes when a distributed solution is chosen.

Dependable communication mechanisms have been identified as the key issue for providing reliable services for process fail-over strategies and dynamic reconfiguration.

It has been shown, how on basis of standard hardware and software, by adding some additional software using Ethernet multicasts to provide the valid configuration status information, and simulated multicasts within the ISO stack framework, a reasonable dependable system with reasonable real-time characteristics has been implemented. Some figures and field data as well as relevant implementation details have been presented.

Until the end of the year, the CSS system will be installed at ten sites, mainly large banks and museums (including WAN-networks connecting several buildings and branches).

References

1. Schoitsch E, Kuhn W, Herzner W, Thuswald M. Experience in Design and Development of a Highly Dependable and Scaleable Distributed Security, Alarm and Control System. Proc. of the IFAC/IFIP/EWICS/SRE Symposium, Trondheim, Norway, 1991, pp 141-147

2. Redmill FJ (ed). Dependability of Critical Computer Systems, Vol. 2. Elsevier Applied Science, London, New York, 1989

3. Bishop PG (ed). Dependability of Critical Computer Systems, Vol. 3, Technical Directory. Elsevier Applied Science, London, New York, 1990

4. ISO. Information Processing Systems - Computer Graphics - Graphical Kernel System (GKS) Functional Description. Document no. ISO 7942, First edition 1985-08-15
5. Powell D, Veríssimo P. Distributed Fault-Tolerance. In: [15], pp 89-124
6. Veríssimo P. Real-Time Communication. In: [16], pp 335-351
7. Pâris JF. Evaluating the Impacts of Network Partitions on Replicated Data Availability. In: [17], pp 49-65
8. Schoitsch E. The Interaction between Practical Experience, Standardization and the Application of Standards. Proc. of the IFAC/IFIP Workshop, Vienna, Austria, 1989, pp 17-24
9. Mullender SJ. Operating System Support for Distributed Computing. In [16], pp 233-260
10. Kopetz H, Fohler G, Grünsteidl G, et al. The Programmer's View of MARS. In: [16], pp 443-458
11. Veríssimo P, Rodrigues L, Rufino J. The Atomic Multicast protocol (AMp). In: [15], pp 267-294
12. Wybranietz D. Multicast-Kommunikation in verteilten Systemen (Informatik-Fachberichte 242). Springer-Verlag, Berlin, Heidelberg, 1990
13. Martin J, Leben J. DECnet Phase V: An OSI Implementation. Digital Press, Bedford MA, 1992
14. Di Vito BL, Butler RW. Formal Techniques for Synchronized Fault-Tolerant Systems. Preprints of the 3rd IFIP International Working Conference on DCCA, Mondello, Italy, 1992, pp 85-97
15. Powell D (ed). Delta-4: A Generic Architecture for Dependable Distributed Computing. Springer-Verlag, 1991
16. Handouts of Lisboa '92. An Advanced Course on Distributed Systems. Estoril, Portugal, 29th June - 8th July 1992
17. Meyer JF, Schlichtinger RD (eds). Dependable Computing for Critical Applications 2 (Dependable Computing and Fault-Tolerant Systems, Vol. 6). Springer-Verlag, Wien, 1992

Real-Time Detection of Failures of Reactive Systems

Rudolph E. Seviora
Department of Electrical and Computer Engineering
University of Waterloo, Waterloo, Ont., Canada   N2L 3G1

1. Introduction

This paper addresses some issues involved in real-time detection of failures of reactive systems. The system architecture considered is shown in Figure 1. External behavior of the reactive system is monitored by a supervisor, which may execute on a separate platform. The supervisor monitors the inputs and outputs of the system and reports the failures that occur.

Real-time detection of failures has a number of benefits. Consider an application such as telecom switching:

a) Early reporting of failures gives the operating company an opportunity to repair the underlying fault before the users start filing complaints.

b) Certain kinds of failures, such as those due to loss of shared resource units, are visible only to an entity with global perspective. Early notification of such failures makes it possible to take corrective steps before its accumulated effects result in major service disruptions.

c) In many reactive systems, failures of control software do not have immediate effect. Because of mechanical inertia etc., a long time interval may elapse before a software failure has detrimental impact on the controlled hardware. Real-time detection of failures provides a basis for subsequent retraction of their effects.

Supervision-based approaches to failure detection and retraction are becoming more important as systems and their control programs are constructed from off-the-shelf components.

In applications in which the external behavior of the reactive system is specified formally, it is attractive to have the supervisor execute (or interpret) a model derived from system specification.

The paper considers the case when the external behavior of the target system is specified by a model based on communicating, extended finite state machines (specification processes). The formalism used is the CCITT Specification and Description Language (SDL)[1]. SDL is an international standard used in the telecommunication industry. SDL specification of external behavior is supplemented by the specification of response times. The focus of the paper is on event-driven applications whose processing is relatively simple. A typical application is telecom switching.

Supervision-based failure detection has some similarities to automated test oracles [4]. However, automated oracles do not detect failures as they occur and usually assume a particular resolution of specification nondeterminisms. Real-time monitors

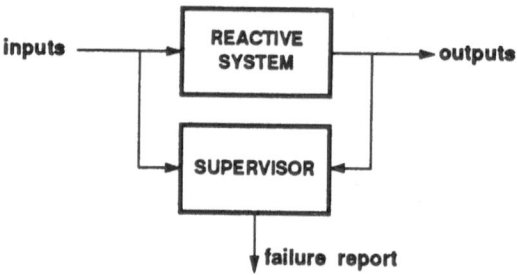

Figure 1. Real-Time System Supervisor

(see, e.g. [5]) work in real-time, but are typically closely coupled with the program being monitored. Supervision-based failure detection also resembles approaches such as the safety bag[6]. However, SB checks for violations of safety regulations and aims to prevent failures from occurring in the first place. Specialized techniques developed to detect certain kinds of telephone exchange failures in real-time can be found in articles describing their maintenance software (see, e.g. [7]). The theory of beliefs, introduced in Section 3, was inspired by the truth maintenance systems and nonmonotonic reasoning[8].

This paper is organized as follows. Section 2 overviews the CCITT SDL. Section 3 discusses the two basic strategies for failure detection in real-time (input and output-driven) and presents formulas that estimate their processing and memory requirements. Section 4 describes experience with input-driven supervisor which was developed to automatically collect failure data of a small exchange. Section 5 offers concluding remarks.

2. Specification Formalism and Issues

Structurally, an SDL specification consists of a hierarchy of blocks. Blocks are interconnected by channels. Channels carry signals between blocks. A leaf block contains one or more SDL processes, whose behavior is specified by an extended finite state machine. Specification processes may contain local variables, which may be updated and tested. Processes within a block communicate by exchanging signals over signalroutes. SDL semantics is defined operationally, by the Abstract SDL Machine [1].

SDL is illustrated in Figure 2. This figure shows partial, SDL-based specification of call processing for a small (and simplified) telephone exchange. Part (a) of the figure shows the block diagram and part (b) gives a fragment of behavioral specification for the Line Handler, one of the processes in the block diagram.

Part (a) shows that the specification consists of two major blocks. One contains the LineHandler processes, which are responsible for the external behavior of the exchange seen by individual phones. The other contains a resource manager process. This process controls the sharing of exchange hardware resources needed to process a call. One resource class may be the touchtone receivers, which decode the digit from the tones sent by the phone when a key is pressed. For simplicity, Figure 2 shows only one resource manager process.

Part (b) states that when the telephone is idle and goes offhook, a request signal for the resources needed to handle the origination will be sent to the Resource Man-

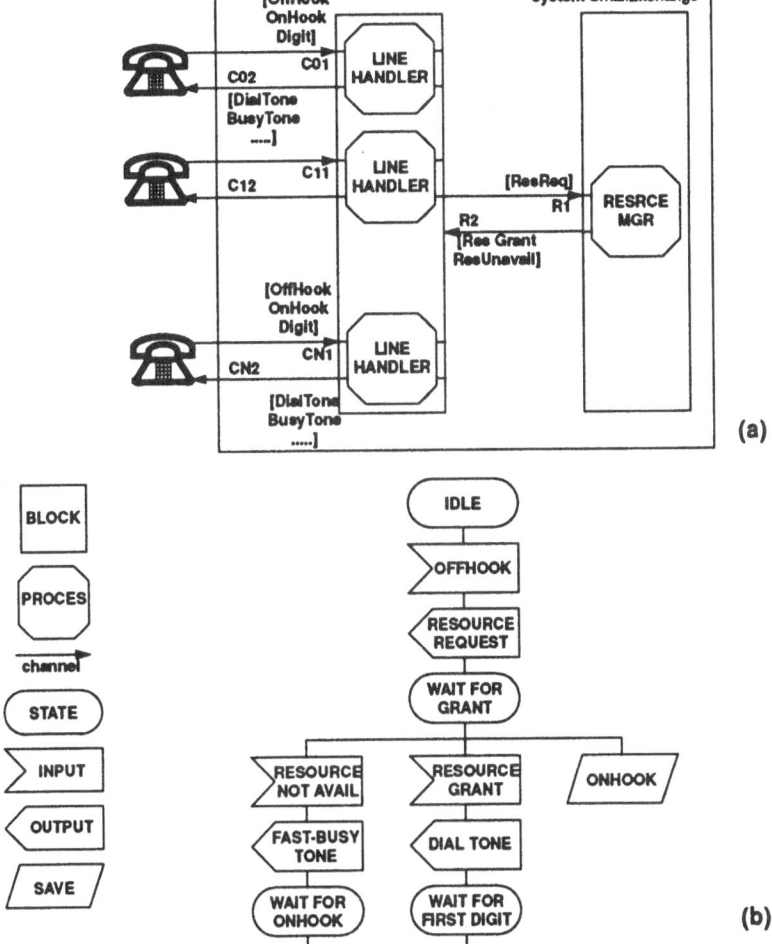

Figure 2. Illustration of SDL Specifications

ager. The Resource Manager may grant the resources by sending a Grant signal to the Line Handler, in which case dial tone is applied to the phone. If the resources are not granted (signal Resource NotAvailable), the phone gets the fast busy tone. If the phone goes onhook while its Line Handler is waiting for response, the OnHook signal is kept (Save'd) until a subsequent state.

SDL specification of external behavior are complemented by performance specification which give the maximum permissible time intervals between an input signal and the response(s) it triggers (for example, the time from OffHook to DialTone). Furthermore, some external signals may be communicated only indirectly, by a signal carrier. For example, OffHook and OnHook are encoded in changes of loop current. The specification includes the definition of the minimum and maximum time for which the signal carrier change must be present in order for the encoded signal to be recognized.

Several issues must be considered in the development of the supervisor. One is the incorporation of specification of response times into the supervisor. A major issue arises out of the nondeterminisms permissible under the specification formalism used. SDL nondeterminisms fall into two major categories:
- indeterminate delays in communication of signals over channels;
- nondeterminisms in the specification of behavior of individual processes (spontaneous transition NONE and nondeterministic path selection ANY [3]).

These nondeterminisms give rise to different but legitimate external behaviors. The supervisor must be able to properly deal with such behavioral alternatives; it should not have a preconceived idea about how the nondeterminism should be resolved in the target system and consider any other alternative as failure.

The supervisor must also be able to properly handle uncertainties arising out of the encoding of external signals in signal carriers. Over a short interval of time (between the min and the max permissible signal recognition time), a state change of signal carrier may but need not be recognized as a valid signal.

3. Supervisor Strategies for Failure Detection

In principle, there are two basic strategies for supervisor-based detection of failures of reactive systems - the input-driven and the output-driven. These two strategies are discussed below.

3.1. Input-Driven Failure Detection

In the input-driven strategy, when an input is observed, the supervisor precomputes the possible system outputs triggered by it and stores them (Figure 3). Because of specification nondeterminisms, there may be more than one legitimate output. When an output from the target system is observed, the supervisor compares it to those on the list. If a match is found, the supervisor removes from the list the alternatives not pursued by the target system and updates the supervisor model state. If no match is found, the supervisor concludes that a failure has occurred and reports it. The supervisor then attempt to re-synchronize with the target system so that does not report the subsequent legitimate behavior of the target system as failures.

To properly handle the nondeterminisms present in the specification model, the supervisor must be able to consider several behavioral alternatives simultaneously. The *theory of beliefs* has been developed for this purpose[9]. In this theory, a sepa-

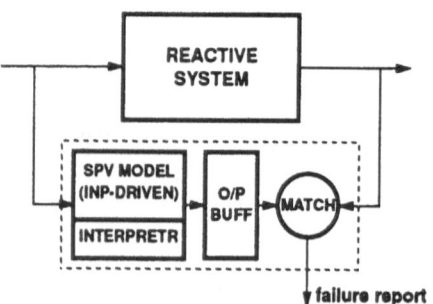

Figure 3. Input-Driven Supervision

rate thread of a specification process (a belief about its behavior) is created to represent a behavioral alternative. In the case of SDL, a major source of nondeterminism is the indeterminate propagation delay of signals over channels. In the belief theory, when a process sends a signal over a channel, the destination process is split into two threads. One thread represents the alternative that the destination process has received the signal and the other that the signal is still in transit. The former thread will process the signal and, if appropriate, produce signals to other specification processes or to the external world. The latter thread stores the signal in transit. This thread is needed to properly handle the case when another process sends a signal to the destination process at about the same time. Due to the indeterminate delays over channels, the second signal might actually have arrived to the destination process earlier than the first. The signal-in-transit thread is used to generate all possible signal arrival sequences at the destination process. The threads representing consistent behavioral alteratives of specification processes are linked into sets. Note that in the scenario discussed, the two threads of the destination process stand for mutually exclusive behavioral alternatives.

When an output from the system is observed, the behavioral alternatives (thread sets) disproved by it are terminated and their constituent threads deleted.

If the specification of behavior of a process includes a nondeterministic construct in the transition being executed, a separate thread must be created for each possible transition path. As before, the alternatives invalidated by the subsequent, actually observed external behavior are terminated.

Figure 4 presents a high level model of the processing involved in propagating an input signal through D communicating processes before the output(s) it triggers are produced. The small rectangles attached to processes represent the signals in transit.

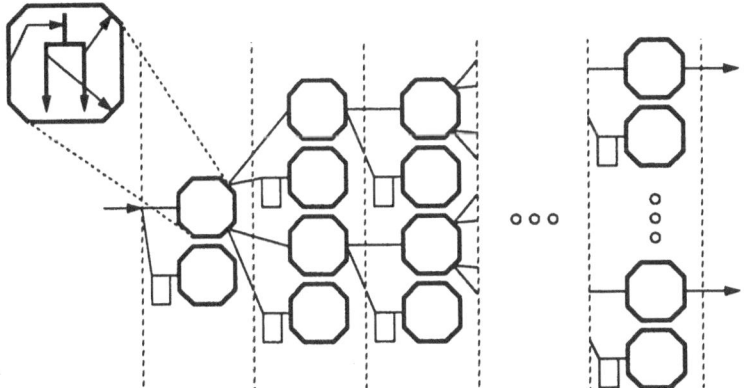

Figure 4. Processing of Inputs in Input-Driven Strategy

The processing time requirements of input-driven strategy can be estimated from Figure 4. The processing time needed to pass a signal through a specification process P is

$$T_P = N_f t_t + N_f(t_s + t_c) \tag{3.1}$$

where

N_f = the mean forward nondeterminism factor (the number of transitions that may potentially be executed as a result of a given incoming signal),

t_t = the mean transition processing time,

t_s = the mean interprocess communication and context switch time,

t_c = the mean cost of creating a new thread (a clone) of a specification process and storing its signal in transit.

The output signal(s) produced by P are going to be processed by N_f specification processes. (Note that the signal-in-transit thread does not process the signal; the signal is merely stored.) This will repeat itself, until a specification process is reached which generates an external output signal. If D is the mean number of processes the external input passes through before an external output is generated, the total cost of processing of an input signal can be approximated as

$$C_i = (1 + N_f + N_f^2 + \ldots + N_f^{D-1}) T_P \qquad (3.2)$$

Note that the cost of matching (and of ensuing termination of invalidated threads) was not included in the above formula.

The additional memory needed for signal-in-transit threads can be estimated as

$$M = (1 + N_f + N_f^2 + \ldots + N_f^{D-1}) m_r \qquad (3.3)$$

where m_r is the memory required for a thread (including its input port).

In the model considered, each process along the input signal propagation path had only one thread. However, under some circumstances, more than one thread may temporarily co-exist. This is, for example, the case with the resource manager process of Figure 2, which is on several input-output paths. Consider the case when a request signal R_P from process P is sent to a resource management process (M). Two threads of M will co-exist for a brief time, until an external output is observed which will cause one to terminate. If another process, Q, sends request R_Q to M before the termination occurs, five threads will have to be created reflecting all signal arrival possibilities at M - $R_P R_Q$, $R_Q R_P$, R_P received and R_Q in transit, R_Q received and R_P in transit, and both R_Q and R_P in transit. In general, if r is the number of requests to M the effects of which have not yet been confirmed through external output, the number of threads of M is [10]

$$\sum_{i=0}^{r} \frac{r!}{i!} \qquad (3.4)$$

Even for small r, the number of additional threads may be large. As a consequence, the processing costs and memory requirements in the input-driven approach may be subject to sudden surges.

To detect response-time failures, the input-driven strategy may take advantage of the form of response-time specifications, which are stated in terms of maximum time interval between a stimulus (external input) and a response (external output). Two cases are possible. When the same specification process receives the stimulus *and* produces the response, it is sufficient for it to set up a timer upon the receipt of the stimulus. If the response arrives before the timer expires, it is canceled. If not, the timer times out and performance failure is reported. This approach must be extended in cases when the response is generated by a specification process different from the one that received the stimulus. Note that the cost of setting up and cancelation of timers was not included in the above formulas.

3.2. Output-Driven Failure Detection

The output-driven strategy is an opposite of the input-driven one. It is feasible when the processing done in state transitions can be easily reversed. In this strategy, the inputs to the target system are kept in a buffer (Figure 5). When an output from the target system is observed, it is propagated backward through the specification model. The input signal(s) that caused it are determined. The input buffer is searched for the signal(s) expected. If a match is found, the supervisor updates the state of the specification model and removes the input signals whose effects have been fully accounted for from the input buffer. If there is no match, the supervisor concludes that a failure must have occurred (there is no cause for the output observed). Note that the backward tracing of an output signal may not necessarily reach a specification process that takes input from the environment. It may cease at an internal process which is in a state that cannot produce the needed signal.

Figure 6 presents a high level model of the processing involved. The model takes into account the possibility that the signal traced might have been produced by several transitions emanating from the current specification state in the sending process and that there might have been several possible sources for the triggering signal.

The processing time requirements of output-driven strategy can be estimated from Figure 6. The processing time needed to trace a signal backward through a specification process P is

$$T_P = (N_t t_t + N_e N_t t_s)$$ (3.5)

where

$t_t =$ the mean cost of (backward) transition processing,

$t_s =$ the mean cost of backward signal propagation,

$N_t =$ the mean number of transitions in the current state of the specification process that could have emitted the signal traced,

$N_e =$ the mean number of processes that could have emitted the triggering signal.

The number of specification processes that must be visited after the trace-back through one specification process is $N_t N_e$. The overall cost of tracing back the output

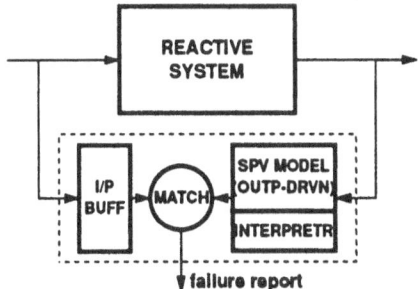

Figure 5. Output-Driven Supervision

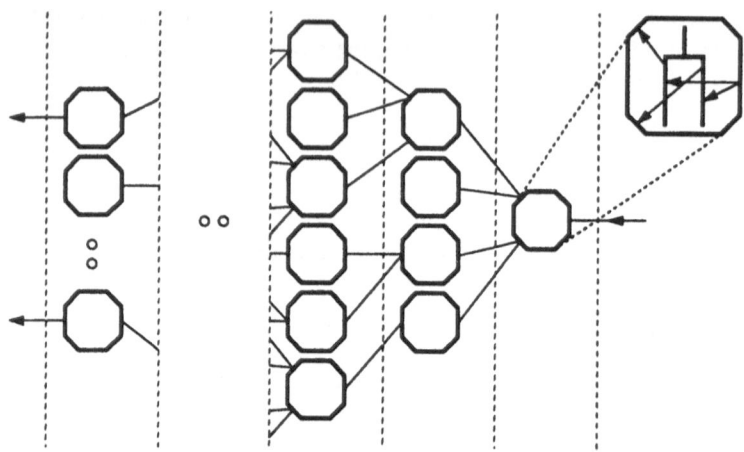

Figure 6. Processing of Outputs in Output-Drive Strategy

produced through D process levels can then be expressed as

$$C_o = (1 + N_t N_e + (N_t N_e)^2 + \dots + (N_t N_e)^{D-1}) T_P \qquad (3.6)$$

In the processing model of Figure 6, no additional memory is required to process an output signal.

As observed earlier, a major advantage of output-driven strategy with respect to the input-driven one is that it does not have to directly enumerate all possible behavioral alternatives. It waits to see which one will actually happen. However, the specification nondeterminisms must be taken into account in explaining what has happened. In particular, the indeterminate channel delays must be considered to correctly explain the outputs observed. As in illustration, consider the scenario when phone A and B call phone C almost simultaneously. Assume that B is going to be successful. If the ringing on phone C is the first output signal detected, it would be incorrect for the supervisor to stop its search as soon as it discovers that A has dialed C. For a brief interval of time, it has to consider both A and B. It is only when the external outputs on phones A and B (i.e. busy and ring tone) are observed that the supervisor may eliminate the alternatives invalidated. The theory of beliefs can handle such scenarios by creating two Line Handler threads for each of A and B. However, at least in the application domain considered, such scenarios appeared to be relatively rare and the cost of thread creation was not included in the formulas given above.

The detection of response time failures in output-driven supervisor is rather difficult, if it is to be done in real time (i.e. as soon as the response interval expires). This is because of the nature of output-driven approach, in which the work is deferred until until the output (the response) appears. If it is imperative that the detection of such failures be carried out in real time, it is usually necessary to separate the detection of behavioral and response time failures and use a separate checker for the latter.

4. Illustration and Experience

A supervisor based on the ideas discussed in this paper was implemented for detection of failures in a small exchange. Real-time detection of failures was required for automatic acquisition of failure data needed in the development and validation of new software reliability prediction models[12]. The exchange and its telephones were emulated on a Unix workstation. Programmable telephone traffic generators were employed to generate random telephone traffic with the specified distributions.

The specification of the exchange had the general form of Figure 2. Only POTS calls (plain, ordinary telephone service) were supported. The exchange served 60 telephones. Call origination rates ranged from 8 to 15 originations/phone/hour. Instead of monitoring the signals between the exchange and the telephones as shown in Figure 1, the supervisor was monitoring the hardware interface memory through which the exchange control program sensed and controlled the exchange hardware. This eliminated uncertainties in output signal recognition (the detection of output signals did not suffer from signal recognition latencies), but it left input signal recognition uncertainties in place.

The analysis given in Section 3 was used to evaluate the tradeoffs involved and to select the supervision strategy. For the input-driven strategy, the N_f factor was 1 and the processing cost was dominated by t_c. D ranged from 1 to 3. For the output-driven, the N_t factor was close to 1. However, for some output signals, the N_e factor was large. This was the case whenever more than one Line Handler is involved in backward propagation of outputs. For example, when ring tone to a phone is observed, the tone must be traced back to the Line Handler for the called phone and from there back again to the Line Handler for the caller. For these signals, N_e is the number of telephones served by the exchange. Even for a small exchange, N_e^2 is a very large number. Although some heuristics could be built into the backward search, this alternative was rejected because of concern of ending up with an ad-hoc, difficult to maintain supervisor.

Based on these considerations, the input-driven strategy was chosen for the supervisor. To reduce the cost of implementation, the matcher of Figure 3 was combined with the processes that produce external outputs (i.e. Line Handlers). For example, the bottom half of the FSM of Figure 2b was converted into the segment of supervisor Line Handler process shown in Figure 7. (To reduce the size of this figure, the treatment of the OnHook signal is not shown.)

This figure contains two extensions to the standard SDL [11]. *O stands for 'any

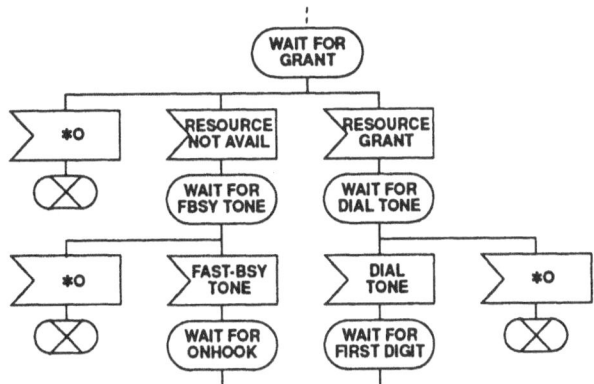

Figure 7. Segment of LineHandler Supervisor

output from the target system pertaining to the line being supervised by this process instance'. The crossed oval denotes the termination of the process thread. The thread is terminated when an external output indicate that the behavioral alternative represented by the thread has not been pursued by the target system. (This is the case, for example, when a different tone is sent to the phone).

Note that such combination of functionality is possible only if it can be guaranteed that the external inputs are propagated through the supervisor model faster than they propagate through the target system. This was the case with the emulated exchange. Alternative approaches are available for non-emulated applications.

Subsequent experience with the purely input-driven version of the supervisor has shown that, during some intervals of operation of the target system, the supervisor ran out of memory. This turned out to be due to the rapid growth in the number of threads of resource management processes. This occurred when the random variations in telephone traffic resulted in a large number of almost simultaneous call originations. In retrospect, this is not surprising in light of equation (3.4), but this point was only realized at a later time. What made this phenomenon worse was the positive feedback in its dynamics - the heavier the load on the exchange, the longer the time interval between an OffHook and the response (dial tone or fast-busy tone) and the larger the number of threads that coexist before they can be terminated.

To resolve this difficulty, it was noted that the processing of call originating off-hooks typically results in dial tone output to the telephone. The product $N_i N_e$ for the dial tone output signal was 1 in the exchange considered. This has led to the decision to introduce a degree of output-driven processing into the input-driven supervisor. The output-driven processing only applied to the output signals that indicate what the outcome of resource request was. The boundary at which the input and output driven processing for these signals (dial and fast-busy tone) tone met was moved into the resource manager. The idea of combining the functionality of input-driven supervisor and matcher was retained. In the implementation of the mixed strategy supervisor, the originating OffHooks were propagated only to the Resource Manager. When dial tone is observed on a phone, the Line Handler sends a notification to the resource manager. As a consequence, the number of behavioral alternatives that had to be considered for the resource manager had become substantially smaller. The memory overflows no longer occurred. The underlying theory is described in [10].

5. Concluding Remarks

The paper considered real-time detection of failures of reactive systems. Failures are detected by the supervisor, a unit that monitors the inputs and outputs from the target system. The supervisor executes a model obtained from the specification of the target system. The paper dealt with the case when the target system is specified in CCITT SDL, a language based on communicating extended finite state machines. The focus was on event-driven applications such as telecom switching.

A major issue in specification-based detection of failures are the nondeterminisms intrinsic to the specification formalism. The supervisor should have no preconceived idea about how the nondeterminisms should be resolved and consider any other alternative as failure. The paper briefly overviewed the theory of beliefs which permits the supervisor to keep track of simultaneous behavioral alternatives.

The paper discussed two basic strategies for real-time failure detection, the input and the output-driven one. In the former, when an input is observed, the supervisor determines what may happen in the future at system outputs. In the latter, the supervisor tries to explain the system outputs from past inputs. The paper presented formulas that estimate the processing and memory requirements for the two strategies. The formulas were based on high-level model of the processing involved and give only a rough estimate of the quantities estimated. They are principally useful in determining the tradeoffs involved and in the choice of supervisor mode of operation.

The paper described an application of real-time failure detection to automatic collection of call processing failure data in a small telephone exchange. The exchange and its phones were emulated on a workstation. A purely input-driven strategy was initially implemented. However, subsequent experience showed that this implementation was subject to excessive surges in processing and memory requirements under certain input scenarios. To gain insight, the models and formulas presented above were developed. A hybrid approach based on partly output-driven processing of certain output signals was implemented. This implementation no longer exhibited the large surges in processing and memory requirements.

Acknowledgments

D. Hay, J. Li and F. Chan were the main contributors to the ideas underlying this paper. The work presented was funded by Bell Canada and by the University Research Incentive Fund of Ontario.

References

[1] International Telegraph and Telephone Consultative Committee, *Functional Specification and Description Language, Recommendations Z.100-Z.104*. Geneva: ITU, 1989.

[2] International Telegraph and Telephone Consultative Committee, *Annex F.1 to Recommendation Z.100: SDL Formal Definition*. Geneva: ITU, 1989.

[3] F. Belina, D. Hogreffe and A. Sarma, *SDL with Applications from Protocol Specification*. Prentice-Hall, 1991.

[4] D. B. Brown et al., "An Automated Oracle for Software Testing", *IEEE Trans. Reliability*, vol. 41, no. 2, pp. 272-280, June 1992.

[5] S. Sankar and M. Mandal, "Concurrent Runtime Monitoring of Formally Specified Programs," *IEEE Computer*, vol.26, no.3, pp. 32-41, March 1993.

[6] A. Erb, "Safety Measures of the Electronic Interlocking System ELEKTRA", *Safety of Computer Control Systems 1989*, Pergamon Press, London, pp. 49-52.

[7] M. N. Myers, W. A. Routt and K. W. Yoder, "Maintenance Software," *The Bell System Technical Journal*, vol. 56, No. 7, pp. 1139-1167, September 1977.

[8] E. Rich, *Artificial Intelligence*, McGraw-Hill, 1983.

[9] D. B. Hay, *A Belief Method for Detecting Operational Failures in Soft Real-Time Systems*, MASc Thesis, Dept. Elect. and Comp. Engg., University of Waterloo, 1991.

[10] J. Li and R. E. Seviora, "Real-Time Supervisor with Reduced Space and Time Requirements," to appear, *Proc. 1993 IEE System Engineering for Real-Time Applications*, London, UK, 1993.

[11] J. Li and R. E. Seviora, "An Extension to SDL," submitted, 1993.

[12] P.Lam, R.E.Seviora and F.C.L.Chan, "Invocation-Count Based Structural Prediction Models," *Proc. Second Bellcore Symposium on Issues in Software Reliability Estimation*, pp.113-129, Oct. 1992.

Reliability and Safety Analysis
of
Hierarchical Voting Schemes

Henryk Krawczyk and Saleh Al-Karaawy
Faculty of Electronics,Technical University of Gdansk
Gdansk/Poland

Abstract

To improve dependability various voting schemes are implemented in computer systems. The paper analyses reliability and safety of elementary and composed majority voting systems. The tool for evaluation of such systems is also proposed. It allows to choose an architecture suitable for given reliability and safety requirements.

1 Introduction

To achieve higher reliability and fault tolerance of computer systems, high-quality components and strict quality control procedure during the assembly phase can be used, or some form of redundancy techniques can be implemented [1]. Both of these complementary techniques lead to an increase in system cost. This is a price which we pay to satisfy dependability requirements that are needed. Moreover, another main point that faces the designers of computer systems is to detect errors at the same time when the real-operations are performed. This means that a system does not need to be stopped to find out which resources are faulty. To satisfy this dependability and time requirements, the *majority voting schemes* are implemented in computer systems. This means that a system must be composed of at least three nodes (modules) [2] which are performing the same job and are establishing the valid result by majority voting. In general, we have two types of elementary voting schemes which will be named Centralized and Distributed Voting Architectures or briefly CVA and DVA, respectively. Moreover, compositions of these fundamental schemes can create more complex (hierarchical) systems which satisfy the highest dependability requirements.

In the literature only centralized voting systems were analyzed very attentively [1,3]. Presently the importance of distributed systems is growing rapidly, so the decentralized voting strategies should be considered and compared. Some ideas referring to the hierarchical systems are given in [4], where some rollback recovery strategies are analyzed. In [5] the matrix and channel voter based architectures are considered. Note that, the latter corresponds to DVAs defined above. In the paper we concentrate on the dependability analysis of the basic architectures. We also propose the systematic approach to evaluate the reliability and safety of more complex voting architectures that are the composition of such elementary systems.

System reliability R(t) is the probability of the correct system work (success) during a certain period of time. System safety S(t) is the probability that the system will survive for a certain period of time. To estimate these parameters, Markov models are used [3]. Based on these models, reliability and safety of system nodes are calculated in Section 2. Then elementary centralized and distributed voting schemes are analyzed and compared in Sections 3 and 4, respectively. Section 5 introduces compositions of the elementary systems. The package program evaluating dependability of different voting systems is presented in Section 6 and its functions are given and discussed.

2 Dependability of System Nodes

Let consider that a computer system consists of n nodes (processing elements). Each node can communicate with some other nodes by interconnect lines. Most of studies in dependability of computer voting system assume that a system remains operable as long as there exist suitable number of fault-free nodes. In consequence, the dependence of system nodes have direct impact on the total system dependability. In this section we concentrate on the dependability estimation of no-repairable and repairable system nodes.

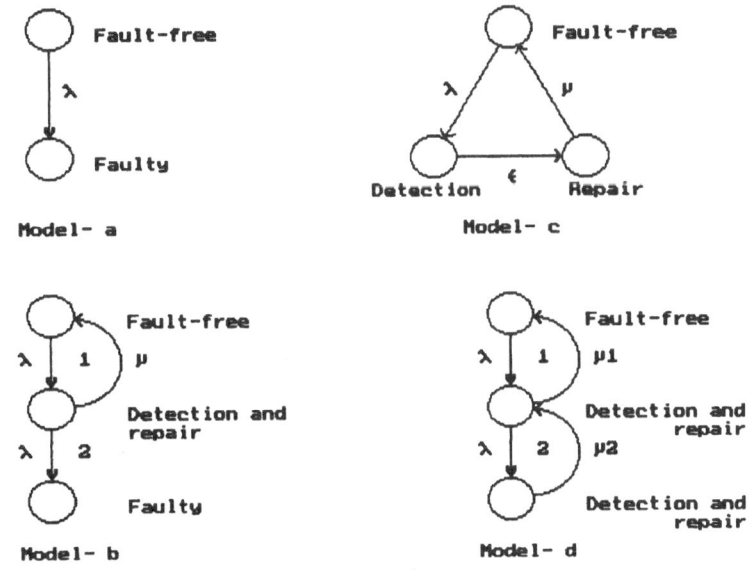

Fig 1 Reliability Models of:
a) Non-repairable node   b) Partial self-repairable node
c) Self-direct repairable node d) Self-indirect repairable node

Let nodes consist of processing unit, memory unit(s), and input/output unit. In case of no-repairable node, each fault of these units can cause the failure of the whole node (see Fig. 1-a). Thus we can assume that:

$R_N(t) = e^{(-\lambda N t)}$, where $\lambda_N = \lambda_P + \lambda_M + \lambda_{i/o}$ (1)

Moreover, it is highly unreasonable to assume that each single node should totally fail. Therefore we consider three types of self-repairable nodes. It is assumed that either a node can survive some faults (see Fig 1-b) or full-success repair can take place (Fig. 1-c,d). These models can be described by the differential equations. Using Laplace transforms, the problem is reduced from a set of differential equation to a set of simultaneous linear equations. For the model given in Fig. 1-c, they are as follows:

$SP_1(S) - 1 = -\lambda P_1(S) + \mu P_3(S)$

$SP_2(s) \quad = \quad \lambda P_1(S) - \epsilon P_2(S)$

$S p_3(s) \quad = \quad \epsilon P_2(S) - \mu P_3(S)$

where:

$P_1(0) + P_2(0) + P_3(0) = 1,$ and $P_1(0) = 1, \; P_2(0) = P_3(0) = 0.$

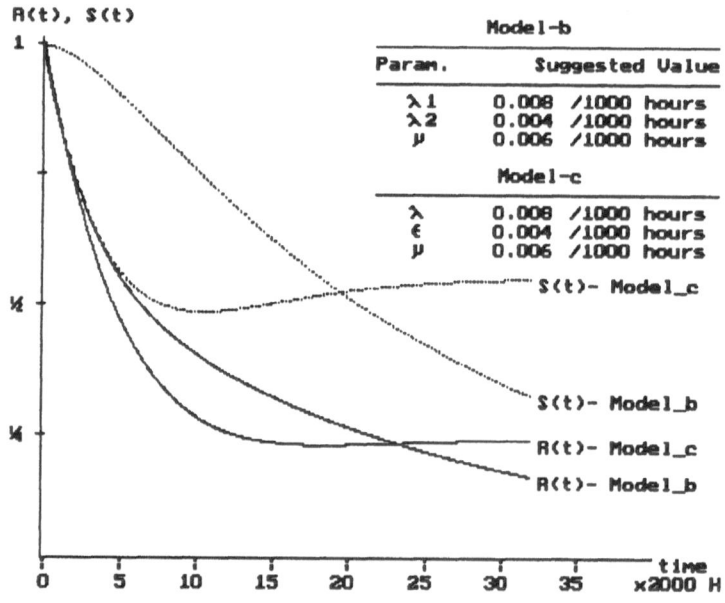

Fig 2 Reliability and Safety Curves For Models b and c

Solving these equations we obtain expressions with variables $P_1(S)$, $P_2(S)$ and $P_3(S)$ which can be transformed directly to the time domain. Then we obtain:

$P_1(t) = A_1 \, e^{at} \sin(\omega t + \alpha_1) + k_1$ (2)

$P_2(t) = A_2 \, e^{at} \sin(\omega t + \alpha_2) + k_2$ (3)

where:

$\omega = 0.5[2(\lambda\mu + \epsilon\mu + \lambda\epsilon) - (\lambda^2 + \mu^2 + \epsilon^2)]^{\frac{1}{4}}$

$a = -0.5(\lambda + \mu + \epsilon)$

$g = \mu + \epsilon$

$d_1 = \mu\epsilon$

$d_2 = \mu$

$k_1 = d1/(a^2 + \omega^2)$

$k_2 = d2/(a^2 + \omega^2)$

$b = (a^2 - \omega^2 + ag + d1)^2$

$A_1 = (1/\omega) [\{b^2 + \omega^2(2a+g)^2\}/\{a^2 + \omega^2\}]^{\frac{1}{4}}$

$A_2 = (1/\omega) [\{(a+d_2)^2 + \omega^2\}/\{a^2 + \omega^2\}]^{\frac{1}{4}}$

$\alpha_1 = \arctan \omega(2a+g)/(a^2 - \omega^2 + ag + d_1) - \arctan (\omega/a)$

$\alpha_2 = \arctan \omega/(a+d_2) - \arctan (\omega/a)$

Then the reliability and safety of a node can be expressed as follows:

$R_N = P_1(t)$ and $S_N = 1 - P_2(t)$.

Fig. 2 shows some graphs for models b and c given in Fig. 1.

3 Centralized and Distributed Voting Architectures

CVA is the classical voting architecture (Fig 3) where each computing node (CN) is described by one of the reliability models shown in Fig 1. The Centralized Voter (CV) is made up of the Bus Interfacing Unit (BIU), which receives and sends some information to the CN, and Voting Unit (VU) which in turn performs a majority voting algorithm to establish valid results. It is assumed that the voter must be a hard-core unit to achieve reliable work of the whole system. Then the CVA may tolerate of maximum of f faulty CNs, where $n \geq 2f + 1$. K out of n system is a generalization of the voting architecture in which k of n nodes must work correctly to perform system functions [3].

We assume that the reliability of the CVA can be determined as a function of the reliability of the CN - $R_{CN}(t)$ and the reliability of the CV - $R_{CV}(t)$.

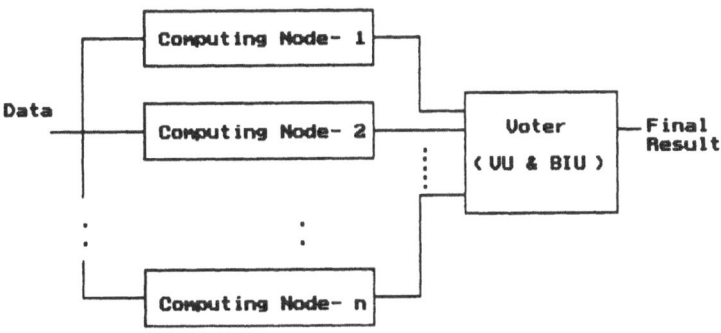

Fig 3 Centralized Voting Architecture (CVA)

Let us assume that all CNs are identical and have the same failure rates λ_{CN}, and the system can tolerate up to $f \leq n-k$ faulty CNs, where k is the minimum number of CNs that must be operational for a system to be reliable. The structure-based reliability assessment of the system can be assessed by assuming a parallel-series structural configuration. Then we define [1,3]:

$$R_{CVA}(t) = R_{CV}(t) \sum_{i=0}^{p-k} \binom{p}{i} R_{CN}(t)^{(p-i)} (1-R_{CN}(t))^i \tag{4}$$

where:

λ_{CV} - is the failure rate of CV $(\lambda_{CV} = \lambda_{BIU} + \lambda_{VU})$

The safety of the system $S_{CVA}(t)$ can be is determined similarly as follows:

$$S_{CVA}(t) = S_{CV}(t) \sum_{i=0}^{p-k} \binom{p}{i} S_{CN}(t)^{(p-i)} (1-S_{CN}(t))^i \tag{5}$$

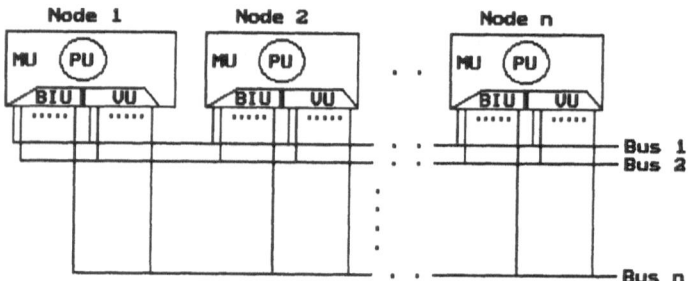

Fig 4   Distributed Voting Architecture (DVA)

DVA is (as shown in Fig 4) composed of computing nodes (CNs) and redundant buses. The nodes transmit data over the buses and each node can receive data from all nodes including itself. The main difference between CVA and DVA architectures is the type of voting. The nodes of DVAs contain BIU and VU, which are used for communication and distributed voting respectively. An example of such architecture is described in [5,6]. BIU modules arbitrate between two CNs when both of them want to access the voter. It is also responsible for reconfiguring a faulty CN out of the system. Then each VU is programmed to perform voting, which is a more complex operation than a simple comparison of the received data from all of the nodes, and also to perform the error logging. When voting is performed, final result is transferred to each node by its respective BIU. Whenever data from a particular node do not agree with the data from the other nodes, an error condition is latched in each node which detects such an error. The latched information specifies the faulty node as well as the source of the faulty data. No hard-core is necessary because voting function is distributed among the nodes and to eliminate f faulty CNs, $n \geq 2f+1$. In case of Byzantine faults very popular in case of the distributed voting $n \geq 3f+1$ [7]. Reliability of DVA is evaluated as follows:

$$R_{DVA}(t) = \sum_{i=0}^{p-k} \binom{p}{i} R_{CN}(t)^{(p-i)} (1-R_{CN}(t))^i \qquad (6)$$

The Safety of the system $S_{DVA}(t)$ is determined in the same way as follows:

$$S_{DVA}(t) = \sum_{i=0}^{p-k} \binom{p}{i} S_{CN}(t)^{(p-i)} (1-S_{CN}(t))^i \qquad (7)$$

4 Comparison of CVAs and DVAs

Let consider the reliability of the architectures presented in Sections 2 and 3. We assume that a computing system consists of homogeneous nodes. This means that all reliability parameters are the same for each node. Reliability of buses is described by the reliability of BIU. Fig. 5 plots the reliability curves for a system of $n=5$ (then $k=3$) and for two different schemes (CVA and DVA). We assume that the reliability of the voting unit and bus interfacing unit are higher than the reliability of other units, i.e., the failure rate of a CN = the failure rate of the MU < the failure rate of the VU < the failure rate of the BIU.

We restrict our analysis to mission time less than 5 years. The S-shaped curves are obtained which are typical for redundant systems [1,3]. Above the knee, then CVA and DVA have spare components that tolerate failures and keep the probability of system access high. Once the system has exhausted its redundancy, however, there is merely more hardware to fail. Because distributed voting systems have more redundant components (except CNs) their reliability for the first period of time (nearly for one year) is higher, then is lower in comparison to the CVA. This tendency is also true for safety of system.

5 Hierarchical Compositions of Voting Schemes

Complex systems are typically structured hierarchically in multi-levels organization. This means that such a system consists of smaller subsystems (each of which is either CVA or DVA) and shared buses. We introduce a new class of voting architectures named *composed voting architectures*. They are defined on the base of the composition operation [8]. The simplest composition is a cascaded series of CVAs or DVAs [3]. Other examples are shown in Fig 6. The scheme made up of

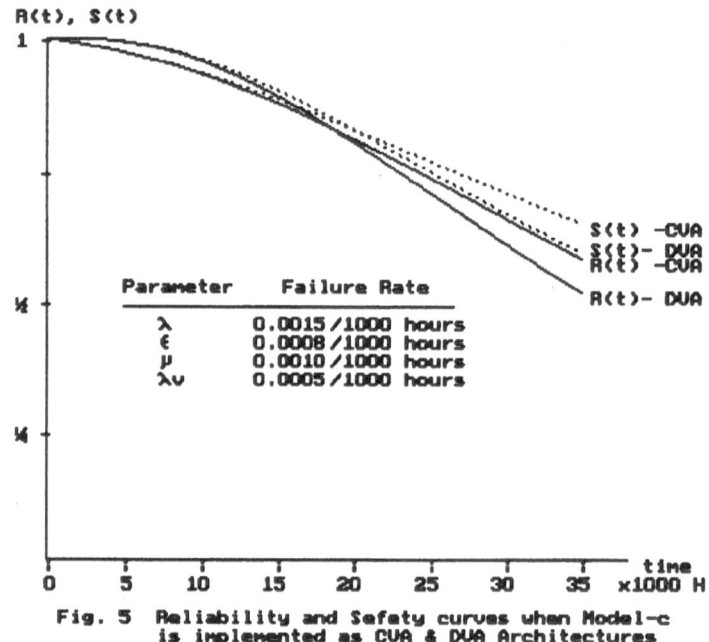

Fig. 5 Reliability and Safety curves when Model-c
is implemented as CVA & DVA Architectures

r distributed voting subsystems, where their signals are coupled by a centralized voter, is named Centralized - Distributed Voting Architecture and is denoted by CDVA. Another architecture, is made up of a set of r centralized voting architectures connected by redundant buses, is named Distributed - Centralized Voting Architecture and denoted by DCVA. We assume that such a complex system works correctly if at least m of r subsystems are fault-free (m out of r).

It is easy to note that for analysis of composed majority voting systems, we may use the formulas given in Section 2 and 3 provided the node reliabilities formulas are replaced by the subsystem reliabilities formulas. Fig 7 shows the curves for the first two types of voting architectures discussed above and for model-c of node reliability. In general, DCVAs are more reliable than CDVAs. The utilization of self-repairing nodes leads to an increase of system reliability more significantly.

6 Reliability and Safety Estimation Package

Below, the newly developed program named "RASEP" is described. RASEP (Reliability And Safety Estimation Package) is dedicated to analysis and comparison of hierarchical voting architectures. The main modelling objective is to provide the estimates of reliability and safety of complex computer systems.

The prototype version of RASEP has been implemented in C programming language and destined for a single processor environment of an IBM PC AT computer running under the DOS. The system structure of RASEP is given in Fig.

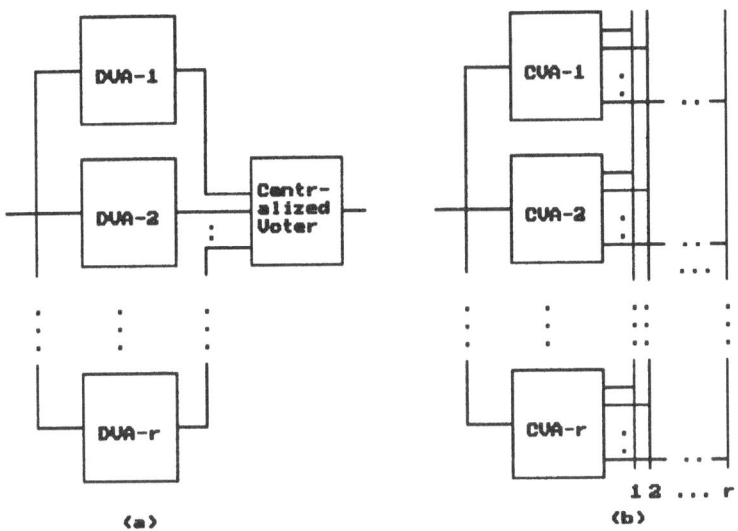

Fig 6   Examples of Composed Voting Architectures
a) CDVA   b) DCVA

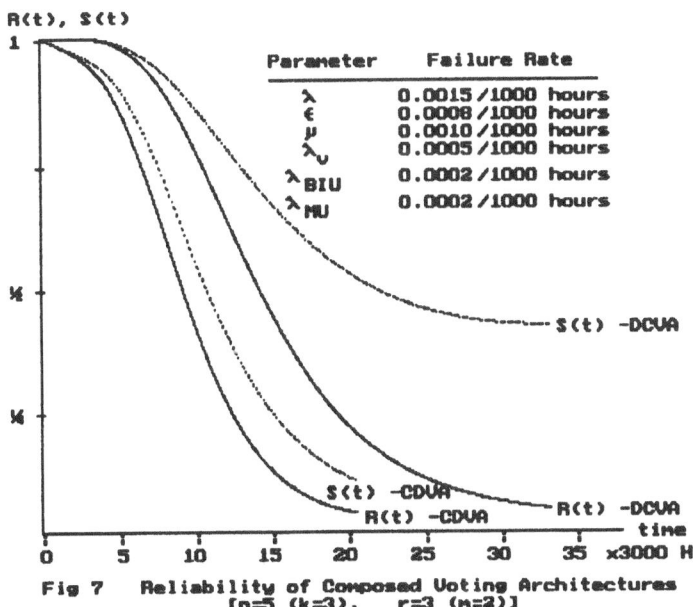

Parameter	Failure Rate
λ	0.0015/1000 hours
ϵ	0.0008/1000 hours
μ	0.0010/1000 hours
λ_v	0.0005/1000 hours
λ_{BIU}	0.0002/1000 hours
λ_{MU}	0.0002/1000 hours

Fig 7   Reliability of Composed Voting Architectures
[n=5 (k=3),   r=3 (m=2)]

8. Presently, the program is working for reliability models of system units presented in Fig 1. However, new models can be added, because the choice of a given model is pointed by unique name. The hierarchical system architecture is described by the following formula:

$$X\,[Y(U_1),\ Y(U_2),\ \ldots\ldots,\ Y(U_r)] \tag{8}$$

284

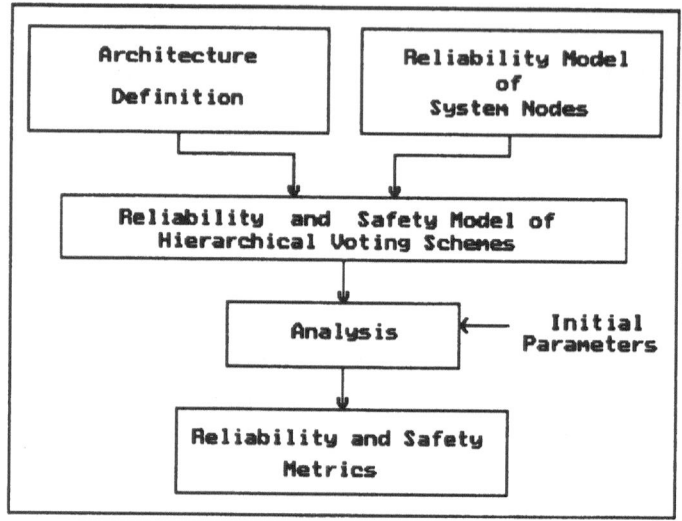

Fig 8   The Structure of RASEP

where:

X, Y - determine the type of architecture, e.g. CVA, DVA,

U_i, i=1,2,...,r - denotes either the kind of a system node or recursively a formula like (8) describing architecture of the subsystem,

r - is the number of elements and it can be different on each level of the description.

For example architectures of Fig. 6 can be described in the following way:

CVA(DVA_1, DVA_2, ..., DVA_r), DVA(CVA_1, CVA_2, ..., CVA_r).

It is easy to note that different types of architectures can be used on the same level e.g.;

DVA(CVA_1, DVA_2(CVA_3, DVA_4)).

Based on the formula (8) and the expressions (1÷7), reliability and safety models are generated and concrete metrics are determined for given parameters of failure and repair rates. The all figures presented in the paper are obtained by the RASEP.

7 Conclusions

There are some real computing systems where majority voting schemes are implemented [1,5,6,7]. The aim of this paper is to pay attention to a new possibility of generation of various voting architectures and to show some methods of their evaluation. In order to support modelling and evaluation of those architectures, a program "RASEP" has been built. Using this package, for a given reliability requirements, we may choose the suitable architecture. Presently, the package is still under development. The fault coverage and error latency will be included into the program as well.

References

1. Holt C, Smith J. Fault tolerant & fault testable hardware design. Prentice Hall International, Inc., London, 1985
2. Nelson V. Safety in numbers: Redundancy lets a system perform its intended functions despite some number of faults. Byte, 1991; 16:175-184
3. Siewiorek D, Schwarz B. Theory and practice of reliability system design. Digital Press, 1982
4. Shieh Y-B, Ghosal D, Tripathi S. Modelling of fault tolerant techniques in hierarchical systems. FTCS-19, 1989, pp 167-174
5. Somani A, Sarnaik T. Reliability analysis and comparison of two fail-op/fail-op/ fail-safe architecture. FTCS-19, 1989, pp 566-573.
6. Kanekawa N, Maejima H, Kato H, Ihara H. Dependability onboard computer systems with a new method-stepwise negotiating voting. FTCS-19, 1989, pp 13-20
7. Laha J. A Byzantine resilient fault tolerant computer for nuclear power plant applications. FTCS-16, 1986, pp 338-343
8. Krawczyk H, Kozlowski W. Fault diagnosis in distributed systems with incomplete test. Microprocessing and Microprogramming, 1987; 20:39-44

Session 9

LANGUAGES

Chair: W.M. Turski,
Warsaw University, PL

Designing a High–level Language for Safety Systems

G. Sen[1], J. Brummer[2]

[1]Reactor Control Division, Bhabha Atomic Research Centre
Bombay, India

[2]Institute for Safety Technology(IST), Gesellschaft für Anlagen– und Reaktorsicherheit (GRS) mbH
Garching, Germany

Abstract

As an alternative to the classical approach for system specification on the basis of a formalised general purpose language a graphical and specialised language for application to safety critical systems is outlined. The architecture of the language is constructed in accordance with the functional and timing requirements typically for operationality in safety systems. The fundamental and generic elements of the language are presented: the syntax and semantics of function and net diagrams. A wide range of operational behaviour (functional and timing) can be determined by this graphical specification technique, several ways of specification analysis are opened. Some examples show how to benefit from the combination of illustrative graphical demonstration and strictly defined rules for their interpretation.

1 Introduction: Universal versus Special Language

A main task in software technology is the computer based and (as far as possible) automatic development of complex software systems. The essential basis for that is settled in the early stages of the development process: The system's work has to be specified in a way the computer can understand and operate with. For that reason and in order to avoid severe misunderstandings, the elimination of which often requires enormous efforts, a precise formulation of the intended system and its design is required. Formal methods for system definition have been suggested which fulfil these requirements to some extent.

The classical approach for a high–level system specification is the formal language representation of its functionality. There are several language concepts [1,2,3], almost all based on the data type description of the system properties: The idea is that a data type is not just a definition or enumeration of its admissible values, but the concept of types also comprises of all operations that are meaningful for these data objects. The way to the system's behaviour is opened by axiomatic rules regulating the relationship between values and the admissible state transfers. The system properties result from rewriting sequences according to the stated rules and according to algebraic transformation principles [4].

Most of these languages do not have the feature to project operational behaviour with respect to time constraints and synchronisation of different processes which constitute the total system. Another difficulty with these formal languages arises because of their universality: To cover a wide range of applications, the vocabulary of these languages has to be very elementary (set–theoretic notation and predicate calculus terminology), and a system description normally consists of a complex set of relational rules. Because only the relational structure of a system is formulated, even for relatively simple systems the consequences of the stated rules and the final behaviour cannot be realised immediately. Therefore, for proving that the specified system meets the original requirements, extensive verification procedures have to be carried out [4].

The situation changes if the universality principle for the specification language is dropped and a formalisation for a restricted application area is taken into consideration: The language vocabulary for a special technical field can be adjusted to the particular subjects, the combination rules (the grammar) can be arranged according to the requirements of the special field.

The safety system in nuclear power plants for example (or similar plant protection systems) fulfills the prerequisites of structurally and conceptually restricted operational technique. The input quantities are regulated and the operational logics for the safety functions are constituted from elementary functional units, because the safety functions follow simple operational patterns: Data acquisition and preparation, accident control by evaluation and comparison of measurements and a few normed reaction schemes. Usually this operational procedure runs simultaneously and redundantly on different computers and is cyclically repeated. Therefore, a synchronisation mechanism has to be implemented and timing constraints have to be considered.

Along this operational paradigm a graphical language for specification and design of such type of safety system is outlined in the next sections. For that, the proposals made in [5,6] for designing a graphical, high–level language are taken up and modified for special applications. In section 2 the architectural concept of the language is discussed: The two essential constituents of the language, functional and net constructors, are established. In section 3 a more detailed description of these two language features is given: The combination of functional units to function diagrams is defined, and a special class of time Petri nets is introduced for managing synchronisation and regarding timing aspects. In section 4 the system's analysis on this very early stage of the development and the possibilities of a direct implementation of the graphically specified system is discussed. The main results of the report are summarised and a short outlook on the future work is given.

2 Global Architecture of a High–level Language for Safety Systems

A language for the specification of control and safety systems should be designed in accordance with the operational and behavioural patterns of these systems. The language features have to be adapted to describe the functional and performance specialities: The vocabulary and the rules for their combination has to reflect the properties of the systems.

For that purpose, the basic functional elements for safety systems have to be isolated, the basic arrangments for building up complexer functions have to be identified and the main rules for integration and synchronisation of the usually distributed computations have to be fixed.

As the main characteristics of typical safety system one can state:

– Mostly simple functional lines, starting with data acquisition, data examination and plausibility checks, and then performing data comparison and evaluation and a few reaction schemes

– Standardised boolean or arithmetical operations with the data values (e.g. selection, sort, threshold comparison)

– A straightforward execution of control functions with few branches and careful looping

– Execution of the same or of similar functionality on different computers (caused by the requirements of redundancy/diversity)

– Rigorous requirements for integration and synchronisation of the distributed computer system

– Strict timing requirements and cyclical execution routines.

The specification language has to take care of these aspects: It should facilitate the formulation of linear functionality and provide constructs for the determination of timing and performance constraints. In correspondence to these demands the language include two graphical forms:

The <u>function form</u>, represented in functional diagrams (see section 3.1).
The functional vocabulary consists of a set of predefined units (bricks) to the designer's disposal enabling the construction of a large class of complexer control functions. In addition to that, the open character of the language allows the integration of new, user defined units for special applications.

The <u>net form</u>, represented in net diagrams (see section 3.2),
is used for the arrangement and management of distributed functionality, for expression of synchronisation and timing requirements.

With these language constructors a program for a distributed safety system is built up: For the different computers their tasks are formulated in separated function diagrams.

Depending on the complexity of these functions the functional specification can be divided into several hierarchically ordered diagrams. The co–operation of the computers is co–ordinated by net diagrams which manage the right ordering and timing of the global computation. Out of the net process other functions may be called and a cyclical execution procedure may be established.

3 Basic Elements of a High–level Language

3.1 The Function Form

The vocabulary of a safety language is a collection of simple functional units, from which the more complex safety functions can be built up. For the definition of the proper words of the specification language one has to make a compromise between simplicity and complexity: The items should not be too elementary because then the more complex constructs loose their transparency; on the other hand they should not be too extensive and should not contain too much information because then the language looses its flexibility and power.

A language for safety systems in nuclear power plants has to provide items

– for manipulation of safety indicators, for example

- simple Boolean operators

- simple numerical evaluators (threshold comparison, sort routines, max/min checkers)

– for special technical features, as they may be

- particular controllers

- drivers for control equipment

– for operational strategies, as they are required for

- process synchronisation and communication

- process priority and access regulation to commonly used resources.

The single words of the language are introduced as graphical symbols, the combination of which is ruled by the inherent properties of these functional units (number and type of input–output relations). The semantics of the language are given by small generic algorithms or models connected to the single words (see below), the language power is determined by these functional bricks and their combination effects.

To specify a function, the input–output relation of that function has to be analyzed and divided into smaller functional blocks (depending on the complexity of the system) revealing the constructive elements of the specific function. Then the functional diagram is built up using the functional units of the language as constituents and connecting them in correspondence to the structure of the function.

This analysis and partition procedure aims at the reconstitution of the function from well–known and predefined, indivisible functional units. The verification process later on is based on the complete verification of these constituents and is going along the composition rules used for the construction of the complexer function (To cover a larger range of applications, there is the possibility to enhance the basic vocabulary by new generic, user defined elements).

A small (data acquisition and check) example should illustrate the construction process for functional diagrams:

The values of two signals (received from different sensors) have to be checked for plausibility. Only if both of the values are realized as valid (not less than a fixed threshold a and not larger than b), the larger one is selected for further evaluation; otherwise the failure of data acquisition process has to be announced.

A specification of this acquisition process is presented in the function diagram of figure 1 (where ω stands for invalid value/failure in the acquisition process).

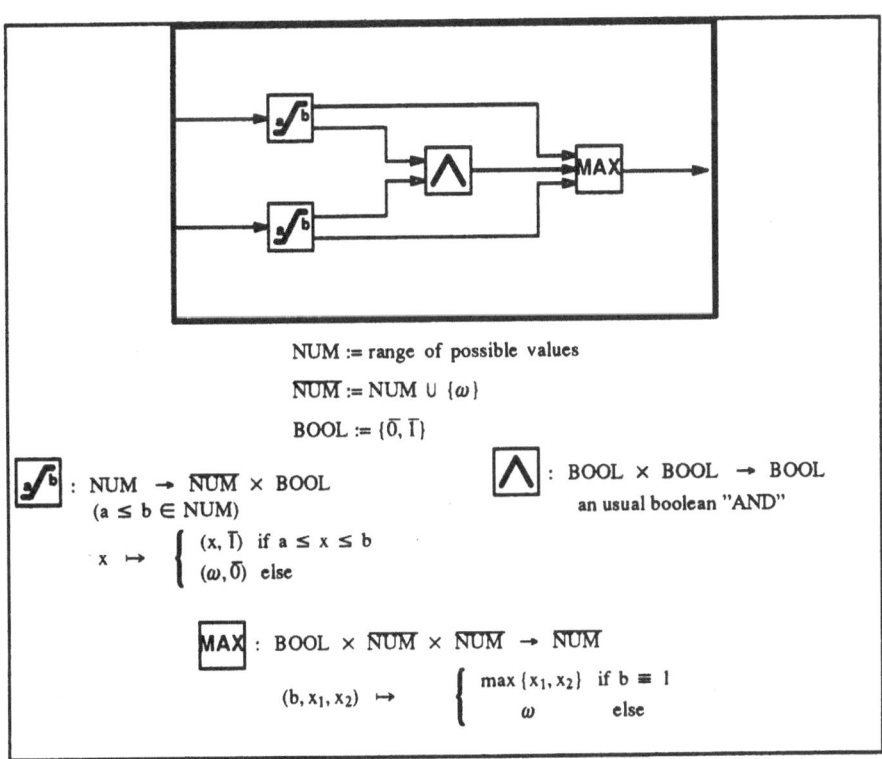

Figure 1: Function diagram example

3.2 The Net Form

Concurrency, synchronisation and time constraints

In general, the function diagrams are distributed on different processors or computers and executed independantly (according to the requirement of redundant/diverse measurement acquisition and evaluation in safety critical systems). At special places the results of these calculations have to be assembled, compared and an adequate reaction has to be caused.

For the purpose of combination and synchronisation of originally separated activities, special language features have to be prepared enabling the integration and organisation of different computational procedures. So, there is a need for language constructs which allow one to express in a formal and unique way

- which procedures (functional diagrams) should be combined, integrated and worked up together
- where and how a synchronisation of the independantly operating units has to be achieved
- what timing requirements have to be fulfilled (duration of calculation, cyclical constraints)
- when and how interrupts are allowed and how to continue afterwards

To meet with the requirements of assembling individual procedures and of timing constraints a special feature is introduced into the language: the net form. The syntax and semantics of these net constructs are borrowed from the well–known Petri net models adding timing facilities and modifying the rules for the dynamical behaviour of the classical Petri nets. These extended Petri nets, so–called time Petri nets, allow the exact specification of action sequencing in accordance with logical and timing constraints (see below).

The arrangement and combination of function and net forms is roughly sketched in figure 2.

For synchronisation and combination of functions' results the outputs of the different function diagrams are collected in a net diagram. The actual initiation of the net (leading to an initial state, see example in figure 3 below) is caused by these function outputs. To the single transitions of the net there may be allocated "actions" (other function diagrams) which are executed in the case of transition's firing and which also may use output values of connected function diagrams (dashed line in figure 2). The sequencing of transitions' firing, i.e. the development of the system's activities, is determined by the structure of the net.

In the following a detailed description of time Petri nets and their use within the language is given.

Specification with time Petri nets (TPN)

In the following a formal definition of time Petri nets is given according to [7,8]. For treating the special requirements stated above some essential modifications to the firing rules and to the net behaviour are introduced.

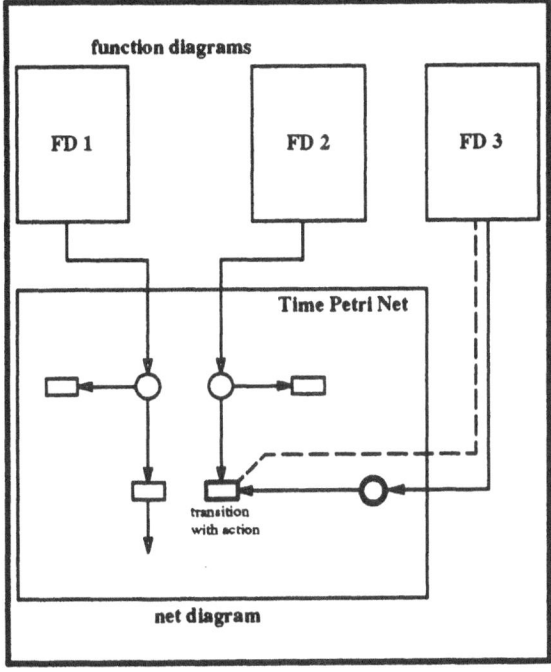

Figure 2: Integration of function and net diagrams

A TPN is a tuple (P, T(A), F, M_0, SIM, DIM), where

- P is a finite set of places, symbolized as circles

- T(A) is a finite set of transitions with possible actions, symbolized as rectangles with action description

- $F \subset P \times T \cup T \times P$ is a relation between places and transitions, symbolized as arcs

- $M_0 : P \rightarrow \mathbb{N}$ is an initial marking, symbolized as black tokens on the corresponding net places

- SIM: T $\rightarrow$ (time)intervals

- DIM: T $\rightarrow$ (time)intervals

The SIM(t)–interval (α_t, β_t) characterizes the time delay for transition firing and may be used for the specification of synchronisation and time sequencing of transitions (and actions). These time values α_t and β_t are relative time indicators, relative to the moment at which the corresponding transition is enabled: Assume that transition t is enabled at absolute time T_{abs}, then t may fire as soon as possible, but not before $T_{abs} + \alpha_t$ and not later than $T_{abs} + \beta_t$. These values are dynamically updated in correspondence to the behaviour of the net (see firing rule below).

If there is an action, added to a specific transition and the duration cannot be neglected, the DIM(t)–interval characterizes the minimal/maximal duration of this action. By that duration value the net activities and the general behaviour may be influenced to some extent (see firing rules below).

Whereas in ordinary Petri nets the behaviour is ruled only by the markings, in TPNs additional conditions concerning the time constraints have to be fulfilled for enabledness and firing of transitions. These conditions constitute, together with the usual marking, the state of the net and determine the next possible activity. In general, a state S of a TPN consists of a pair

S = (M,I), where

- M is a marking of the net
- I, the firing domain, is the set of enabled transitions (by marking M) together with the corresponding firing intervals (The number of entries of I will vary during the behaviour of the net according to the number of transitions enabled by the current marking).

Now, let us assume the current state of the TPN be S = (M,I). The transitions in I are enabled by the marking M, but not all of them are allowed to fire immediately due to the timing constraints (of the firing intervals). The "firability condition" for a transition contains a time parameter and expresses the fact that an enabled transition may not fire before its left time interval value and must not fire after its right time interval value (unless another transition fires before and modifies marking M and so state S).

Accordingly, transition t is firable from state S = (M,I) at time $T_{abs} + \tau$ iff

i) t is enabled by marking M at time T_{abs}

ii) the relative time τ is within the actual time interval for t

The firing rule itself is introduced by the definition of the follower state S' = (M',I') reached by firing an enabled transition t at time $T_{abs} + \tau$:

i) M' is computed according to the firing rule in ordinary Petri nets.

ii) The new firing domain I' is computed from I in three steps:

- All transitions, which are disabled when t is fired, are removed from I
- For a transition t which remains enabled by M' the new firing interval is calculated as

$$\alpha' = \max\{0, \alpha - \tau - d\}$$
$$\beta' = \beta - \tau - d$$
$$(\text{for } \beta' < 0 : \text{SIM(t)})$$

- For a newly enabled transition t introduce as firing interval SIM(t)

In other terms, the first step corresponds to projecting the firing domain to the transitions that remain enabled after t has fired. In the second step the time consumption by the firing delay (and, eventually, a time duration of an action connected to t) is regarded: Time is incremented by the value $\tau + d$, which appears as an interval shifting operation. If the time interval has completely passed ($\beta' < 0$), the transition interval is set to the original delay SIM(t).In the third step the domain of the new state is supplemented by the original firing intervals of the newly enabled transitions.

With the rule "firing transition t from state S at time τ" a relation on the set of possible states

STATES = {(M,I):M marking,I firing domain }

is introduced and is denoted as

$$S \to^{(t,\tau)} S'$$

A firing schedule, that means a sequence of pairs $(t_1, \tau_1), ..., (t_n, \tau_n)$ (t_i transitions, τ_i times), is feasible from a state S iff there exist states S1, S2, ..., Sn such that:

$$S \to^{(t_1, \tau_1)} S1 \to^{(t_2, \tau_2)} S2 ... \to^{(t_n, \tau_n)} Sn$$

A state S' is reachable from state S within time T' iff there is a firing schedule $(t_1, \tau_1), ..., (t_m, \tau_m)$ leading from S to S' and $\sum_{i=1}^{m} \tau_i < T'$

Therefore, the firing rule permits one to compute states and a reachability relation among them. The set of states that are reachable from the initial state or the set of firing schedules feasible from the initial state characterize the behaviour of the TPN, in the same way as the set of reachable markings and the firing sequences characterize the behaviour of ordinary Petri nets.

For instance, let us consider a small example in order to illustrate the firing conditions and the behaviour of a TPN (see figure 3a):
The net is initiated providing a certain value and allowing an interruption demand (the net is marked at the places 1 and 2 from an outside function). The further treatment of the value is delayed some time (θ, starting at the moment the net is initiated).
If no interrupt is demanded within this time period θ, the transition 1 will fire after that time.
If an interrupt is demanded within this time period (a token appears at place 3), the interrupt transition 2 is fired and a possibly connected action is started immediately. In this (interrupt) case the ongoing behaviour of the net depends on the duration of the interrupt action:
If the interrupt action is finished before 2θ, the transition 1 is still enabled and will fire, otherwise the token is removed from place 1 by transition 3 and the execution (transition 1) is omitted.
In figure 3b the states are precisely noted and their relational combination is graphically presented.

298

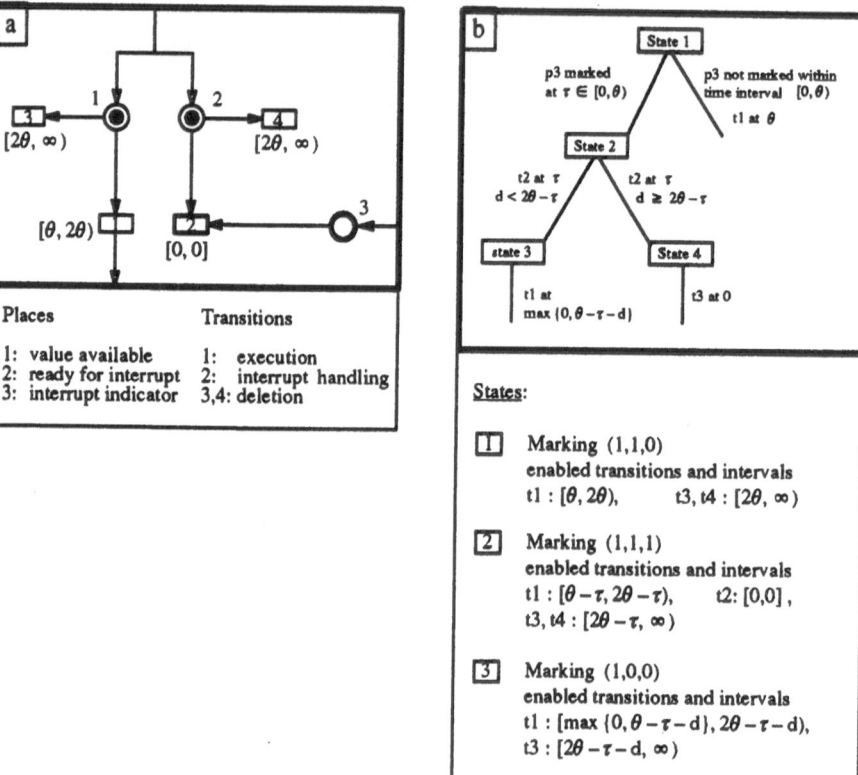

Figure 3: Interrupt example (a: state graph, b: net specification)

In the following a more sophisticated example should give an impression about the facilities of synchronisation and memory management by time Petri net integration of distributed systems:

Two values, received from different computers (output of different function diagrams), have to be compared, exactly at the time points θ, 2θ, 3θ ,... (cyclical synchronisation). If the comparison indicates a critical situation, an exception handler has to be started, if the comparison indicates normal behaviour, the cyclical execution is started again. The exception handler also has to be called, if one of the values (or both) are not available right in time twice (in subsequent runs).

The net specification of this example is shown in figure 4. There, the synchronisation is established by the transition 5, which fires exactly at the moments θ, 2θ, 3θ ,... After that firing the evaluation transition 4 is enabled for firing immediately, if both of the values are available (places 1 and 2 are marked in the meantime). If only one of the values is missing at that moment, transition 6 (permanently enabled) gets the priority for firing: In that case it depends on the memory token, how to exit the net form (token at place 9: there was no missing value before, token at place 10: there was already a missing value in the last computation cycle).

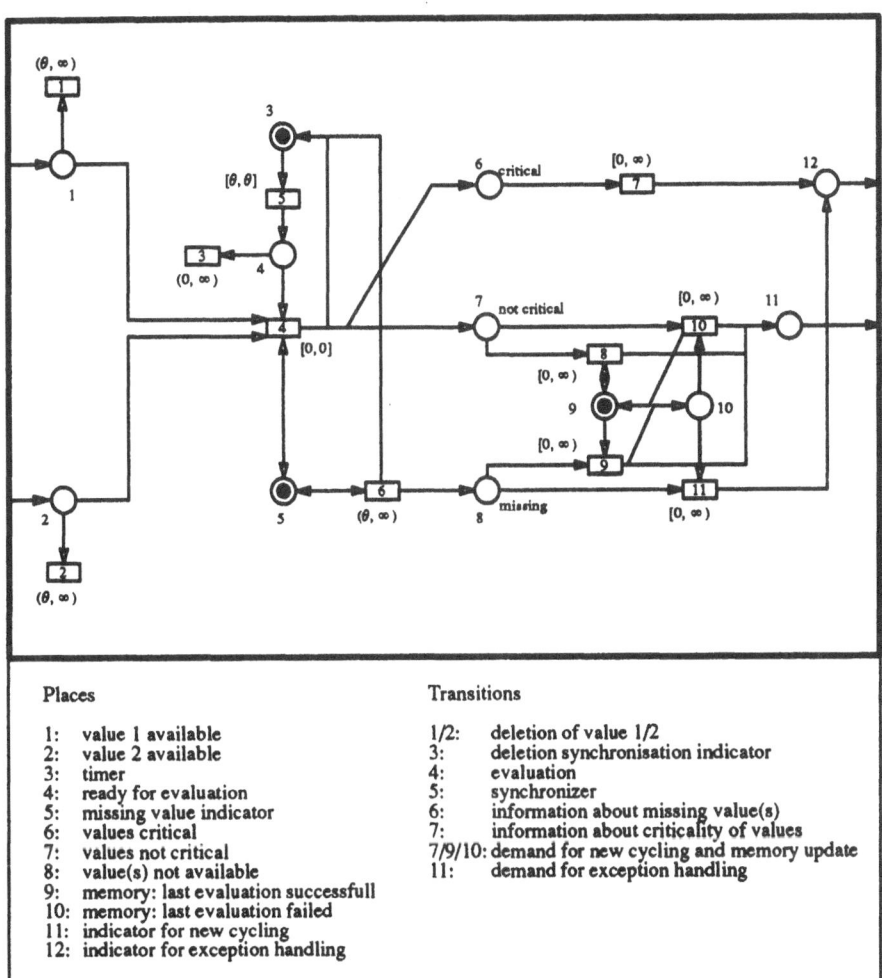

Figure 4: Synchronisation and evaluation example

4. Conclusions: Verification and Implementaion of the Safety Language

This paper presented an approach to formally specify software systems in a graphical language – as an alternative to the classical formal approach where the relational structure of a system is given in elememtary mathematical terms. The architectural concept for designing the graphical language by combining functional form of representation along with the net form (time Petri nets) has been described in some detail. This composition feature allows the description of the functional evaluations as well as the consideration of the operational behaviour of the distributed computation (concurrency, synchronisation, timing constraints).

Function forms are built up by the combination of a fixed set of basic functional units using only simple composition rules. Some investigations revealed that about 50 units (most of them relatively short modules) are sufficient for the specification of the large majority of the occuring automation problems in nuclear safety systems [9]. Because of the limited size of the basic units it should be possible to prove formally their correctness using for example the rules of the Hoare calculus [10]. For the verification of complete function diagrams one has to consider following:

– Loops and condition branches are packed into the single functional units which can be proven separately and in advance.

– The function diagrams themselves are constructed without any backward arcs.

With these prerequisites in general a function diagram can be translated into a (semantically) equivalent computation sequence consisting of the single functional constituents (linear ordering of the functional units). On the premises that the individual units are proven to be correct, the verification process for function diagrams is limited to the checking of the adequate transformation of the results along a linear computational sequence.

The net form enforces a careful construction of the system's concurrent parts and demands an explicitly formulated schedule for the temporal patterns. Using and extending classical Petri nets or place–transition nets by adding time features allows one to simultaneously model the behaviour and analyze the properties of timed systems. This technique is related to the reachability analysis method for usual Petri nets: The firing rule, modified with a timing condition, permits one to compute states and a reachability relation among them. The set of states that are reachable (along firing schedules feasible from the initial state) characterize the behaviour of the time Petri net and determine in a rigorous way the actions of the system.

Using this set of states for analysis purposes, reasonable computational limits may be exceeded for complexer systems. Therefore, to find exhaustive and computational analysis methods for time Petri nets will be one of the main tasks for the future. Nevertheless, the examples presented in the figures 3 and 4 show that in many practical situations the TPN approach enables an exact definition and verification of the intended system's activities.

Work carried out in this direction is in its early stage of development. Up to now the vocabulary of the graphical language is restricted to issues relevant to the operational behaviour of reactor safety systems. A graphical editor tool will be developed whereby it will be possible to specify a system by generating functional diagrams and integrating them into suitable net forms (on a graphical screen by selecting the different symbols displayed on a separate window, connecting them by arcs).

For the purpose of analysis and simulation of the system specification the semantical information about the constructed function and net forms are stored in the background. Finally, the ultimate goal will be to automatise code generation directly from the system specification.

References

1. Futatsugi K, Goguen J.A, Jouannaud J–P, Meseguer J. Principles of OBJ2. In: Proceedings ACM Princ. of Prog. Lang., 1985.

2. Jones C.B. Systematic Software Development Using VDM. Prentice Hall, 1985

3. Spivey J.M. An Introduction to Z and Formal Specifications. Software Engineering Journal 1989; 1

4. Brummer J. Representation and Verification of Discrete–Event Systems by Means of Petri Nets. In: Proceedings of the Third International Workshop on Software Engineering and its Application. Toulouse, France, 1990

5. Kaufmamn F, Schillinger D. Functional Language as User–Friendly Programming Aid. In: Brown Boveri Review, 1984

6. IEC SC65A/WG6/TF3(Coordinator)4. Discontinous Process Control, Working Draft: Standards for Programmable Controllers, Part 3: Programming Languages, 1988

7. Merlin P, Faber D.J. Recoverability of communication protocols. IEEE Trans. Commun. 1976; 24:9

8. Berthomieu B, Diaz M. Modelling and Verification of Time Dependent Systems Using Time Petri Nets. IEEE Trans. Software Engineering 1991; 17:3

9. Siemens AG. Sicherheitsleittechnik, Konzeptbeschreibung Teil 1. KWU E451, 1991 (in German)

10. Hoare C.A.R. An axiomatic basis for computer programming. Communications of the ACM 1969, 12

Oreste : a Reliable Reactive Real-Time Language

Pierre Molinaro et Olivier H. Roux
Ecole Centrale de Nantes / Université de Nantes
Laboratoire d'Automatique de Nantes, URA 823
Nantes, France

Abstract

The behavior during execution of an Oreste program is driven by the application. To perform reliability, this behavior has to be always defined. The failure of an Oreste's software component execution is either explicitly recovered, either implicitly propagated to the caller of the component. This is performed by a multi-tasking extension of programming by contract, organized panic and/or resumption proposed for the Eiffel sequential language by B. Meyer.

1 Introduction

This work is based upon the French contribution to the Programming Language for Robots (PLR) [1] developed by the ISO working group 4 of TC 184 SC 2. From this contribution, further work led to the definition of a general purpose reactive real time language called Oreste for specifying distributed concurrence, i.e the behavior during execution is driven by the application; a program can be executed on a monoprocessor calculator or on calculator network without shared memory. It requires a real time executive on every station and an inter-station communication manager which provides a point-to-point communication with on-receive synchronization.

The main design goal of Oreste is security :

 • the compiler should detect as many errors as possible : this is achieved through explicit declaration of every program entity, and strong typing ; furthermore, the behavior of programs written with a subset of the language can be expressed by a finite states automaton ;
 • always defined behavior during execution : based on a multi-tasking extension of Eiffel's programming by contract, an on-line deadlock detection, and a clean termination of concurrent execution.

An Oreste program is described by a hierarchical composition of software components ; Oreste defines the following software components : function, procedure, task type and statements that include usual ones, and those dedicated to multi-tasking : fork, accept, reply, request, wait and loopwait statements.

Every software component either succeeds, either fails and when its fails, the failure is either recovered, either propagated to its caller. This is performed by the programming by contract proposed in the Eiffel language by B. Meyer [2], [3]. In this paper, we describe in a first part the exception mechanism and the programming by contract adopted by Oreste, then the Oreste's task and fork and the propagation of a failure in a multi-tasking Oreste program, then in the last part the failure of the communication statement.

2 Failure, Exception and Contract

2.1 Introduction

Several languages introduce exception mechanisms for dealing with abnormal cases. Most of them (as in Ada [4], CLU [5]) are not safe, i.e. a computation can fail without propagating the failure to the caller. By contrast, B. Meyer has defined for Eiffel language [2,3] a clean and safe mechanism. In the following, we summarize its main features.

2.2 Eiffel's Exceptions Mechanism

2.2.1 Default Mechanism

Eiffel defines the following software components : statements, routine (i.e procedure or function), and proposes the following definitions of exception and failure :

An **exception** is the occurrence of an abnormal condition during the execution of a software component ;

A **failure** is the inability of a software component to satisfy its purpose.

Every software component either succeeds or fails. Failure of a routine statement execution implies the failure of the routine call statement execution, i.e. the execution of the routine sequence of statements is aborted, and control is returned to the caller. Routine call statement execution failure is just a particular case of statement execution failure : so, by this way, failure is implicitly propagated to the caller, until the failure of the top-level routine that aborts execution. With this default mechanism, the occurrence of an exception led to abort execution.

2.2.2 Rescue Mechanism

Eiffel introduces two ways for providing fault tolerance :
- organized panic ;
- resumption.

Syntactically, fault tolerance is expressed at routine level by a rescue clause which contains a sequence of statements. Eiffel allows using a particular statement, the retry statement, only in rescue clauses. When a routine statement execution fails, the rescue clause sequence of statement is executed ; if this execution fails, the routine call statement execution fails ; if it succeeds and the retry statement is invoked, the routine is executed again from its first statement (resumption) ; if it succeeds without invoking any retry statement, the routine call statement execution fails (organized panic).

Note that routine local variables are always set to their default values during routine invocation, and resumption does not modify any variable : so re-execution is started with current values.

Organized panic is useful just for restoring a clean state, by example for allowing resumption at a higher level. In either case, note that the rescue clause does not perform the purpose of the routine.

2.2.3 Programming by Contract : Preconditions and Postconditions

An Eiffel routine can be completed by an optional precondition and an optional postcondition.

A precondition expresses condition that must be satisfied by the context and/or the values of effective routine call parameters before the execution of routine first statement. Syntactically, a precondition is defined by a set of boolean expressions, which have to be evaluated true ; if evaluation fails, or if one of them is false, the routine is not executed, and the routine call statement execution fails.

A postcondition expresses conditions that must be satisfied by the context and/or local variables values after the execution of the routine sequence of statements has succeeded. Syntactically, a postcondition is defined by a set of boolean expressions, which have to be evaluated true ; if evaluation fails, or if one of them is false, the rescue clause is executed as defined before.

Preconditions and postconditions are the basis of B. Meyer's programming by contract [2]. It is essential to express this contract precisely. A contract represents the service that a software component has to perform, and necessary conditions to execute it.

Oreste has adopted Eiffel's exception mechanism and programming by contract ; however, as Eiffel is a sequential language, the exception mechanism has to be extended.

3. Concurrency in Oreste

As many other concurrent languages, concurrency in Oreste is expressed through parallel execution of sequential tasks ; we have adopted the Ada definition [4] : *a program without any task is executed sequentially on a single logical processor. Tasks are entities whose executions proceed in parallel in the following sense : each task has its own logical processor; different tasks (different logical processors) proceed independently, except at point they synchronize.* However, there are several differences between Ada and Oreste tasks :

> • Oreste tasks can be described only through task types ;
> • an Oreste task has its own private variables (there is no shared data between tasks) ;
> • tasks declaration, instantiation, and execution is handled by a new dedicated statement, the fork statement.

3.1 Task types

In Oreste, we cannot describe directly a processing task but only a model of task we call a task type. Every task type can be instanced in order to obtain a processing task. A task type is described by a set of private variables, a set of ports and a set of private procedures and functions ; ports defines communication/synchronization potentiality with other tasks ; one particular procedure is the main one ; execution of an instance consists of executing the main procedure. An *Oreste application* is a set of task types and its execution consists of instantiation of a particular task type, the root task type and to execute it.

As execution of a task is sequential, Eiffel's exception mechanism applies : so a task execution either succeeds, either fails.

3.2 Basic Fork Statement

3.2.1 Presentation

The only way to express concurrency in Oreste is using the fork statement, which declares a set of tasks, links between the tasks ports, and tasks scheduling through the scheduling clause. The scope of declared tasks is limited to the current fork statement, so a concurrent execution is completely embedded within a fork statement.

In the following example (figure 1), the fork statement declares three tasks : the tasks T1 and T2, instances of elsewhere defined task type tasktype1, and the task T3 instance of elsewhere defined task type tasktype2. Theses task types define no port (i.e. there is no communication/synchronization between the tasks T1, T2 and T3) ; the scheduling clause specifies parallel execution between the execution of T1 and the sequential execution of T2 and T3.

```
fork
        task T1,T2 : typetask1 ;
        task T3 : typetask2 ;
begin
        T1 || (T2 ; T3)
end  fork ;
```

Figure 1 : Example of Fork Statement

More generally, the scheduling clause is an expression whose operands are the declared tasks and whose operators are sequential and parallel operators. Parenthesis can be used to show grouping. Each declared task must appear exactly once in the scheduling clause.

The sequential execution is expressed by the " ; " operator which is associative :

$$A; (B;C) \Leftrightarrow (A;B);C \Leftrightarrow A;B;C$$

The parallel execution is expressed by the " | | " operator which is associative and commutative :

$$A||(B||C) \Leftrightarrow (A||B)||C \Leftrightarrow A||B||C$$

$$A||B \Leftrightarrow B||A$$

The parallel operator has higher priority than the sequential operator :

$$A||B;C \Leftrightarrow (A||B);C$$

The execution of the basic fork statement consists of the following steps :
        • instancing all declared tasks, including setting the task private variables to their default values ;
        • concurrent execution, as specified by the scheduling clause ;
        • destructing all declared tasks.

3.2.2 Execution and Failure of Fork Scheduling Clause

Depending from the execution of the fork scheduling clause, a fork statement execution either succeeds, either fails.

Scheduling clause execution is recursively described by the execution of sequential execution of its terms and parallel execution of its factors. Sequential execution of A; B, where A and B are scheduling clause terms, is defined as follows :
        • A is first started ; when A has succeeded, B is started ; when B succeeds, the execution of A; B succeeds ;

• if the execution of A fails, B is not executed, and the execution of A ; B

fails ;

    • if the execution of B fails, the execution of A ; B fails.

Parallel execution of A | | B, where A and B are scheduling clause factors, is defined as follows :
    • A and B are started in an order that is not defined to the language ; when both have succeeded, the execution of A | | B succeeds ;
    • if the execution of A fails, the execution of A | | B will fail when B will terminate its execution (by a success or by a failure) ;
    • by symmetry, if the execution of B fails, the execution of A | | B will fail when A will terminate its execution.

Note that failing of one task of a parallel expression does not directly influence the other one, because *"different tasks proceed independently, except at point they synchronize"* ; so failing of a task affects the behavior of other tasks only when they wish to synchronize with it.

The propagation of the failure of a task can be abstracted as follows : failure of a component of a sequence causes the failure of the sequence ; failure of a component of a parallel expression does not affect directly other components, however, when all other components have succeeded or failed, the parallel expression will fail.

3.2.3 Example

We go back to the scheduling clause of the figure 1 :

$$T1 \; || \; (T2 \; ; \; T3)$$

The propagation of the failure of a task to the failure of the fork scheduling clause is as following :
    • if the execution of the task T1 fails, the fork scheduling clause will fail when the execution of T3 will terminate ;
    • if the execution of the task T2 fails, the fork scheduling clause will fail when the execution of the task T1 will terminate. The task T3 is not executed ;
    • if the execution of the task T3 fails, the fork scheduling clause will fail when the execution of T1 will terminate.

3.3 Fork Statement with Post Conditions and/or Rescue Clauses

A fork statement can be completed with an optional postcondition, and an optional rescue clause.

The post condition is evaluated only if the scheduling clause execution has succeeded.

The rescue clause is executed when the scheduling clause has failed, or when the post condition has failed. Oreste does not defines a Eiffel's retry statement (it acts as a goto), but uses a predefined boolean local variable, called DoRetry. At the end of the execution of the rescue clause, if the boolean variable DoRetry is true, the scheduling clause is executed again.

As for an Eiffel routine, execution of a rescue clause leads to organized panic or resumption, which consists of executing again the fork declared tasks. Note that, as the tasks are not destructing when the scheduling clause execution terminates, they are not re-instanced, and the task private variables are not resetting to their default values.

Note that although a fork statement execution includes concurrent execution, post condition evaluation and rescue clause execution are sequential.

The programming by contract and its multi-tasking extension provides both a debugging mechanism and a tool for fault tolerance and failure recovery.

Figure 2 shows a fork statement with a post condition and a rescue clause ; figure 3 illustrates the effect of retrying execution : T1 and T2 are first instanced, and then started ; both tasks succeed ; but evaluation of post condition fails, leading execution of the rescue clause, that sets the 'DoRetry' variable to true ; So T1 and T2 are re-executed ; both tasks succeed, also does post condition evaluation ; both tasks are then destructed : fork statement execution has succeeded. If the second evaluation of the post condition should have failed, the second execution of the rescue clause should set the 'DoRetry' variable to false, leading to the failure of the fork statement execution.

```
      FirstAttempt := true ;
      fork
            task T1 : typetask1 ;
            task T2 : typetask2 ;
      begin
            T1 || T2
      ensure
            -- boolean expressions
      rescue
            DoRetry := FirstAttempt ; FirstAttempt := false ;
      end  fork ;
      -- statement b
```

Figure 2 : Example of Fork Statement with post condition and rescue clause

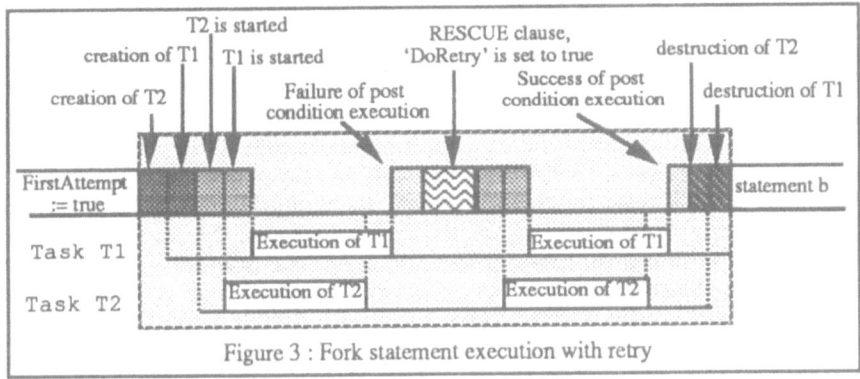

Figure 3 : Fork statement execution with retry

4. Failure of Communication Statements

Synchronization and communication in Oreste comes from the concept of "port" introduced by Silberschatz [6]. It achieves synchronous message passing, without explicit naming of the correspondent. Every port has one and only one owner, and one or more users. An Oreste port performs [7] :

• for users, synchronous sending-and-receiving by the request statement ;

• for the owner, synchronous receiving by the accept statement and asynchronous sending by the reply statement ; every accept statement execution must be balanced with a reply statement execution ; one or more accept statements can appear in wait and loop wait statements, corresponding respectively to the alternative and repetitive commands of CSP [8, 9].

A task declares all the ports it owns and it uses ; theses declarations introduce local names and specify the type of the transmitted information. In a fork statement, link clauses define the links between the different ports of the tasks.

Note that using local names allows the writing of general type tasks : by contrast, Ada entry calls either explicitly names the corresponding task, either implicit naming is resolved through visibility rules. Furthermore, as all potential interactions are declared in the fork statement, all potential correspondents are always known.

As other statements, request, accept and reply statements either succeed or fail.

If all users of a port are terminated when the owner invokes an accept or a reply statement, the communication statement fails. Also, if the owner of a port is terminated when an user invokes a request statement, the request statement fails.

As for other statements, the communication statement execution failure causes the rescue clause of current procedure to be executed. This ensures that a task never waits for a communication when all its potential correspondents are terminated and allows a clean and automatic termination of tasks.

5 Conclusions and Perspectives

The multi-tasking extension of programming by contract, organized panic and resumption provide both a debbugging mechanism and a mechanism for fault tolerance and failure recovery for real time application programs. Failure are recovered or propagated on run-time, it allows an always defined behavior during execution of an Oreste program. Also, the communication lock, when all the potential correspondent are terminated can be avoided by the failure of the communication statement. However, the problem remains in the case of deadlock between several tasks. The behavior of programs written in a subset of the language can be expressed by a finite states automaton, so that off-line deadlock detection can be performed. For the full language, current work is in progress for providing an on-line deadlock detection mechanism.

References

1. ISO /TC 184 / SC 2 / WG 4 / N106 ISO /WD 11513.1 Manipuling Industrial Robots, "Programming Languages for Robots" (PLR) Sept. 91.

2. B. Meyer, 'Applying "Design by Contract ", IEEE Computer, pp 40-51, october 1992.

3. B. Meyer, "Object-Oriented Software Construction", Prentice Hall, Englewood Cliffs, N.J., 534p., 1988.

4. Le Langage de programmation ADA, Norme AFNOR NF EN 28652, 1989.

5. Barbara Liskov, John Guttag, "Abstraction and Specification in Program Development", MIT Press.

6. A. Silberschatz, "Port Directed Communication", The Computer Journal, Vol. 24, n°1, pp 78-82, 1981.

7. O.H. Roux, P. Molinaro, "Mécanismes de communication et de synchronisation du langage Temps Reel Oreste", JJCSIR, Grenoble, France, pp 5-10, 14-16 April 1993.

8. C.A.R. Hoare, "Communicating Sequential Processes", Comm. of the ACM, Vol. 21, n°8, pp 666-677, 1978.

9. E.W. Dijkstra, "Guarded Commands, Nondeterminacy and Formal Derivation of Programs", Comm. of the ACM, Vol. 18, n°8, pp 453-457, 1975.

INVITED PAPER

How Far Can You Trust A Computer?

Carl E. Landwehr
Center for High Assurance Computing Systems,
Naval Research Laboratory
Washington, D.C., U.S.A

Abstract

The history of attempts to secure computer systems against threats to confidentiality, integrity, and availability of data is briefly surveyed, and the danger of repeating a portion of that history is noted. Areas needing research attention are highlighted, and a new approach to developing certified systems is described.

1 Introduction

Concerns about the security of data processed by or stored in computers are probably as old as computing, at least in the sense that some of the earliest modern computing machines were built and used in sensitive applications -- for example, the Polish "Bombe" and its British descendants that were used to attack German ciphers during World War II [1]. But it was only with the development of large-scale, shared multiprocessing systems that computer security, in the sense that term is used today, began to be an issue of general concern. The advent of time-sharing systems in the late 1960's and early 1970's brought the difficulties of protecting users from each other within a single computing environment into sharper focus, because people expected to store data for long periods in such systems, as well as to receive a fair share of interactive computing services. This paper will review briefly some of the history of computer security work starting from that time, summarize some of the lessons we have learned, sketch a recently developed approach to developing and certifying computer systems with security requirements, and suggest some research directions.

Because words like "security" and "trust" are commonly, but imprecisely, used, we introduce a few definitions before proceeding. We say a computer system is *secure* if it can preserve the *confidentiality, integrity,* and *availability* of the data it processes and stores against some anticipated set of threats. Preserving confidentiality has typically denoted protecting the data against unauthorized disclosure; preserving integrity has denoted preventing its unauthorized modification; and preserving availability has denoted preventing its unauthorized withholding. These definitions are perhaps narrower than the casual reader might expect, and indeed they have provoked some debate even within the computer security community. Integrity, in particular, continues to be a much-debated term [2]. A computer system or component is *trusted* if we rely on it to perform some critical function or preserve some critical property (such as security); it is only *trustworthy* if we have evidence to justify the trust we place in it. A computer system is called multilevel secure (MLS) if it is trusted to separate users with different clearances from data with different classifications.

2 Penetrate and Patch

When, in the late 1960's and early 1970's, operating system developers (and their customers) began to discover that their operating systems were somewhat less secure than they had thought, they treated security flaws like any other bugs, and installed fixes. Customers who were particularly interested in the security of their systems sometimes hired "tiger teams" to try to penetrate them, so that all holes might be found and patched. One product of such efforts was the flaw hypothesis methodology, which suggested an informal but systematic approach to this activity [3].

This approach to computer security was a victim of its own success -- not its success in achieving secure computer systems, but its success in penetrating insecure ones. Every time a new person or group attempted to penetrate a system, even one that had previously been penetrated and patched, new holes were found [4]. Although records were sometimes collected concerning the holes found, they were not widely circulated, and many of them have since been lost [5]. A forthcoming research report collects fifty surviving examples and proposes a taxonomy for organizing this kind of data [6]. Figure 1 reproduces a chart from that report characterizing the genesis, location, and time in the system life cycle where these flaws were introduced. The taxonomy and charts in that report are intended to provide a helpful method for abstracting current flaw data so that future attempts to improve system security can build on a stronger empirical base.

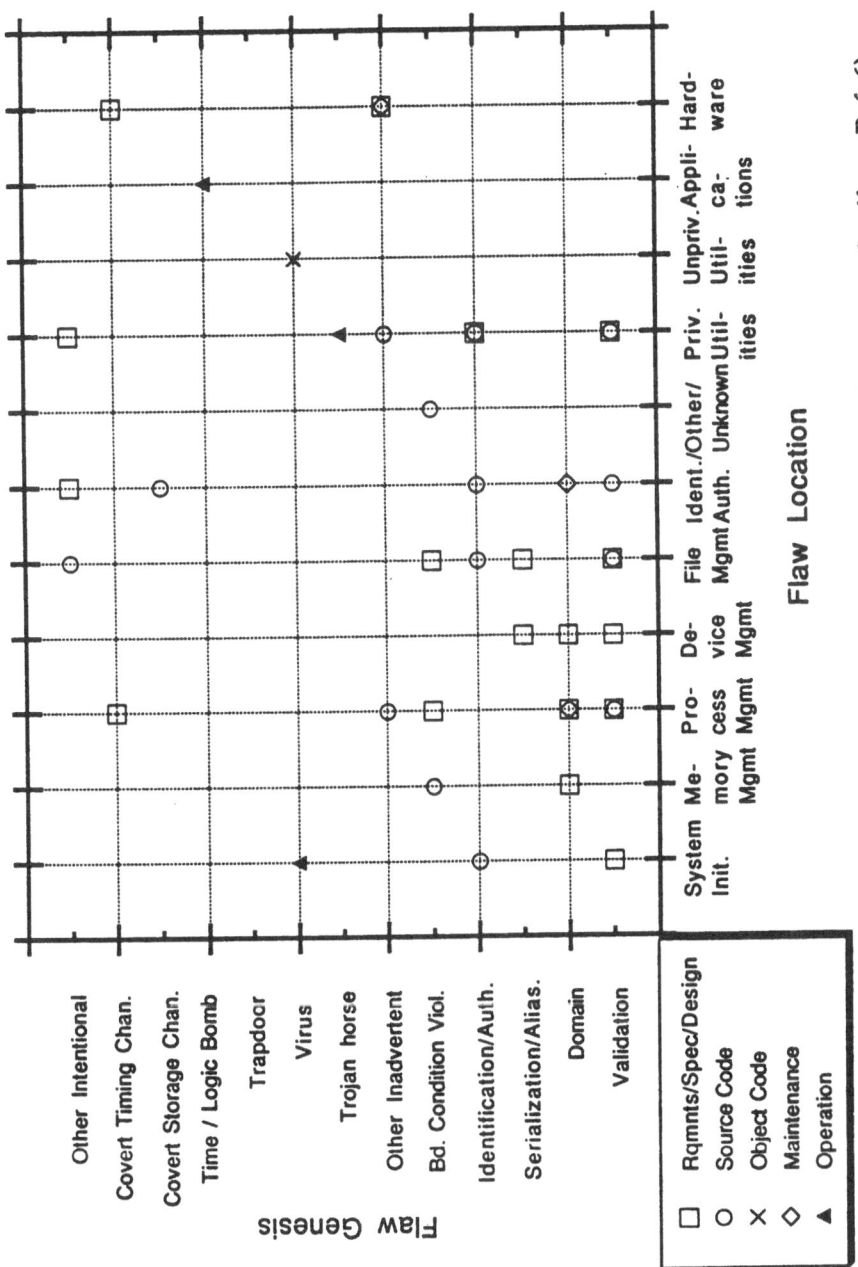

Figure 1. Characteristics of computer security flaw examples (from Ref. 6).

3 Technology for Trustworthy Operating Systems

The discouraging results of penetrate and patch activities led to the realization that a more systematic approach to building secure systems was needed. From the operating system notion of *monitors* as software components that controlled access to critical system resources, computer security researchers developed the concept of a *reference monitor* that would validate that all references issued by subjects (executing programs) to objects (memory and files) were consistent with an access control policy [7]. If the hardware and software needed to perform the reference monitor functions could be isolated and encapsulated in a small and simple enough part of a system so that high confidence in its correctness could be established, that component would be called a *security kernel* [8]. For some time, researchers labored to demonstrate prototype security kernel implementations. The difficulty of actually isolating all of the code on which security-enforcement depended proved greater than had been supposed, but it nevertheless seemed clear that much more trustworthy systems could be built using this approach than by the penetrate-and-patch method.

To persuade vendors to build more trustworthy systems and to make it easier for users to purchase such systems, the U.S. government established a National Computer Security Evaluation Center in 1981. It was to produce a set of computer security evaluation criteria and then evaluate products submitted voluntarily by vendors. Results would be maintained on an Evaluated Products List that could be used to qualify products for government purchase; this qualification provided the incentive for vendors to submit their products for evaluation.

The *Trusted Computer System Evaluation Criteria,* better known as "the Orange Book," which defined seven different evaluation levels, appeared officially in 1983. It defined a *trusted computing base* as the "totality of protection mechanisms within a computer system ... that is responsible for enforcing a security policy." Thus a system that had its security enforcement mechanisms distributed throughout its software and hardware could be said to have a trusted computing base, but not a security kernel. Only the highest two levels defined by the TCSEC (B3 and A1) require that the TCB be structured to exclude code not essential to security policy enforcement -- that is, to have a security kernel.

The Orange Book was tacitly based on an abstraction of the dominant computing model of the 1970's: a shared, central-server timesharing system. As the computing world shifted toward workstations and networks, it became clear that, at the very least, some additional thought was required to see how to apply the Orange Book in this context. One result of this re-thinking was the *Trusted Network Interpretation of the TCSEC* (the "Red Book"), published in 1987.

It also became clear that having an operating system that could separate differently classified information from differently cleared users was not the same thing as having an application that provided a useful MLS interface. The operating system might support a variety of applications, each operating at single security levels or it might support several instances of the same application, with each instance operating at a different (single) level. But to support a single application that operated across a range of security levels required that some of the security enforcement be provided by the application -- the product TCB would have to be modified or extended to incorporate parts of the application in this case. To address this issue, particularly in the context of database management systems, the *Trusted Database Interpretation* was published in 1991.

The rest of the world did not sit still during this period. Toward the end of the 1980's, France, Germany, the Netherlands, and the United Kingdom produced the *Information Technology Security Evaluation Criteria* (the ITSEC) -- a "harmonised" version of evaluation criteria created jointly by representatives from all four countries. The ITSEC permit greater flexibility than the TCSEC, in that vendors and customers can separately specify functional requirements and assurance requirements, which are joined in the Orange Book classes. The ITSEC also attempt to define a structure that supports both product and system evaluation, although their utility for the latter role is not universally accepted. The *Canadian Trusted Product Evaluation Criteria* (CTCPEC) also permit separating function and assurance, but (as their title indicates) is restricted to product evaluations. However, the CTCPEC have gone farthest in explicitly addressing integrity and denial-of-service (availability) issues.

4 What We Can Do Today

Many products have been built and submitted for evaluation in the decade since the Orange Book first appeared. As of June 1992, the U.S. Evaluated Products List included over a dozen products rated C2, four rated B1, two rated B2, and one at B3; in addition there are two systems that have been evaluated as network components under the TNI; one received a B2 rating and the other an A1. Products have also been evaluated successfully against the ITSEC and the CTCPEC.

On this basis, it seems fair to conclude that today we can specify, design, and build operating system products to meet the requirements reflected by Orange Book levels D through B3. We also know how to evaluate products against those criteria, and there is some evidence that products satisfying the higher levels of the criteria are indeed more difficult to penetrate than those that don't.

These conclusions should be tempered with the understanding that the evaluation process is still typically long, arduous, and, particularly at the higher assurance levels, expensive. This fact is partly attributable to the way that the evaluation process has evolved, as Steve Lipner noted in his insightful paper presented two years ago [9], but it is also attributable to the fact that developing high assurance systems based on security kernels requires greater control and documentation of the system engineering process, and particularly the software engineering process, than most developers customarily provide.

Something else that we can (and do) do today is plug together systems from products, including workstations, local area networks, routers, gateways, and software from a wide variety of sources. Often these systems perform their functions quite effectively, but the security they provide is usually hard to determine and hard to control, even if the security properties of the individual components are known.

5 Problems We Face

There are many problems that need to be solved before we will see the widespread and effective application of trustworthy computing technology. A few of these problems are highlighted in this section; the following section reports some recent advances in addressing the first of them.

We need less costly product evaluation/system certification techniques that can be applied more quickly, but with effectiveness at least equal to current approaches. Problems with the current product evaluation process in the U.S., as described by Lipner [9], have stimulated attempts to improve it. Unquestionably, better methods are needed, but there seems to be an increasing tendency in some quarters to accept commercial off-the-shelf products, together with some form of testing, as sufficient to assure the security of a product. We must be sure that this tendency does not simply lead us back to the discredited "penetrate-and-patch" paradigm. Research areas relevant to this problem include techniques for assessing software development methods, techniques for documenting and assessing software specifications and designs, tools and methods for product testing, and reverse engineering methods.

We need better ways to understand the security provided by composite and distributed systems. The only reason that we can build systems today by plugging them together is that there is a degree of standardization at the level of physical connectors and device protocols. These standards permit the possibility of component-based system engineering. We are not likely to be able to deduce much about the security provided by systems built this way, however, until at least a comparable level of standardization of the security functions and assurances provided by components is achieved. Further, we must take into account the potentially world-wide distribution of modern systems, which, particularly in the commercial sphere, often means that reliable authentication of the originator and recipient of a message, rather than the confidentiality of its contents, is the paramount security concern. Research areas that address this issue encompass abstract, formal work on security modeling techniques that support composition and decomposition, methods and tools for reasoning about and finding flaws in cryptographic protocols, and concrete approaches for standardizing security function and assurance requirements.

We need better ways to control the security functions that current (and future) systems provide. Many actual security problems occur not because security controls are lacking but because the existing controls have not been set up correctly. Steve Kent of BBN Communications has observed that the US is a country of people who are unable to program their videotape recorders, yet we are building interfaces to our security controls that are much more complex than the average VCR control panel. It is difficult to find much research aimed at this problem presently, but work to identify common requirements for application-based security controls and to develop user and administrator interfaces to them that are based on work in the area of human-computer interaction could lead to significantly improved security in practice.

We need to develop practical methods for building high assurance systems. There are definite needs for systems that can provide very high confidence that they will not have security failures. The leading technology for developing high assurance software is to apply formal techniques to its specification and development. Although a recent study shows increased industrial application of formal methods [10], their use is still seen as a significant cost factor, and there is uncertainty as to whether they can be successfully applied in large projects. Further, it is difficult to assess the cost-effectiveness of their application because it is hard to quantify the security provided by the resulting system. Imaginative approaches are needed to organize systems so that requirements for high assurance software are kept to a minimum. We need practical methodologies for exploiting formal methods on those portions of systems that unavoidably require high assurance, and we need methods to estimate or measure the security actually provided.

We need to broaden the scope of "security" and to develop methods for addressing security properties in conjunction with other critical system properties. Few systems are purchased strictly to provide security. Typically, a customer requires a system to perform some function -- communication, record keeping, real-time control, etc. -- and may acknowledge that to perform the function properly, some security requirements must be met as well. In commercial applications, confidentiality may frequently take a back seat to integrity and authenticity, and availability may be the strongest security concern. In control systems, timely delivery of results may be paramount. If we are to build systems that incorporate security as well as the other properties users require, we need techniques for developing designs that can meet a variety of critical requirements and that permit a system designer to make rational trade-offs among them. Research that permits quantification of covert channel bandwidths, for example, is a step in this direction to the extent that it permits us to quantify the rate at which a particular system design permits information to leak [11]. Work to model denial of service protection is similarly relevant [12].

6 Developing and Certifying Trustworthy Systems: A New Approach

As noted above, we cannot at present develop and certify the security properties of integrated systems nearly as well as monolithic products. In this section, we briefly describe an informal, but structured, approach to system development and certification developed recently at the Naval Research Laboratory. This approach has yet to be applied in sufficient detail to a large example to permit us to make strong claims about its effectiveness, but it is based on concepts proven in our earlier work [13,14]. It has strong intuitive appeal, both as a way to address security requirements during system development and as a way to explain to the *accreditor* (the person responsible for deciding whether to permit the system to be operated) what security the system provides and what risks its operation would pose. *Certification* denotes a technical assessment of the ability of a system to meet specified technical standards (e.g. for enforcing security requirements). A more comprehensive description of this approach has recently appeared [15].

The approach is based on recording *assertions* and *assumptions* that capture the system security requirements within the framework of a documented *assurance strategy*. At the beginning of the project this strategy records both an initial, high-level, abstract version of the *assurance argument* for the system as well as the plan for creating the final, more detailed and concrete assurance argument that will form the primary technical basis for the certification decision. As the project progresses, the assurance strategy is elaborated to reveal the increasingly detailed outline of the assurance argument, which demonstrates that the system as designed and built actually satisfies its security requirements. This argument will not exist as a separate document; the final assurance strategy will in effect be an index to other parts of system documentation (software specification and design documents, test plans and results, etc.) that provide the "nuts and bolts" of the assurance argument. With this approach, certification of a trusted system can largely be accomplished as an audit of the development process.

For a given system, assertions are predicates that are enforced by the system, and assumptions are predicates that must be enforced in the system's environment. The system itself is unable to enforce its assumptions, but must rely upon them. Together, assumptions and assertions represent what must be true of the system and its environment to satisfy the security policy. If an assumption or an assertion is false, a security violation may occur.

For example, consider a medical information system used by physicians, nurses, and pharmacists within a single hospital to record the current symptoms, diagnosis, treatment plan, and billing information for each patient. Suppose that the system's security policy requires that (1) only an administrator can create a new patient record or enter authorizations for doctors, nurses, and pharmacists to use the system, (2) only physicians may update the recorded diagnosis and treatment plan, (3) nurses may update the record of symptoms and medication administered, and (4) pharmacists may read the treatment plan but can only update the billing information. Finally, (5) patients are prohibited from any access to the system.

The architect of such a system has a number of security disciplines available to help satisfy system security requirements, including personnel security, physical security, procedural security, communications security, computer security, and others. The system architect typically seeks the most cost-effective combination of methods drawn from these disciplines that will satisfy the overall system security policy in the face of anticipated threats.

If the system architect decided to rely primarily on the discipline of computer security to enforce the security policy, most of the predicates would be enforced by the medical information system software, and they would be assertions about that software system. If the architect chose to rely on personnel and procedural security measures (e.g., by training the users in their roles and relying on them to invoke only the system functions

322

appropriate to those roles) then, from the standpoint of the information system, all of the predicates would be assumptions about the environment in which it operates.

The assurance strategy provides a framework for recording the assertions and assumptions according to a chosen system security architecture. Suppose a design is developed that requires the administrator to use, and to write down on paper, a password for authentication purposes. Figure 2

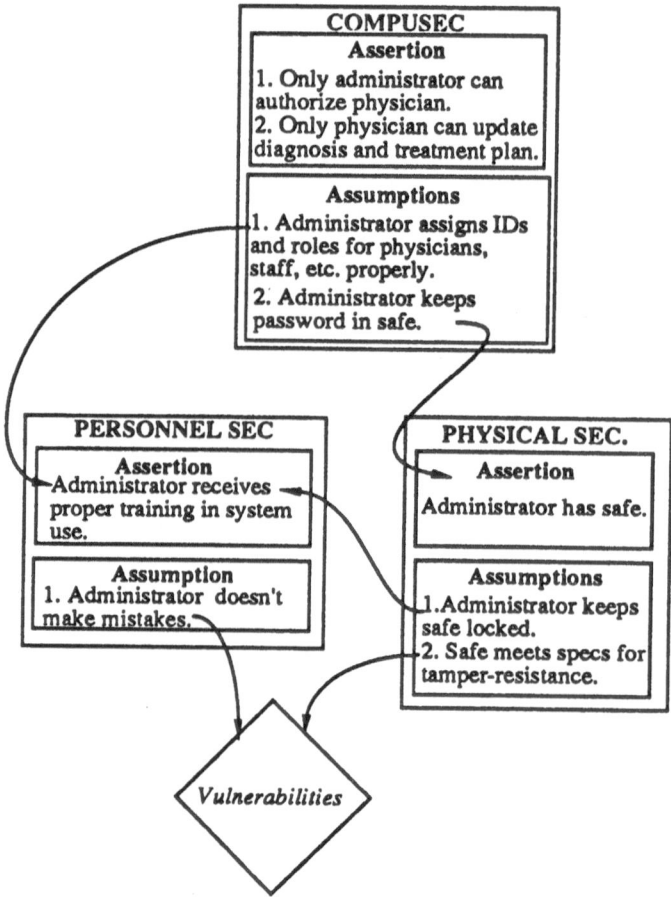

Figure 2. Part of an assertion strategy.

illustrates part of an initial assurance strategy for the medical information system example. Notice that each assumption is supported by an assertion from another discipline or else maps to the "vulnerabilities" symbol. In this representation, assertions do not map to other parts of the framework; they form a set of requirements to be satisfied by the discipline in which they occur. The full assurance strategy for

COMPUSEC, for example, would have to explain what kind of assurance would be provided that the COMPUSEC assertions are enforced (e.g., through use of a trusted computing base supporting access controls).

A certifier can review the assurance strategy at the beginning of the development and decide at that time whether this strategy (if followed) is likely to produce an acceptable assurance argument at delivery. The assurance strategy may be modified during the development, since as design trade-offs are made it may be appropriate to modify the kind or degree of assurance (e.g., code reviews, formal verification, testing) required to support particular parts of the assurance argument that is being created. It allows certifiers to assess the role various design decisions play in the overall assurance argument and to determine whether the proposed assurance techniques are effective for demonstrating the validity of the decision.

This approach is not a panacea; correctly defining the security policy and creating a complete assurance argument will continue to be a challenging task. But it does promote the integration of system and security engineering, it can reduce the risk that unforeseen certification issues will impede system development and delivery, and it can make the risks of operating the system more clearly visible to the accreditor.

7 Summary and Conclusion

We have reviewed briefly the development of trustworthy computing technology. Readers should particularly note the lessons of the "penetrate and patch" approach, so that we do not repeat the experiences of that era. We have noted needs for improvements in system certification methods, in understanding security implications of composite and distributed systems, in the control interfaces for security functions, in methods for developing high assurance systems, and in integrating security and other critical properties. Finally, we have glimpsed a new approach to system certification.

How far can we trust computers? We already trust them to fly our airplanes and rockets, and it is certainly easier to purchase a computer system and know its security properties now than it was ten years ago. But we still have far to go to before we can make rigorous statements about the *trustworthiness* of the software in our trusted systems.

References

1. Hodges A. Alan Turing: the enigma. Simon and Schuster, New York, 1983.

2. Sandhu R. On four definitions of data integrity. In: Keefe, T (ed) Proc. IFIP WG11.3 seventh working conf. on database security, Sept., 1993 (to appear as Database Security VII: Status and Prospects, Elsevier, 1994).

3. Linde R. Operating system penetration. In: Proc. National Computer Conference, 1975. AFIPS Press, Montvale, N.J., 1975, pp 361-368.

4. Neumann P G. Computer security evaluation. In: Proc. National Computer Conference, 1978. AFIPS Press, Montvale, N.J., 1978, pp 1087-1095.

5. Bisbey R. Personal communication. 26 July 1990.

6. Landwehr C E, Bull A R, McDermott J P, Choi W S. A taxonomy of computer program security flaws with examples. NRL Report (forthcoming), Naval Research Laboratory, Washington DC, 1993.

7. Anderson J P. Computer security technology planning study (vols I and II). ESD-TR-73-51, Hanscom Field, Bedford MA; NTIS AD 758 206, 1972.

8. Gasser M. Building a secure computer system. Van Nostrand Reinhold, New York, 1988.

9. Lipner S B. Criteria, evaluation, and the international environment: where have we been, where are we going? In: Lindsay and Price (ed), Proc. IFIP-SEC 91, Brighton, England. Elsevier - North Holland, 1991.

10. Craigen D, Gerhart S, Ralston T. An international survey of industrial applications of formal methods. NRL Report 9554, Naval Research Laboratory, Washington DC, 1993.

11. Gray J W. On introducing noise into the bus-contention channel. In: Proc. 1993 IEEE CS Symp. on Research in Security and Privacy. IEEE Computer Society Press, 1993, pp 90-99.

12. Millen J K. A resource allocation model for denial of service. In: Proc. 1992 IEEE CS Symp. on Research in Security and Privacy. IEEE Computer Society Press, 1992, pp 137-147.

13. Landwehr C E, Heitmeyer C L, McLean J. A security model for military message systems. ACM Trans. on Computer Systems 1984; 2(3):198-222.

14. Froscher J N, Carroll J M. Security requirements of Navy embedded computers. NRL Memorandum Report 5425, Naval Research Laboratory, Washington DC, 1984.

15. Payne C N, Froscher J N, Landwehr C E. Toward a comprehensive INFOSEC certification methodology. In: Proc. 16th National Computer Security Conference. National Institutes of Standards and Technology / National Computer Security Center, Baltimore, MD, Sept. 1993.

Session 10

SECURITY

Chair: P. Daniel
GEC Marconi Secure Systems, UK

Security Audit Trail Analysis Using Genetic Algorithms[*]

Ludovic Mé
Laboratoire d'informatique, SUPÉLEC
Avenue de la Boulaie
B.P. 28, F-35511 Cesson Sévigné Cedex
France

Abstract

We propose a security audit trail analysis approach based on predefined attack scenarios and using genetic algorithms. This paper shows the validity of this approach and presents some of its problems.

1. Introduction

Several approaches exist in computer security: (i) enforcing users' respect to particular rules when they are using the system, (ii) identifying threats to the system (COPS [1] is the best known tool in the UNIX world for such an identification) and (iii) recording some or all actions performed on the system in order to analyze the trail to detect some attacks (it is an after-the-events detective control).

The third approach is known as "**security audit trail analysis**" (SATA) and seems especially important because [2] (i) even the most secure systems are vulnerable to legal users' misuses (audit trails may be the only means of detecting authorized but abusive user activity), (ii) existing systems have security flaws and thus are vulnerable to attacks, (iii) substitution of existing systems by secure ones is not easy to achieve for economic reasons and (iv) development, installation and management of secure systems is not an easy task.

So we need to record events in audit trails. However there are some problems: (i) this approach is very expensive in terms of disk space, (ii) its impact on the systems performance is noticeable and (iii) its efficiency is low because the **security officer** (**SO**) has to manage such a huge amount of data recorded that it is not humanely possible.

Our objective is to design an automatic tool to increase the SATA efficiency. This paper presents our work with the following framework. Section 2 exposes two classical approaches for intrusion detection. Section 3 introduces an alternative approach based on attack scenarios. In section 4 we give our own vision of the security audit trail analysis problem which is proved to be NP-complete. Section 5 proposes a brief survey of **genetic algorithms** (**GA**). In section 6 we derive a simplified but still NP-complete version of SATA and show how to apply GAs to it. Section 7 discusses our experiments which exhibit fairly good results. Section 8 highlights some remaining problems of our approach and section 9 concludes the paper.

---

[*] Pierre Rolin (Télécom Bretagne) supervises this PhD work.

2. Two Classical Approaches

Intruders can attack a system using either unknown or known techniques. In the first case, a possible strategy for the SO to detect intrusion is the use of a statistical approach (reference [3] uses neural networks). The SO has to respond to the question "is the users behavior normal according to the past?". This approach is known as "the comportemental model" [2]. In the second case, it is possible to provide attack detection rules which enables the SO to use expert systems tools. The question is "does the user behavior correspond to a known attack?". Some tools, and especially the most famous, IDES (Intrusion Detection Expert System) [4] [5], implement both approaches. Others rely on expert systems exclusively. Some tools already provide quasi-real-time intrusion detection.

The statistical approach leads to some problems: (i) the choice of the parameters of the statistical model is tricky, (ii) the statistic model leads to a flow of alarms in the case of a noticeable systems environment modification and (iii) a user can slowly change his behavior in order to cheat the system.

In the expert system approach, the SOs knowledge is encoded in a set of rules used to analyze the audit log. In practice, however, the SO has only gained limited expertise, essentially because the huge amounts of data recorded leads to an untractable duty.

3. An Alternative Approach

A third approach, "model-based reasoning", is recommended as an additional intrusion detection system by Teresa Lunt and Thomas Garvey [6]. It consists of designing attack scenarios as sequences of user behavior. These behavior sequences are then translated (depending on the audit system used) into audit events' sequences (for example the password copy activity is translated into an execute access to /bin/cp and a read access to /etc/passwd). Using attack scenarios has some advantages:

1. the SO can design the attack scenarios himself according to the threats he is afraid of,
2. the modification of an existing scenario or the addition of a new scenario (e.g. after its detection by a statistical method) is easy,
3. the events to be stored are only those present in at least one scenario,
4. the SO can attribute to each scenario a weight, according to the consequences of the corresponding attack.

We use this approach for our work. Let us consider the attack scenarios as sequences of audit events. A method to simplify the design of the scenarios must be found. We propose the following requirements for this method:

1. it should spare the designer the trouble of enumerating all the possible variants of the same scenario
2. it should allow the shortest expression of any scenario, especially when it contains repetitive patterns (e.g. denial of service)
3. it should be as general as possible in order to allow the design of any scenario.

It is useful to refer to the works done in the field of pattern matching: the set of audit events can be seen as an alphabet, each audit event as a character, the audit trail as a main string and the scenarios as sub-strings to locate in this main string. To reach the previous requirements, we chose **regular expressions with back referencing (rewbr)** [7] as a language to design the attack scenario. The referencing operator allows the design of any scenario.

In this way, the **"security audit trail analysis problem"** (SATAP) becomes similar to a "finding a rewbr in a string" problem (**FRSP**). The NP-Completeness of FRSP [7] makes classical algorithms quite impossible to apply to real audit logs (recall the huge amount of data recorded).

4. A More Precise Expression of the SATA Problem

If the SATA was made attack by attack (i.e. rewbr by rewbr), exclusive attacks could be declared present at the same time. To avoid this problem, we have to consider the whole set of attack scenarios in a single analysis. We have to determine, among all the possible attacks' sub-sets, the one which presents the greatest risk to the system. For this, we suppose that the SO is able to evaluate the risk inherent in each scenario. Consequently we attribute to each scenario a weight proportional to that risk. By default, these weights are all equal to 1 and we simply look for the biggest possible sub-set.

More formally, our approach of SATAP can be expressed by the following statement: A is an alphabet whose letters are auditable events, S is a set of attack scenarios expressed by rewbr made with A's letters (each scenario S_i is associated to a weight W_i), T is the audit trail which is to be analyzed and viewed as a string of A's letters; We have to find the sub-set S' of S so that the total weight is maximized and so that each rewbr of S matches a different sub-string of T.

Theorem 1. SATAP is NP-complete.
Proof. If there are n different rewbr in S, there are M possible non-empty sub-sets of S. If we ignore the order of the successive attacks, we have:

$$M = \sum_{i=1}^{n} C_n^i = 2^n - 1 \qquad (1)$$

The number of possible sub-sets increases exponentially with the number of potential attacks. Each sub-set relates to a hypothesis corresponding to the attacks which would be actually present in the audit trail. However, not all the hypotheses are realistic. So for each of the 2^n-1 possible sub-sets, we must make a decision about the realism of the corresponding hypothesis. For this, we have to solve from 1 to n FRSP problems (one for each element of the sub-set until there is no match in T for a particular element of the sub-set). To solve SATAP, it is necessary to solve at least 2^n-1 FRSPs. We have here linked NP-complete problems.

5. A Search Algorithm for the SATA Problem

The NP-Completeness of SATAP makes it quite impossible classical algorithms to apply to real audit logs (recall the huge amount of data recorded). So we propose to use an heuristic method, the so called "genetic algorithm".

Genetic algorithms (GA) [8], proposed by Holland (1975) [9], are optimum search algorithms based on the mechanism of natural selection in a population.

A population is a set of artificial creatures (individuals or chromosomes). These creatures are strings of length l coding a potential solution to the problem to be solved, most often with a binary alphabet. The size L of the population is constant. The population is nothing but a set of points in a search space.

The population is randomly generated and then evolves: in every generation, a new set of artificial creatures is created using the fittest or pieces of the fittest individuals of the previous one. The fitness of each individual is simply the value of the function to be optimized (the fitness function) for the point corresponding to the individual. The iterative process of population creation is achieved by three basic genetic operators [9]: **selection** (selects the fittest individuals), **reproduction** (promotes exploration of new regions of the search space by crossing over parts of individuals) and **mutation** (protects the population against an irrecoverable loss of information). The general structure of a GA is thus the following:

```
Random generation of the first generation
Repeat
        Individual Selection
        Reproduction
        Mutation
Until an individual outclasses others
```

Genetic operators are randomized ones but genetic algorithms are no simple random walk: they efficiently exploit historical information to speculate on new search points with expected improved performance.

Because of space constraint, we cannot expand this section. For more developments, please refer to the literature [8] [9] [10] [11] [12].

6. Security Audit Trail Analysis Using Genetic Algorithms

Two sub-problems arise when applying GAs to a particular problem: (i) coding a solution for that problem with a string of bits and (ii) finding a fitness function to evaluate each individual of the population. To satisfy these two requirements, we had to simplify our vision of SATAP.

6.1 A Simplified Vision of SATAP

As seen previously, we have linked NP-complete problems: for each possible sub-set among 2^n-1 and for each element of the sub-set find a different matched string in the audit trail. Our goal is to determine, among all the possible attacks' sub-sets, the one which presents the greatest risk to the system: we adopt a simplified scheme which consists of bypassing the matching problem.

To make a decision about the realism of the hypothesis corresponding to a particular sub-set, we propose to work on the events rather than on the attacks. This means that we count, for a particular attacks' sub-set, the number of events of each type generated by all the attacks. If, for one or more types, this number is less than or equal to the number of recorded events of that type, then the hypothesis corresponding to the sub-set is realistic. It is a way to achieve a hypothetico-deductive scheme [13], just like a human expert would probably do for SATA: a hypothesis is made (e.g. among the set of 18 possible attacks (figure 1), the attacks 3, 7 and 12 are present) and the deduction involves an evaluation of the hypothesis (in our approach, deduction is enforced through the fitness function, as presented in the remainder of this paper). According to this evaluation, an improved hypothesis is tried, until a solution is found.

This **simplified vision of SATAP** (SSATAP) implies the translation (with a linear one-pass algorithm) of the audit trail into an observed audit vector O (a real-time construction of O could be considered). Basically, Oi counts the number of i type events present in the audit trail. This translation results in the loss of the time sequence. This presents some difficulties but, in addition, avoids the problem of events reordering which is not easy when timing information gained from the audit sub-system is not precise enough. In the case of network audit trail, where a global time does not exist, this could be an advantage. In some cases, it could be useful to apply other building rules for O. Oi then counts the number of sequences of particular events.

6.2 Proof of the NP-completeness of SSATAP

Formally, SSATAP can be expressed by the following statement:
- let N_e be the number of audit events and N_a the number of potential attacks
- let AE be an $N_e \times N_a$ attacks-events matrix which gives the set of events generated by each attack. AE_{ij} is the number of audit events of type i generated by the scenario j ($AE_{ij} \geq 0$). (See section 7.2 for an example of such a matrix)
- let W be a N_a-dimensional weight vector, where W_i ($W_i > 0$) is the weight associated with the attack i (W_i is proportional to the risk inherent in the attack scenario i)
- let O be the N_e-dimensional observed audit vector defined in the previous section
- let H be a N_a-dimensional hypothesis vector, where H_i equals 1 if the attack i is present according to the hypothesis and H_i equals 0 otherwise (H describes a particular attacks' sub-set).

SSATAP consists in finding the H vector which maximizes the W.H product, subject to the constraint $(AE.H)_i \leq O_i$ $(1 \leq i \leq N_a)$.

Theorem 2. SSATAP is NP-complete.
Proof. SSATAP can be polynomially reduced to the **zero-one integer programming problem** (ZOIP). ZOIP can be expressed by the following statement [14]:
> *Instance*: A finite set S of pairs (X,b), where X is an m-tuple of integers and b is an integer, an m-tuple C of integers and an integer B.
> *Question*: Is there an m-tuple Y of integers such that $X.Y \leq b$ for all pairs (X,b) and such that $C.Y \geq B$?

SSATAP can be directely reduced to ZOIP by writing:
> number of pairs (X,b) in the S set = N_e
> $m = N_a$
> $X = (ae_{i1}, ae_{i2}, \dots ac_{iNa})$ (a line of the AE matrix)
> $b = O_i$
> $C = W$

B = the extremum of the fitness function

Y = H

ZOIP is NP-complete [14]. Therefore SSATAP is NP-complete.

6.3 GASSATA, a Genetic Algorithm for Simplified Security Audit Trail Analysis

Reference [15] gives an example of a genetic algorithm solving an NP-complete problem similar to ZOIP, the **set covering problem (SCP)**. Liepins and Potter studied in 1991 a genetic approach for multiple-fault diagnosis [16] based on SCP. We used these two papers as a starting point for GASSATA.

6.3.1 Coding a solution to SSATA with a binary string

Recall that an individual is a 1 length string coding a potential solution to the problem to be solved. In our case, the coding is straightforward: the length of an individual is N_a and each individual in the population corresponds to a particular H vector as defined in the previous section.

6.3.2 The Fitness Function

We have to search, among all the possible attacks' sub-sets, for the one which presents the greatest risk to the system. This results in the maximization of the product W.H. As GAs are optimum search algorithms for finding the maximum of a so-called fitness function, we can easily conclude that in our case this function should be made equal to the product W.H. So we have:

$$\text{Fitness} = \sum_{i=1}^{N_a} W_i \cdot I_i \qquad (2)$$

where I is an individual.

This fitness function does not, however, pay attention to the constraint feature of SSATA which implies that some hypotheses (i.e. some individuals) among the $2^{N_a}-1$ possible ones are not realistic. This is the case for some i type of events when $(AE.H)_i > O_i$. There are several ways to take a constraint into account with GAs: (i) modifying the genetic operators so that they only generate "good" individuals (i.e. with respect to the constraint), (ii) repeating each crossover or mutation process until a "good" individual is generated, (iii) penalizing the "bad" individuals by reducing their fitness value. When a large number of individuals do not respect the constraint (this is precisely our case) the third solution is the best one.

To reduce the fitness value for a "bad" individual, we compute a penalty function (P) which increases as the realism of this individual decreases: let Te be the number of types of events for which $(AE.H)_i > O_i$, the penalty function applied to such an H individual is then:

$$P = Te^p \qquad (3)$$

A quadratic penalty function (i.e. p=2) allows a good discrimination among the individuals. The proposed fitness function is thus the following:

$$F(I_i) \quad = \quad \alpha \; + \; \left[\; \sum_{i=1}^{N_a} W_i . I_i \; - \; \beta . Te^p \; \right] \qquad (4)$$

The β parameter makes it possible to modify the slope of the penalty function and α sets a threshold making the fitness positive. If a negative fitness value is found, it is equaled to 0 and the corresponding individual will die. So the α parameter allows the elimination of unrealistic hypotheses.

6.3.3 The genetic operators

We are using, at the moment, the three basic operators defined by Holland [9].

7. Experimental Results

We give in this section a brief survey of the IBM[1] AIX[2] security audit system. Then, we describe the attacks-events matrix. Lastly, we present our experiments and their results.

7.1 IBM AIX security audit system

The AIX security audit system [17] [18] [19] is designed to satisfy the C2 security requirements of the orange book [20]. Some audit events are generated by the AIX kernel subroutines. Other events can be defined by the SO (some are proposed by default). These later events can be associated with objects (files) for a write, a read or an execute access. Both types of events can be associated with subjects (users). For our experiments, we use 25 kernel or self-defined events [21].
Events are recorded in a protected trail. The audit trail records format is the following: event name, status (OK or FAIL), real user id, login user id, program name (the command which generated the event), process id, parent process id, time in seconds, record tail (contains information depending on the event such as the file name in case of a file opening event).

7.2 The Attacks-Events Matrix

In practice, we encounter problems designing consistent attacks and thus we consider rather short ones. We add in the attacks set some "suspicious actions" such as repeted use of the "who" command.
An attack is a non ordering set of audit events which happened during an audit session. We work with 30 minutes audit sessions. This represents about 85 kilo-octets for each audited user (users are software developers for the moment). We translate the audit trail into user-by-user audit vectors with a linear one-pass algorithm. Successive audit trails and audit vectors should be archived on tapes for possible future investigations.
Figure 1 shows the attacks-events matrix that we use for our experiments ("." taking the place of "0" for lisibility reasons). In that

---

1, 2. IBM and AIX are trademarks of International Business Machines Corporation.

matrix each column corresponds to an attack. For example, column 8 corresponds to a browsing attack which can be characterized by a high rate of use for "ls" and "more" or "pg" commands (in our case 10 of each during the 30 minutes of the audit session) [21].

	1	2	3	4	5	6	7	8	9	10	11	12	13	14	15	16	17	18
passwd read	.	.	1	.	.	.	.	.	.	.	.	.	.	.	.	.	.	.
group read	.	.	1	.	.	.	.	.	.	.	.	.	.	.	.	.	.	.
hosts read	.	.	1	.	.	.	.	.	.	.	.	.	.	.	.	.	.	.
opasswd read	.	.	.	1	.	.	.	.	.	.	.	.	.	.	.	.	.	.
ogroup read	.	.	.	1	.	.	.	.	.	.	.	.	.	.	.	.	.	.
fail sensitive files write	.	.	.	.	1	.	.	.	.	.	.	.	.	.	.	.	.	.
fail day login	3	.	.	.	.	.	.	.	.	.	.	.	.	.	.	.	.	.
night login	.	.	.	.	.	.	.	.	.	.	.	.	.	.	.	.	.	1
su command	.	.	.	.	.	.	.	.	.	.	.	.	.	3	.	.	.	.
who, w, df, ...	.	10	.	.	.	.	.	.	.	.	.	.	.	.	.	.	.	.
fail ls command	.	.	.	.	.	5	.	.	.	.	.	.	.	.	.	.	.	.
ls command	.	.	.	.	.	.	.	10	.	.	.	.	.	.	.	.	.	.
cp command	.	.	.	.	.	.	.	.	.	.	30	.	.	.	.	.	.	.
rm command	.	.	.	.	.	.	.	.	.	.	.	.	10	.	.	.	.	.
whoami command	.	.	.	.	.	.	.	.	.	.	.	.	.	.	.	50	.	.
more or pg cmd	.	.	.	.	.	.	.	10	.	.	.	.	.	.	.	.	.	.
passwd command	.	.	.	.	.	.	.	.	.	.	.	.	.	.	.	3	.	.
fail chmod cmd	.	.	.	.	.	.	.	.	3	.	.	.	.	.	.	.	.	.
fail chown cmd	.	.	.	.	.	.	.	.	.	3	.	.	.	.	.	.	.	.
fail file open	.	.	.	.	.	.	5	.	.	.	.	.	.	.	.	.	.	.
file deletion	.	.	.	.	.	.	.	.	.	.	.	.	.300	.	.	.	.	.
process creation	.	.	.	.	.	.	.	.	.	.	.	.100	.	.	.	.	.	.
process exec.	.	.	.	.	.	.	.	.	.	.	.300	.	.	.	.	.	.	.
process priority modification	.	.	.	.	.	.	.	.	.	.	.	.100	.	.	.	.	.	.

Figure 1. An Attacks-Events Matrix for GASSATA

7.3 Experiments and Results

All our experiments are made using the following parameters for the fitness function: $\alpha=50$, $\beta=1$ and $p=2$. Each experiment can be characterized by a 5-tuple (Pc, Pm, L, g, a) where Pc is the crossover probability, Pm is the mutation probability, L the population size, g the number of generations and a the number of attacks actually present in the audit trail. The default values for these parameters are $Pc=0.6$, $Pm=0.0083$, $L=100$, $g=100$ (they correspond to classical values when using GAs) and $a=2$.

Because of their non-independence, we vary each parameter separately by taking values from the following sets :

Pc ... (0.5, 0.7, 0.8, 0.9, 1.0)
Pm ... (0, 0.00166, 0.00332, 0.00498, 0.00664, 0.00996, 0.01162, 0.01328, 0.01494, 0.0166)
L .... (20, 50, 150)
g .... (1000)
a .... (0, 1, 5, 10, 15, 18)

The 5-tuple (Pc, Pm, L, g, a) thus takes 25 different values. For each of these values, we perform 10 runs (all the following results are averages over the 10 runs):

- for each generation, we compute the minimum, maximum and average whole population fitness values
- when the GA stops, we count, for each bit position i along the strings of the final population, the number of 1 values and compute a rate ti by dividing this number by L. When ti is greater or

equal to a given detection threshold D, the ith attack is declared present.

We define four rates T1, T2, T'1 and T'2 as follow:
- T1 is the number of detected present attacks out of the number of present attacks
- T2 is the number of detected not present attacks out of the number of not present attacks
- T'1 is the number of individuals in which bits corresponding to present attacks are 1 out of the total number L of individuals
- T'2 is the number of individuals in which bits corresponding to not present attacks are 1 out of the total number L of individuals

T1 and T2 are respectively the detection rate and the false alarm rate. They depend on D. Our tool should present a high rate of detection and a low rate of false alarm. T'1 and T'2 qualify the results of the GA. T'1 should be equal to 1 and T'2 to 0.

7.3.1 Minimum, Maximum and Average Fitness Values

The maximum fitness value converges quickly on the optimum (after about 20 generations with the default parameters, figure 2). The remainder of the population follows and after 100 generations, the average fitness is about 9/10 of the maximum fitness. If Pc, Pm or L grows GASSATA converges more slowly but the risk to reach a local optimum (due to a so called "premature convergence" [8]) decreases. We observe that if a>15, the number of generations has to be increased in order to insure GA convergence.

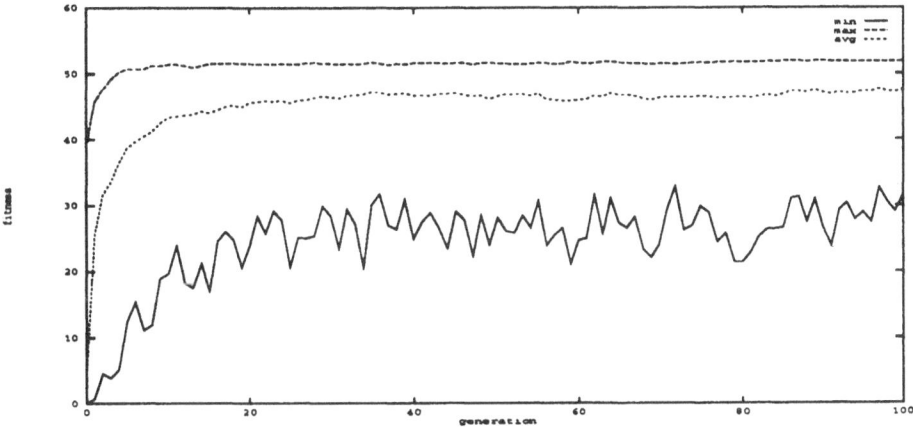

Figure 2. *Average Min, Max and Avg Fitnesses for 10 Runs*
(Pc=0.6, Pm=0.0083, L=100, g=100 and a=2)

7.3.2 T1 and T2

Figure 3 shows the evolution of T1 and T2 versus the detection threshold D for the default parameters. It shows that 0.5 is a good value for D as we then have T1=1 and T2=0. We observe that it is always the case with the default parameters. If Pc or Pm increases, the optimal D value decreases. If L or g increases, the optimal D value increases.

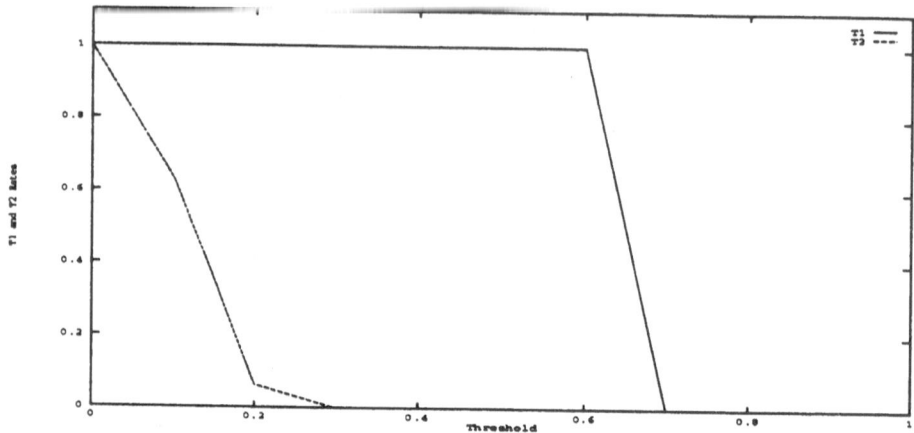

Figure 3. *Average* T1 *and* T2 *for 10 Runs*
(Pc=0.6, Pm=0.0083, L=100, g=100 and a=2)

7.3.3 T'1 and T'2

T'1 is above 0.6 and T'2 below 0.15 (figure 4). This means that more than 60% of the final populations individuals bits corresponding to present attacks are 1 and that less than 15% of the bits corresponding to not present attacks are 1. These results are better for higher values of L or g but decrease when Pc or Pm increase.

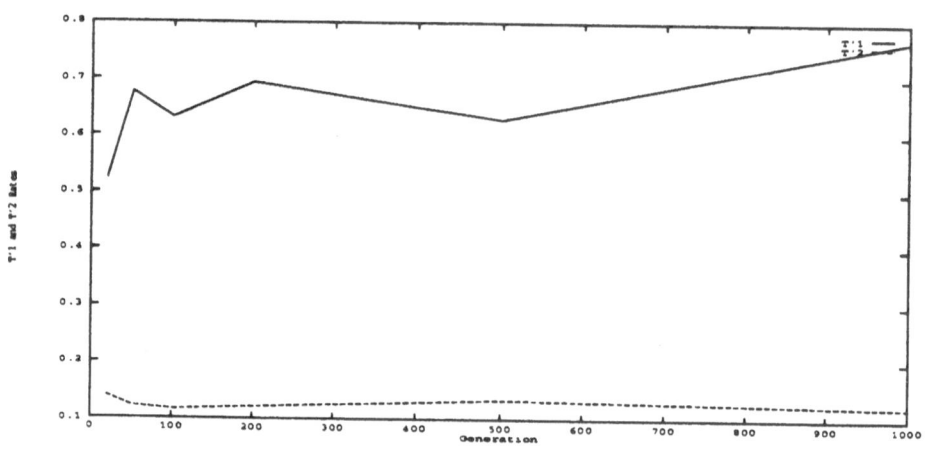

Figure 4. *Average* T'1 *and* T'2 *for 10 Runs*
(Pc=0.6, Pm=0.0083, L=100, g=1000 and a=2)

8. Remaining Problems

We must mention some remaining problems, which arise due to the use of predifined attack scenarios or to the use of our simplified vision of SATAP.

8.1 Problems due to the Predefined Attack Scenarios Approach

In practice, the design of consistent attack scenarios is difficult. For security reasons, information on the subject is often kept secret. Nevertheless, note that our approach allows an easy addition of any scenario, according to the security officer's knowledge.

The translation of intrusive behavior into audit events' sequences or sets requires a very good knowledge of the system kernel. Sometimes, this knowledge is not easy to gain.

8.2 Problems due to the Simplified Vision of SSATA

The translation of the audit trail into an observed audit vector implies the loss of the time sequence. It is a major problem that we must solve. For that we plan to formalize SATAP as a graph contractability problem. No work has been done in that area for the moment.

By using binary coding for the individuals, we cannot detect the multiple realization of a particular attack. As a consequence, we should try non-binary GAs. In that case, the GA execution time should grow because of the search space extension.

GASSATA finds the H vector which maximizes the W.H product, subject to $(AE.H)_i \leq O_i$ $(1 \leq i \leq N_a)$. If the audit session is too long, this constraint is always enforced and GASSATA converges on the N_a-dimensional unit vector. To avoid this problem, the duration of the audit session should be choosen carefully. In our case, we work with 30 minutes audit session.

If the same event or group of events occurs in several attacks, an intruder realizing these attacks simultaneously does not duplicate this event or group of events. In that case, GASSATA fails to find the optimal H vector. We have no solution to that problem for the moment. This means that we only consider independent attacks.

To end this section, it should also be stated that GASSATA does not locate attacks in the audit trail. Just like statistical tools, it only gives a presumptive set of attacks and the audit trail must be investigated latterly by the security officer to precisely locate the attacks.

9. Conclusion

The experiments (section 7) show good results for GASSATA which validates the genetic approach for security audit trail analysis:
- if D=0.5, then T1=1 and T2=0 in every cases: the high rate of detection and the low rate of false alarm requirements are satisfied
- the minimal value for T'1 is about 0.6 and the maximum value for T'2 is about 0.15 with the default configuration: it shows a quite good convergence of the GA which explains the previous results.

If we let the GA run for a given number of generations, the execution time is constant for any audit vector (note that this is not the case for the audit-trail-into-audit-vector translation time). We observe that the GA always converges if we let it run for 100 and if a<10: in this case the execution time is 9 seconds with an IBM RS6000 11.7 Mflops. For higher a values, we must increase the number of generations.

Nevertheless, some problems remain (section 8) which motivate future studies especially for including timing aspects in GASSATA.

Let us finish by noting that we see GASSATA as nothing more than an additional tool in the set of security officers intrusion detection tools.

340

Acknowledgments

The author would like to thank B.Picinbono and R.Baduel (SUPÉLEC) for allowing him to pursue this work, P.Rolin (Télécom Bretagne) for supervising it, V.Alanou (SUPÉLEC) for her help during the experimental phase and V.Cazein and G.Pouëssel (SUPÉLEC) for writing a C results exploitation tool. Thanks to Martin Coy for correcting my english language.

References

1. Salz R. Computer Oracle and Password System (COPS). Internet Newsgroup Article (comp.sources.unix), 1990
2. Denning D.E. An Intrusion-Detection Model. IEEE transaction on Software Engineering 1987; Vol.13, N°2
3. Debar H, Becker M, Siboni D. A Neural Network Component for an Intrusion Detection System. In: Proceedings of the IEEE Symposium of Research in Computer Security and Privacy, 1992
4. Lunt T.F, Tamaru A, Gilham F, Jagannathan R, Jalali C, Javitz H.S, Valdes A, Neumann P.G. A real-Time Intrusion-Detection Expert System. SRI International, Technical Report, 1990
5. Lunt T.F, Tamaru A, Gilham F, Jagannathan R, Neumann P.G, Jalali C. IDES: A Progress Report. In: Computer Security Application, Proceedings, 1990
6. Garvey T.D, Lunt T.F. Model-based Intrusion Detection. In: Proceedings of the 14th National Computer Security Conference, 1991
7. Aho A.V: Algorithms for Finding Patterns in Strings. In: Handbook of Theoretical Computer Science. J. Van Leeuwen, 1990
8. Goldberg D.E. Genetic Algorithms in search, Optimization and Machine Learning. Addison Wesley, 1989
9. Holland J. Adaptation in Natural and Artificial Systems. University of Michigan Press, Ann Arbor, 1975
10. Lawrence Davis and al. Handbook of Genetic Algorithms. Lawrence Davis, 1991
11. Davidor Y. An Intuitive Introduction to Genetic Algorithms as Adaptative Optimizing Procedures. Weizmann Institute of Science, Technical Report CS90-07, 1990
12. Mé L. Algorithmes génétiques. SUPÉLEC, Rapport interne 93-001, 1993
13. Peng Y, Reggia J.E. A Probabilistic Causal Model for Diagnostic Problem Solving - Part 1: Integrating Symbolic Causal Inference with Numeric Probabilistic Inference. In: IEEE transaction on Systems, man and cybernetics, 1987; Vol. 17, N°2
14. Garey M.R, Johnson D.S. Computers and Intractability: A Guide to the Theory of NP-Completeness. W.H. Freeman, 1979
15. Liepings G.E, Hilliard M.R, Richardson J, Palmer M. Genetic algorithms applications to set covering and traveling salesman problems. In: Operations Research and Artificial Intelligence: The Integration of Problem Solving Strategie, 1990
16. Liepins G.E, Potter W.D. A Genetic Algorithm Approach to Multiple-Fault Diagnosis. In: Handbook of genetic algorithms. Lawrence Davis, 1991
17. IBM. Elements of AIX Security. IBM Technical documentation GG24-3622-01, 1991
18. Mé L. Audit de sécurité. SUPÉLEC, Rapport interne 92-002, 1992
19. Alanou V, Mé L. Une Expérience d'audit de sécurité sous AIX R3.1. In: TRIBUNIX, 1992; Vol. 8, N°43
20. U.S. Dep. of Defense. Trusted Computer System Evaluation Criteria. DOD 5200.28-STD, 1985
21. Alanou V, Mé L. Audit de sécurité : Deuxième partie. SUPÉLEC, Rapport interne 93-002, 1993

MRSA - a new public key encryption method

Reinhard Posch

Institute for Applied Information Processin Graz University of Technology
Klosterwiesgasse 32/I A-8010 Graz, Austria
email: rposch@iaik.tu-graz.ac.at

Abstract

This paper focuses on the key generation problem for a modified RSA public key cryptographic system based on the RNS arithmetic. The RNS based modification of the well known RSA algorithm uses highly parallel computation with the restriction that only a subset of key triples $(D, ekey, dkey)$ of a conventional RSA system can be adopted. These restrictions result from the choice of base elements used. The present work shows that the remaining set of possible keys is still large enough to be used in a realistic cryptographic system. The encryption machine under discussion can use parallelism and thus high speed. A rather straight forward algorithm for the generation of keys can be given. The resulting key space can be viewed as satisfactory. The method gives an additional degree of freedom for the implementation on parallel systems avoiding all conversions between number systems.

1 Introduction

Parallel computation of the RSA algorithm is well known to be complicated [1]. This results from the fact that the central operation involved (modulo arithmetic) demands for intensive computation with extremely high communication. From the theoretical point of view some algorithms would outline a solution [2][3]. All these methods are too intensive in communication and not feasible as a parallel chip design since the chip area is limited. Available results therefore usually get stuck around the 50Kbit/sec. performance range with a 512 bit key length, not considering the Chinese remainder theorem [4]. Using top technologies, like silicon on insulator in sub micron, designs can improve these results only by a factor of two. However, as can be seen in presented designs, parallelism is not extensively exploited [5].

The dilemma is twofold. First, highly parallel designs on a single chip are impossible due to limited chip sizes. Second, cutting into many chips cannot be managed due to extensive communication. In previous projects it has been shown that both deficiencies can be fought somehow. Scalable parallelism is one special answer that enables the designer to use up available chip area and to gain encryption rates in the area of 200 to 300 bit/sec. with 1μ CMOS technology [6]. RNS arithmetic could be an answer to the distribution of the processing power over many chips [7]. The second method also reliefs the heat dissipation problem which would also arise on highly parallel single chip designs for RSA. The technique that bases mainly on RNS base extension can be used to construct a public key cryptographic system. This method is equivalent to RSA in terms of security. However it restricts on the key space that can be used.

Restriction on the key space is due to the RNS arithmetic and the restrictions that have to be made on the RNS base elements to ensure efficient computation of the algorithm. These restrictions are discussed in the presented paper and an algorithm to construct RNS base elements and key pairs is presented. The paper also contains a sketch of the algorithm itself and of the underlying register oriented arithmetic.

2 The encryption machine

For the RSA - like algorithm, $cipher = plain^{key}$ MOD D has to be computed [9]. This can be done using Knuth's square and multiply algorithm [10]. This fact turns the modulo exponentiation to break down to modulo multiplication MMUL. Basically this would involve multiply and divide operations. For reasons of performance and uniformity of operations which is of prime importance in an VLSI design it is most desirable to avoid the divide operation. This can be done by the FASTMM fast modulo multiplication as described in [3]. As described in more detail in [7] RNS can load a set of processing elements more equally and thus call for less communication on the single processing element when combined with Mongtomery's reduction [2]. The basic idea is to substitute the radix by a product $N = \prod_{i=1}^{k} p_i$ of RNS base elements. This reduces the modulo multiplication to two base extensions as shown in figure 1.

The method which is described in full detail in [14] leaves an unprocessed factor with each modulo multiplication step. As the number of multiplications is constant for a given key this factor in total is also constant and compensation can be achieved by a single multiplication. This multiplication is done with the scheme shown in figure 1. Again a factor would be introduced. Taking this last fact into account a correcting factor can be precomputed which if multiplied after the encryption process compensates for all errors.

Base extension itself is nontrivial and good methods in log time are known only if control base elements are available. This is definitely not the case for the Z MOD N step of figure 1. In [14] an approximation is presented. This approximation is exact if the initial value and thus the plain text offsets at least by some Δ from zero. This in turn also means that restricting relations between D and the M and N values should hold:

(a) $\qquad D(1+\varepsilon_D) < N < D + \dfrac{D}{3}$

(b) $\qquad 4D < M < 4,44D.$

(c) $\qquad M \cong 4N; \quad 4N < M$

Restriction (a) results from the fact that modulo reduction is achieved with results out of $[0,D(1+\varepsilon_D)]$. Here ε_D can be kept small but depends on the number of bits used in the arithmetic. Typically $\varepsilon_D < 1/1000$ is easy to get. Thus, the assumption $D < N$ can be used for further discussion. Restriction (b) and (c) result from the interval decision process during the reduction described in [14].

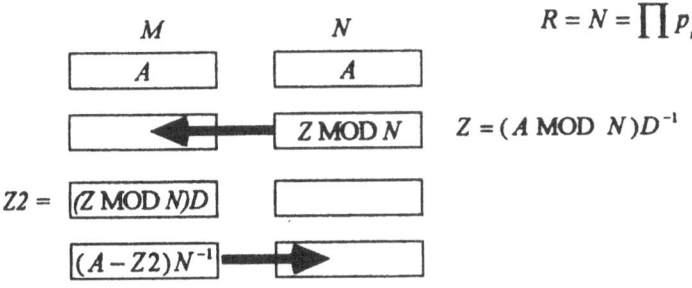

Figure 1: Base extension as substitute for modulo multiplication.

The roughly outlined method can be processed on the register level of the RNS base in all steps.

Building an encryption machine needs connecting these register oriented elements. Optimum parallelism would be reached if a divide and conquer method would be involved when computing the convolution sums. Convolution sums are basically needed in the base extension process used with this work [7]. This however would demand for many processing elements. Therefore a bus connected processor set is assumed to be more practical still giving good results. The result is a technically feasible parallelism. Convolution sums in this context have to be computed with modulo arithmetic within the registers.

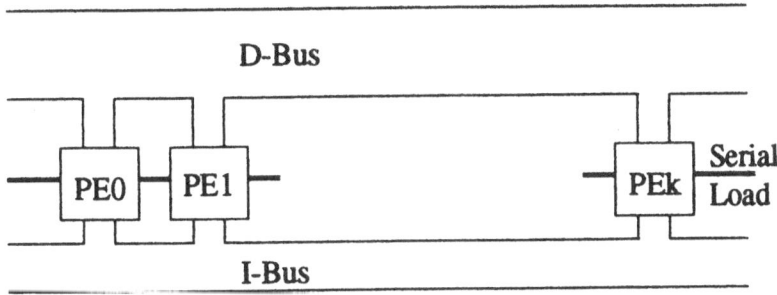

Figure 2: A possible structure for a parallel ciphering machine.

A practical feasible configuration is outlined in figure 2. Serial connection and communication is assumed for distribution of constants and plain text feeding as well as retrieval of the cipher text. The main feature of such a configuration is the possible distribution of processing over a set of chips. The key to this nontrivial issue is low communication as compared with other methods.

3 The modified algorithm with efficient register arithmetic

This paragraph focuses on the $a \cdot b + c$ MOD p_i that has to be computed in each processing element along with the convolution. It is included to point at the necessity to restrict on the set of possible keys. The basic idea is to use only those p_i elements that can easily be managed. Besides the $p_i = 2^r$ where modulo multiplication is just a cut on the binary digits, a p_i that can be written as $p_i = 2^r - \mu_i$, with μ_i having a low 1-bit count, is convenient for processing as well.

Starting with the Wallace scheme the result is a redundant represented number and can be processed as shown in figure 3 using the μ_i assumption in a first phase cutting the $2r$ bits down to $r + log_2(\mu_i) + 2$.

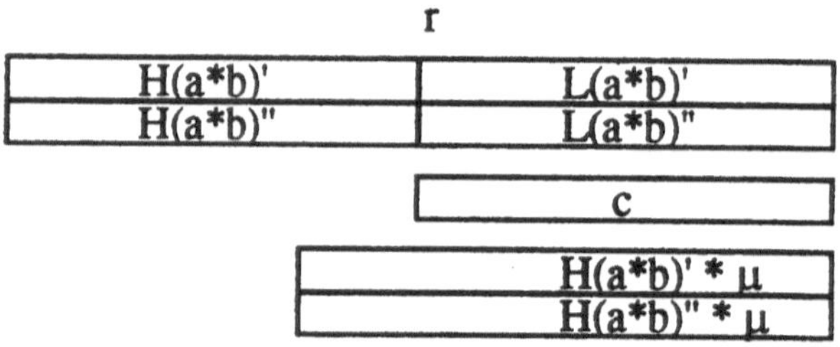

Figure 3: The first phase of $a_i b_i + c_i$ MOD p_i.

A second phase is involved giving the unique result out of the interval $[0, p_i)$. Figure 4 shows the basic idea. At this point it has to be said that a relaxed residuum at the register level will not work as both addition and multiplication will be performed on the results. The idea bases on the same idea as in phase one and uses the fact that μ has only a few bits equal to 1. In addition an evaluation of the two 2^{r-1} bits of the redundant number representation becomes necessary. Taking the constraints on the p_i elements this results in an X out of the interval $[0, 2p_i)$. At this point carry evaluation has to be performed. Since $p_i + \mu_i = 2^r$, a simple decision gives the result X or $X + \mu_i - 2^r$ as the final result at register level.

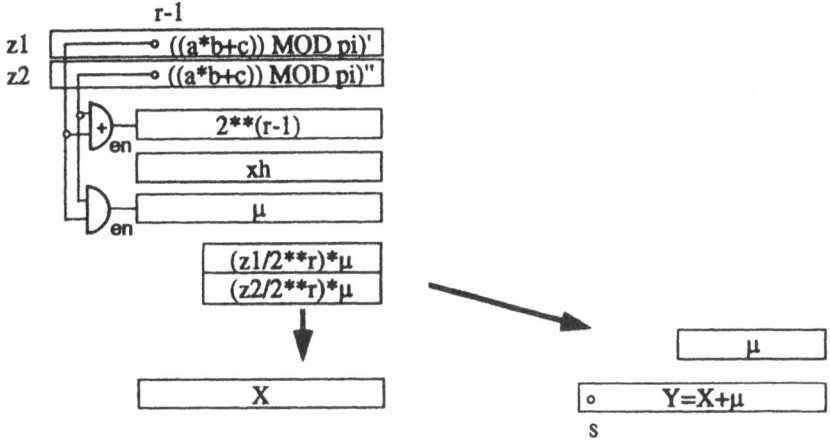

Figure 4: The second phase of $a_i b_i + c_i$ MOD p_i

4 The key generation algorithm

This section focuses on key generation within realistic ranges. As seen from the constraints (a),(b),(c) of figure 1 put mainly on the values N,M and D it is not possible to fulfill them for the arbitrary case. The presented assumption and the proposed algorithm allow only a restricted subset. Only D is key sensible N and M are only relevant for the proocessing phase. However, for realistic relations between the register length n of the individual RNS registers, and the total length of the key $log_2(D)$, keys do exist in a satisfactory variety.

For a given N and M these constraints can be rewritten as :

(d) $\qquad \frac{4N}{4,4} < D < N\left(\frac{1}{1+\varepsilon_D}\right)$

In the special situation, $log_2(D)$ is assumed to range from 512 to 1024 and n is assumed to be equal to 42. This value results from the fact that at the register level a Wallace tree [11] is implemented. Optimal register lengths for the Wallace tree would be 28, 42, 63 etc.

For further considerations detailed calculations are given for $log_2(D)$=672 and r=42. In this case 32 suitable relative primes p_i, q_i are needed. As sketched above a number p is quoted to be suitable if it is close to 2^r and difference 2^r-p has only few ones. Practical tests show that many more possible values for the p_i, q_i would exist. But the restrictions for the register oriented arithmetic show that all these values are very close to 2^r. This means for practical estimations that the existence of many such values does not help to enlarge the interval to choose D of. The previous statements are not true for p_1. This value is chosen $p_1 \cong 2^{r-2}$. This results in the estimation that $4N \cong M$. The relations (e) and (f) show the restrictions on base elements.

(e) $N = \prod\limits_{i=2}^{k} p_i \; : \; p_i = 2^n - \mu_i, \mu_i = \sum\limits_{j=1}^{15} m^i_j * 2^j, \sum\limits_{l=1}^{15} m^i_j \leq 4 : i = 2, k$.

(f) $M = \prod\limits_{i=1}^{k} q_i \; : \; q_i = 2^n - \lambda_i, \lambda_i = \sum\limits_{j=1}^{15} t^i_j * 2^j, \sum\limits_{j=1}^{15} t^i_j \leq 4 : i = 1, k$.

As already pointed out all p and q values are extremely close to 2^r. This means that the following estimations $M = 2^{kr} - \Delta_M$, $\Delta_M << 2^{kr}$, and $N = \dfrac{2^{kr}}{4} - \Delta_N$, with $\Delta_N << 2^{kr}$ can be used.

With RSA D is a product of two large primes $D = prime1 * prime2$. The restriction (e) on D allows the following procedure to select keys:

(1) $\qquad D_{\text{min}} = \dfrac{4N}{4,4}$

(2) $\qquad D_{\text{max}} = \dfrac{N}{(1+\varepsilon_D)}$

(3) $\qquad$ Select $prime1 < \sqrt{N - \dfrac{N}{4}}$

(4) $\qquad$ Select $prime2$ such that D is within the interval $\left[D_{\text{min}}, D_{\text{max}} \right]$.

A short discussion on the availability of MRSA keys as compared to keys in conventional RSA is added. The restrictions on the selection of D are seen in (a),(b), and (c), giving restriction (d). This tells the key finder that for a given number of bits in N only a limited area of the representable values are available for D. From (d) it can be seen that this area covers a little less than 10% of the space. Half of this space can be represented with one bit less so that effectively 18% can be used for keys. This is also shown in figure 5. The number of bits in N can be easily decreased by one by cutting one of the RNS registers. However it should be stated that in the case of efficient register oriented modulo reduction this does mean the use of additional hardware or the predefinition of the key space. As this restriction allows only for twice the amount of the available keys, and as there are sufficiently many keys available, this is not assumed for practical applications.

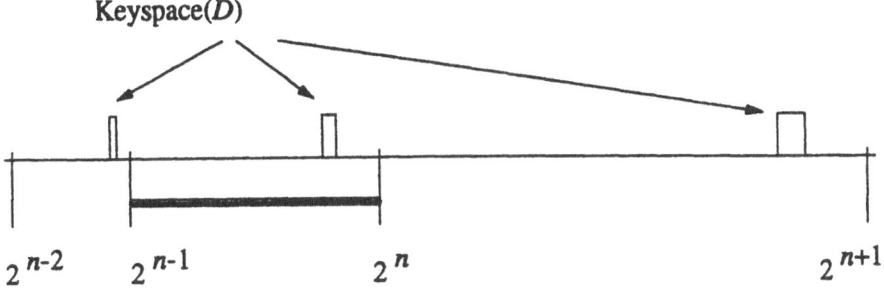

Figure 5: Key space of MRSA in comparison to RSA.

It must also be mentioned that the size of D is adopting to N and not to the block size of the plain text. As this method assumes the plain text block to hold k bits less than N the relative key space as compared to the size of the plain text blocks exceeds by far conventional RSA. Finally it should be stressed that there is no idea that this structure of restricting on the key space could influence the cryptographic quality of the system to any extent. As all the constraints on D just limit the interval whereof D may be chosen this fact does not influence key quality at all as demanded by the RSA encryption system [13].

Due to the nature of the MRSA algorithm *dkey* and *ekey* need not to observe constraints and can therefore be chosen the same way as usual RSA keys. This means that one of the two is a randomly chosen number and the second is computed by application of Berlecamp algorithm [12] to find the multiplicative inverse with respect to $\Phi(D)$.

Reference List

1. Cooper R.H., Patterson W.: *RSA as a Benchmark for Multiprocessor Machines*; Proc. "Advances in Cryptology - AUSCRYPT '90" LNCS, Springer 1990, pp. 356-359

2. Montgomery L.: *Modular Multiplication Without Trial Division*; Mathematics of Computation, Vol. 44, No. 170, April 1985, pp. 519-521.

3. Posch K.C., Posch R.: *Approaching encryption at ISDN speed using partial parallel modulus multiplication*; Microprocessing and Microprogramming, North-Holland, 29, (1990), pp. 177-184.

4. Beeth Th. et al.; *Public key Cryptography, State of the Art and Future Directions*; LNCS 578, Springer 1992

5. Ivey P.A., Walker S.N., Stern J.M., Davidson S.: *An Ultra-High Speed Public Key Encryption Processor*, IEEE Custom Integrated Circuits Conference, 1992, pp. 19.6.1-19.6.4

6. Lippitsch P., Posch K.C., Posch R., Schindler V.: *A scalable RSA design with encryption rates from 200 Kbit/sec to 1,5 Mbit/sec*; Poster at CRYPTO '92

7. Posch R.: *A Parallel Approach to Long Integer Register Oriented Arithmetic*; Fifth International Conference on Parallel and Distributed Computing and Systems; Oct. 1.-3. 1992, Pittsburgh, PA.

8. Schoenfeld L.: *Sharper bounds for the Chebychev functions $\Theta(x)$ ans $\Psi(x)$*; II, Math. Comp. 30, (1976), pp 337-360.

9. Rivest R., Shamir A., Adlemann L.: *A Method for Obtaining Digital Signatures and Public-Key Cryptosystems*; Comm. of the ACM (Feb. 1978),pp. 120-126.

10. Knuth D.E.: *The Art of Computer Programming*; Vol 2, Addison Wesley, Reading, Mass., 1969

11. Wallace C.S.: *A suggestion for a fast multiplier*, IEEE Transaction on Electronic Computers, Vol. EC-13, Feb. 1964, pp. 14-17.

12. Lüneburg, H.: *Vorlesungen über Zahlentheorie*; Elemente der Mathematik vom höheren Standpunkt aus, Band VII, ed. by E. Trost, BirkhäuserVerlag, (basel, 1978).

13. Denning D.E.: *Cryptography and data security*; Addison Wesley, Reading, Mass., 1983

14. Posch K.C, Posch R.: *Residue number systems a key to parallelism in public key cryptography*; Fourth IEEE Symposium on Parallel and Distributed Processing; Dec. 1.-4. 1992, Dallas.

A Virus-Resistant Network Interface

Martin Witte and Wolfgang A. Halang
FernUniversität, Department of Electrical Engineering
D-58084 Hagen, Germany

Abstract

The concept of an interface for electronic mail between a LAN (Local Area Network) within an organisation and an external network is presented. The interface's design renders any external intrusion impossible. The concept is based on a straightforward hardware solution and has already been validated by a prototype implementation.

1   Introduction

The use of computer networks in all fields of business, government, and science is rapidly growing. Information interchange is often a decisive factor for competition. Nevertheless, owing to numerous cases of intrusion via the communication networks, organisations are becoming more and more concerned about security and reluctant to extend their utilisation of external network communication; some are even discontinuing network communication.

Computer "viruses" represent an enormous security risk. They endanger the functionality and integrity of computer systems and the permanent integrity of data by using more and more elaborate ways of manipulation. Apart from intruding a system via exchangeable media such as floppy disks, the most frequent way of entering a computer system is from a computer network. To show how computer "viruses" gain control over a computer system, we first give an overview over the different kinds of "viruses" and their ways of intrusion.

Computer "viruses" are non-self-acting parts of computer programs (host programs) that are copied into memory by starting a host program. There are different kinds of computer viruses, e.g., "*Trojan horses*", "*time bombs*", "*worms*", "boot sector viruses", and "*hybrid viruses*". The variety of kinds of computer "viruses" is expected to grow further in the future. Generally speaking, a computer "virus" consists of two parts:

- Program routines that have to look for further programs on the host system to which the "virus" can copy itself. This is used to increase the number of the "virus" program's copies in a system

• Program routines that have to execute one or more manipulating functions. Most of the manipulation is destruction or change of data which may cause great economical and financial harm.

There are two different ways of "infecting" a program. One way is to overwrite the host program code that is stored on a writable medium. Then, at least parts of the original program are destroyed and the "virus" may easily be detected as the application is unable to run. The other way of infection is that the "virus" adds itself to the body of a program and adjusts the program's entry point and address tables.

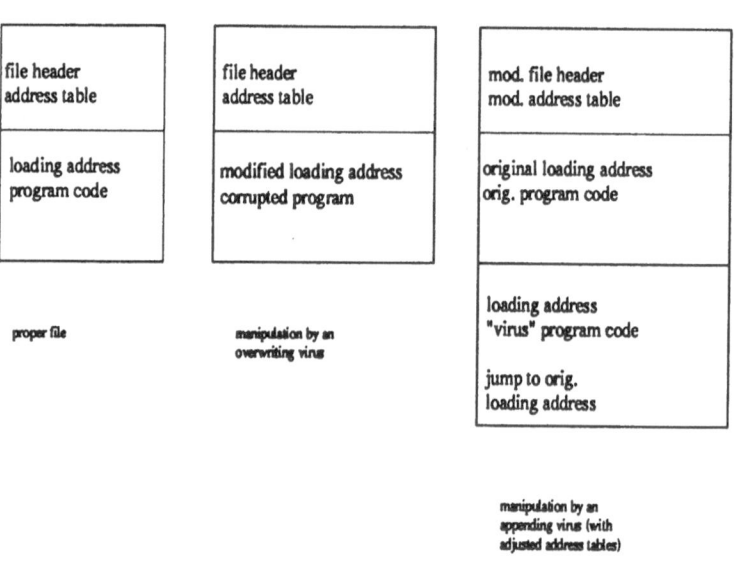

Figure 1: Structure of a proper and of corrupted program files

The "virus" code is loaded into memory together with the host program where it may stay resident and may "contaminate" further programs. A "contamination" may be detected by an extension of the host program's length in the directory. Many viruses keep cross-links to the host program and execute parts of the host program while simultaneously carrying through their manipulations in a way not recognisable by the user.

In the next section we begin with an overview about computer "viruses" outlining the ways of "infecting" computer systems and the different classes of "viruses" which can be distinguished. Next, the existing concepts to protect computer systems against manipulation by "viruses" are discussed. Most of them are software-based, but there are also a few hardware-orientated concepts. After a comparison of the von-Neumann computer architecture with the Harvard architecture, a hardware-based "virus-resistant" network interface is introduced and details of the prototype implementation are presented.

2   Kinds of Computer "Viruses"

Computer "*worms*" can often be found in multi-user systems and computer networks, as they keep busy with copying themselves all over the system, which leads to performance reduction.

"Virus" programs that catch the user's interest and that start their destruction immediately after being executed are called "*Trojan horses*".

"*Time bombs*" or "*logical bombs*" are very similar to "Trojan horses", but they start their manipulation when they detect a particular logical or temporal status. This can be a special time or date. The "*Michelangelo Virus*" is one of the "time bombs" as it starts its destruction at Michelangelo's birthday every year by reading the date from the system clock [2].

"*Boot sector viruses*" are placed in the boot sector of a system disk in order to gain control over a computer system already before the proper operating system is loaded into memory. They can hardly be detected as they can keep control over all protective devices and mechanisms that are offered by the operating system and can bypass protection measures.

The group of "*hybrid viruses*" is the latest variety of "viruses". They are resident in memory (like "boot sector viruses") and, moreover, they exist in files of computer systems. If only the "infected" files are removed from the system, the memory resident part of a "hybrid virus" is still active and may start infecting files anew.

3   Existing Concepts of "Virus" Prevention

3.1   Software-Based Solutions

As already mentioned, most of the protection measures are software-based. They can be classified into:

- Memory-resident "watchers", i.e., programs that guide (operating) system functions and interrupts. They try to secure a system of being corrupted.

- Secondary programs and utilities that have to be executed regularly to scan a system and its storage devices (e.g., floppy disks, hard disks etc.) and search for "infected" data, especially "infected" executable files.

The purpose of memory resident utility programs is to control a system, especially its input and output (I/O) functions, as computer "viruses" have to read and write back the manipulated data. These utilities have to give a message if there is a possible misuse of I/O. The problem is that the permanent function control leads to a performance reduction or that the controlling program cannot decide on its own whether an I/O request is intended or caused by a "virus" manipulation.

The second group of software solutions are characterised by scanning systems for manipulated data and are, therefore, called "virus scanners". Corrupted programs can be detected by sequences of characteristic bytes or string patterns that are supplied with each individual computer "virus". A scanner's quality depends on the number of known viruses that it can search for, and this manifests its disadvantage: as the number of newly developed "viruses" is growing, the scanner has always to be updated. This makes it clear that a scanner can only react to known ways of manipulation and cannot be regarded as a preventive measure.

3.2 Hardware-Based Solutions

Some companies supply additional hardware devices in order to make computer systems secure. This starts with simple mechanical barriers to lock floppy disk drives and to prevent new probably corrupted software from entering a system via exchangeable floppy disks. Furthermore, there are extra boards and smart cards equipped with cryptographic devices. They are used to encode and decode data with user-dependent (e.g., the user identification in computer networks, uid) keys to ensure that only authorised users can get access to security sensitive data. These boards are individually designed and manufactured. Their use increases the economic and administrative costs and efforts for computer support and maintenance.

4 Computer Architectures

4.1 The von Neumann Architecture

The classical von Neumann computer architecture makes it really easy for computer "viruses" to "infect" other programs and to take control over a computer system after a manipulated host program is loaded into memory, since program code and data are stored in a common random access memory (RAM), and since any word in memory can be fetched and executed as an instruction. Abstractly speaking, this architecture consists of the central processing unit (CPU) with only a few internal memory cells (registers), external memory cells in form of RAM, and external, peripheral devices. RAM is used to store program code that is successively loaded in the CPU's instruction register. The communication between the CPU and its RAM is carried out via buses. The von Neumann architecture comprises an address bus and a data bus. There is only a separation between address and data. Whether a binary word in memory is an instruction, a constant, or a data word can at runtime only be perceived from the context. In addition to this, modern operating systems do not offer any support to supervise the instruction fetch cycle of the CPU.

Moreover, also the external storage media, from which programs are loaded into main memory, can generally be rewritten. This means that a proper program once stored into memory cannot prevent it from becoming corrupted.

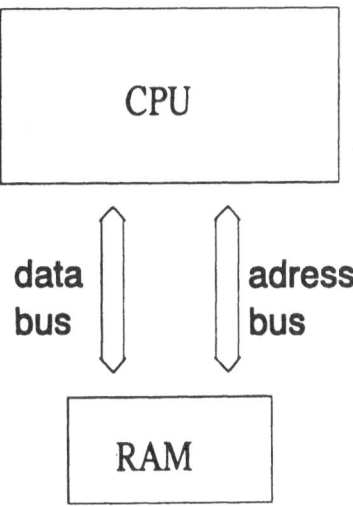

Figure 2: The von Neumann architecture

4.2 The Harvard Architecture

The analysis of the reasons enabling the "virus" problem suggests a straightforward and fully effective solution by hardware, such as the Harvard computer architecture physically separating program and data memories. The above von Neumann architecture is extended by two further buses, viz., one each for instruction addresses and instruction data. This ensures that data and instructions are accessed via two different non-multiplexed buses [3]. The instruction bus (address and data bus) only offers physical access to the instruction register of the CPU, while the data bus exclusively serves the data and operand registers.

Originally, this design was developed to increase performance, as for the independent buses the instruction cycle and the data fetch cycle can be executed simultaneously. With regard to a security sensitive computer application, this architecture represents the better design, because it can protect program code from corruption.

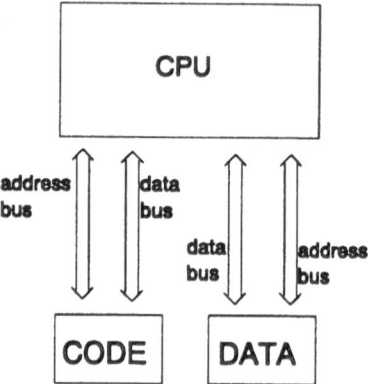

Figure 3: The Harvard architecture

5   Concepts of Protection

In order to prevent a "virus" from entering a system in the first place, and to ensure that (program-) files do not become manipulated and changed afterwards, it is sufficient to implement program memories exclusively as read only memories (ROMs). The possibility of any kind of manipulation and modification of binary program code is thus systematically disabled. Naturally, it has to be made sure that any program code which is placed into ROM is free of any destructive machine instructions.

The prototype for the network interface, which we have built, is structured according to the above reasoning. It is based on the Z80 CPU. Though its architecture corresponds to the von Neumann architecture it is suited for the prototype implementation with small additional feature added. The Z80 CPU works with eight bits for data and a sixteen bits wide address bus. The latter is physically realised by the CPU's sixteen address pins called A0 to A15.

The software required for the purpose of a network interface, i.e., operating system kernel, user interface, editor, and the KERMIT protocol, is provided in a ROM module occupying the lower 32 kB of the CPU's 64 kB physical address space. The upper 32 kB are used for data at runtime. To emulate the Harvard architecture with the Z80 we took advantage of a particular signal called $\overline{M1}$ available at a pin of the CPU chip. $\overline{M1}$ indicates the instruction fetch phase within the instruction execution cycle. We used the disjunction of $\overline{M1}$ and the inverted address-line A15 as input at the $\overline{RESET}$-pin of the Z80 CPU, i.e., trying to read an instruction word in the data space of main memory results in the immediate resetting of the CPU. This is an additional feature coping with errors in the application software.

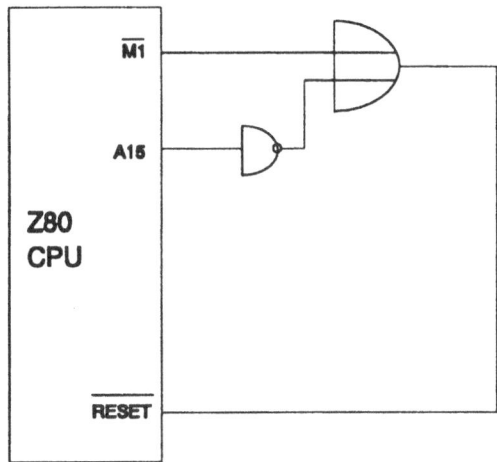

Figure 4: Additional circuitry in the prototype implementation

Usually, private networks are directly and physically connected to public ones and allow for access from outside (remote login, ftp, telnet etc.). This bears the risk that "contaminations" can enter from outside. In contrast to this, our interface is designed to serve as a buffer between two communication networks. Owing to the alternate switches shown in the Figure 5, it can only communicate with one network at a time, making direct links between internal and external networks impossible. Furthermore, any communication is initiated and actively carried through by the interface. Mail data are fetched in file transfer mode from one of the two networks. They are then stored in the interface on disk and may be manipulated with the editor. Not before turning the alternate switch the data can be forwarded to the other network.

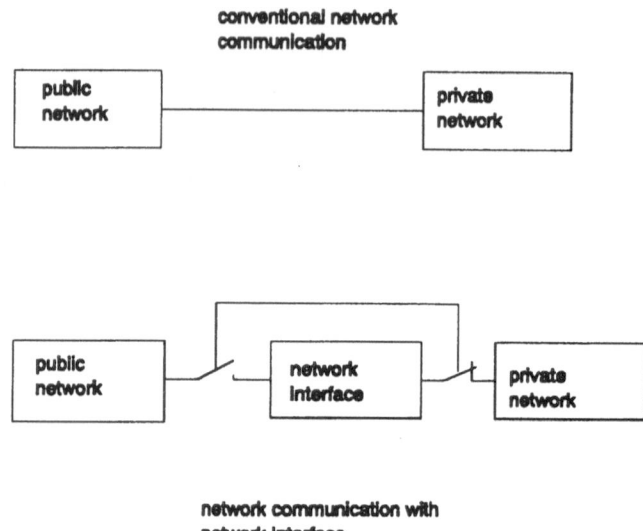

Figure 5: Alternate switch to avoid a direct link in the communication line

6   Conclusion

Although based on the classical von Neumann architecture, with just a few technical modification, the presented prototype implements a fully effective protection against "virus" intrusion. This is achieved by the following features.

- Program code is unchangeably stored in read only memory (ROM)

- An additional circuit is used to reset the CPU in case a program tries to fetch an instruction from RAM.

- A direct link between the local network and external devices is made impossible by means of an alternate switch box

In conclusion, it was constructively shown that the problem of "viruses" entering private networks from public networks can easily be solved by hardware at negligibly low cost in a fully effective way. Thus, the statement "Avoidance of viruses seams to be impossible to attain by technical means since they can be introduced into the system by a properly authorised user" found in [1] has been refuted.

References

1. Gaj K, Górski K, Kossowski R, Sobczyk J. Methods of Protection Against Computer Viruses. In: B. K. Daniels (Ed.): Proc. SAFECOMP '90 Conference, pp. 43 - 48. IFAC Symposia Series, 1990, No. 17, Pergamon Press, Oxford

2. Gleissner W, Grimm R, Herda S, Isselhorst H. Manipulation in Rechnern und Netzen. Addison-Wesley (Germany), 1989

3. Shiva SG. Computer Designs and Architecture. HyperCollins Publishers Inc., New York 991

Session 11

REACTIVE SYSTEMS

Chair: E. Schoitsch
Austrian Research Centre Seibersdorf, A

CIP – Communicating Interacting Processes
A Formal Method for the Development
of Reactive Systems

Hugo Fierz, Hansruedi Müller, Spiros Netos
Institut für Technische Informatik und Kommunikationsnetze, ETH-Zentrum
CH-8092 Zürich, Switzerland, E-mail: fierz@tik.ethz.ch

Abstract

CIP is a formal method for the development of distributed reactive systems. The compositional, real world oriented approach guides the developer from an initial environment modelling step towards the complete definition of the reactive behaviour of a system. The description technique of the method combines graphical and textual notations.

1 Introduction

Although it is well known that the use of formal methods supports the development of robust and reliable systems, there is still a great dislike in applying them in practice. A main reason for the bad acceptance of formal methods is the missing support through constructive development concepts, guiding the user from the informal requirement description to the definition of the system [e.g. 3, 6, 11, 12, 13].

D. Harel and A. Pnueli have characterized reactive systems (process control, embedded and real-time systems) as follows: "A reactive system, in general, does not compute or perform a function, but it is supposed to maintain a certain ongoing relationship with its environment [10]." Many formal methods propose stepwise refinement techniques. Concepts of modularity serve as guidelines for the difficult task of structuring. *Top-down* approaches are suited for the development of transformational systems (functions, algorithms). The structure of a reactive system, however, must reflect the temporal behaviour of its environment, thus an *outside-in* approach is more adequate. A developed reactive structure may then serve as a basis for the specification of functional computation. A similar approach (applicative state transition systems) has been proposed even in 1978 by Backus [1] for purely transformational systems.

CIP bases the development of a system on an explicit model of its environment. The real world modelling approach to system development is well known from the JSD method (Jackson System Development) [4]. CIP differs from JSD mainly in the state oriented view of processes, and by the possibility to specify instantaneous interaction between synchronously cooperating system compenents.

2 The CIP Method
2.1 System Description Concepts
Operational Specification

The operational approach to software engineering [14] has been proposed as an alternative to the conventional approach (Analysis, Design, Coding). An operational specification is a problem oriented system description which is executable by a suitable interpreter.

Instantanteneous Reactivity

A system description which allows the specification of system reactions dependent on the actual state of the environment, must abstract from the non-zero duration of response times. Instantaneous reactivity is therefore the commonly accepted hypothesis for the development of reactive systems [2, 5, 11]. The assumption states that a system responds instantaneously to its inputs. A consumed event and the generated actions compose thus a temporally atomic system reaction.

Sequential Components

We define a system, in contrast to process algebras or petri nets, as static composition of sequential components. Sequential components cooperate synchronously or asynchronously. Synchronous cooperation is usefull for the description of instantaneous system reaction. Distributed environments however require a formalism including asynchronous cooperation.

System description based on statically composed sequential components is suitable for the development of safe critical systems. Due to the time independent state space structure, such systems are much easier to treat than specifications based on more general models of concurrency (non-conservative systems). Furthermore, the constant degree of concurrency simplifies considerably the verification of the instantaneous reactivity assumption for implemented systems.

2.2 Development Steps:
Real World Model, Correspondence and Composition

The general task is always related to an environment composed of real objects: machines, chemical processes, lanes of traffic at an intersection, etc., but the environment may also include previously-developed software components. The developer is asked to construct a specific software interaction between these objects. CIP differs from most other methods in that its development process starts with an environment model and postpones the functional description to a second phase.

In the first phase, the environment objects are modeled in terms of state machines. For every modeled object a corresponding model process is specified. These provisionally incomplete system components define in fact an interaction protocol for the system and its environment. In an implemented system the corresponding synchronisation takes place by means of transmitted events and actions.

The second phase of development involves the introduction of function processes, allowing for instantaneous interaction and asynchronous communication to take place between the established components. Here, the application of classical concepts of modularity strategically supports the composition process. The result is a network of interacting and communicating extended state machines that provide an operational description of the system.

Specifications based on real world models are transparent and easy to understand because all elements are clearly related to the environment. In a running system, the current states of the model processes always correspond to the current states of the associated real objects. This is an important prerequisite for the development of robust and safe systems. The approach also has advantages as regards system maintenance, since a real world model is likely to be more robust than a set of functional requirements.

3 CIP System Description

A CIP specification provides an operational description of how a system reacts to external events. The reactive behaviour is described by extended state machines that can influence each other. Data processing and algorithmic functions are carried out in the transitions of the state machines. Transition structures and data flow networks are graphically specified while functions and conditions are defined through annotations in a functional language.

Systems

A CIP System consists of several concurrent *clusters*. A cluster is a sequential subsystem composed of a set of synchronously cooperating *processes*. A state transition of a cluster represents an instantaneous system transition, defined through the state transitions of its components. The processes of a cluster interact through instantaneous transmission of *pulses* (software events). All clusters may contain processes, which communicate through asynchronous exchange of *messages*.

Processes

A process is an extended state machine which can carry out internal operations through its transitions. We distinguish interaction driven processes (*I-process*) and communication driven processes (*C-process*). The state transitions of an I-process are triggered by external *events* and by *pulses* emitted by other processes. Occurring events must always be accepted while occurring pulses may be ignored. A singular output of an I-process can consist of a *pulse* for other I-processes of the cluster and of an *action* for the system environment. *Messages* for other processes can be transmitted through specified *outports*. A transition of a C-process occurs spontaneously when a message is pending at one of its *inports*. The output of a C-process may also consist of a pulse and of several messages.

The transition graph of a process may contain non-deterministic branchings, i.e. in a given state, several transitions are possible for the same input. In order to resolve this ambiguity, switch conditions must be defined that can be dependent on the states and the variables of the processes within the same cluster. Compared to pulse transmission, state inspection represents a much weaker coupling, because the inspected processes are not affected.

A process' local memory can be extended with variables. In order to support data transmission for each event, action, pulse and message, corresponding record types are declared. Transitions are annotated by sequences of operations, which are ideally described in a functional language.

Moderated Processes

A moderated process consists of several alternative *modes*. The several modes of a process differ in their transition graphs. The state space and the interface, however, are the same for all modes of a given process. The dynamics of mode transitions is defined through a further transition structure called *moderator*, which is built on the modes themselves. A mode transition may trigger a state transition of the new mode.

Instantaneous Interaction

The static pulse flow structure of a cluster is specified trough an interaction net. Through pulse transmission every process which expects external input (events, messages) can cause instantaneous chain reactions. In order to prevent cyclic chains, a *cascade* must be specified for every process with external input. A cascade is a cyclefree partially ordered subnet of the interaction net which defines the possible paths of pulse transmission caused by its top process. The transitions of the processes activated in a cascade define a temporally atomic cluster transition.

Asynchronous Communication

The message flow in a system is defined through networks of processes linked by *datastreams*. A datastream is a FiFo-buffer which stores the messages arriving from the connected process *outports*. When the connected reader process awaits a new message, the oldest message is released to the corresponding *inport*. External datastreams are connected to *devices*.

Denotational Semantics

The meaning of language constructs describing synchronous cooperation has been defined through SCSM-expressions [8]. SCSM is a formalism for the description of synchronous compositions of state machines [7]. The asynchronously cooperating clusters are interpreted as petri net state machines coupled to petri nets which model the behaviour of the datastreams [9].

Extension of the Formalism

The implicitly defined temporal properties of a system can be verified by automated system execution. We are working on an extension of CIP by a logical language for the decription of temporal properties of the controlled environment. The verification of temporal predicates can be based on the common environment model.

4 A Complete Case Study

We specify a simple system which can be described by pure state machines only.

4.1 Requirement Description of the AccessControlSystem

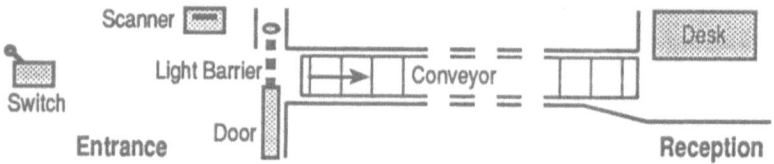

Fig. 1   System Environment

An arriving visitor must insert his badge into the scanner in order to be identified. For an accepted visitor the door is automatically opened and the possibly idle conveyor is started. The opened door closes as soon as the light barrier gets free. The closing door reopens when the light barrier is interrupted again. The scanner is only enabled when the door is fully closed.

Entries are signalized to the reception desk inside the building where entered visitors have to register. If no more visitors are on the way to the reception desk, the conveyor of the entrance is turned off. There are at most three visitors allowed to be on the way for registering.

A switch allows to enable or disable the access control system. If the system is disabled, an already accepted visitor is still allowed to enter. However, when the light barrier is interrupted, the eventually closing door is stopped without reopening. At any time the system can be enabled to work again in its normal mode.

REMARK: We suppose that every accepted visitor really enters the door.

4.2 AccessControlSystem First Phase: Model Processes

The provisionally incomplete model processes behave like their corresponding real world objects of the system environment. They are going to be completed in the second development phase (section 4.3), where asynchronous communication and instantaneous interaction with additional function processes are introduced.

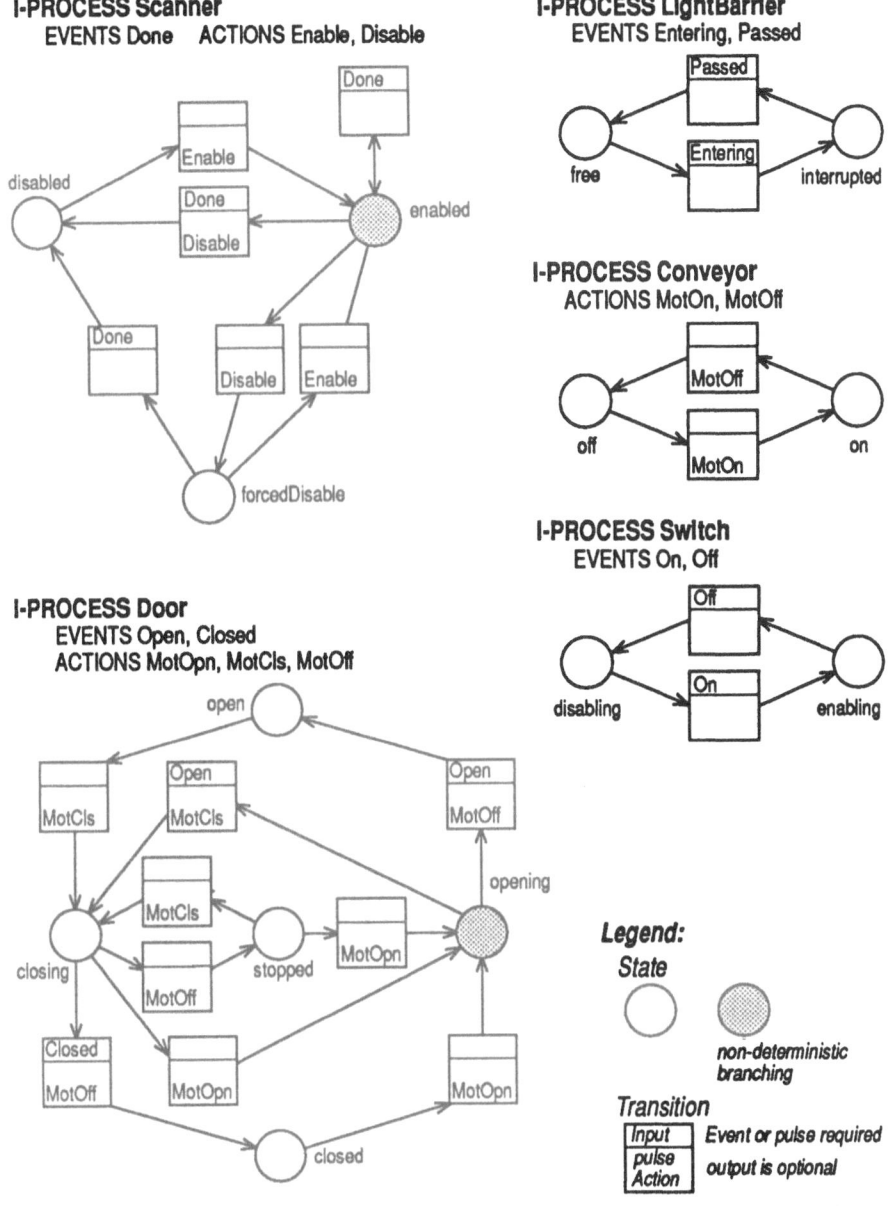

Fig. 2   Incomplete Model Processes

4.3 AccessControlSystem Second Phase: Functions

Comments on the graphical CIP specification below follow in section 4.4.

COMMUNICATION NET OF SYSTEM AccessControlSystem

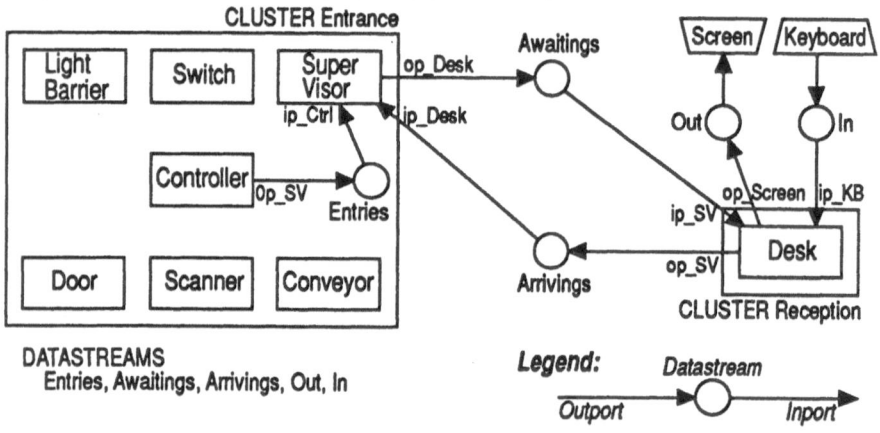

DATASTREAMS
Entries, Awaitings, Arrivings, Out, In

Legend:

INTERACTION NET OF CLUSTER Entrance

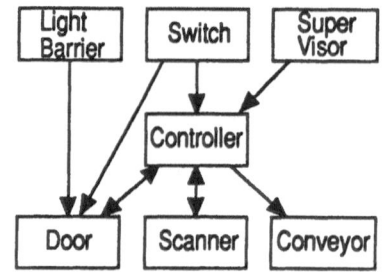

CASCADES OF CLUSTER Entrance

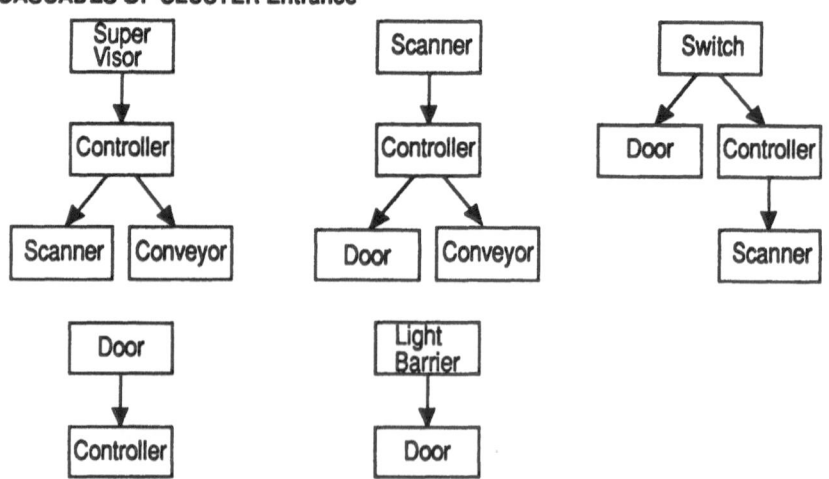

Fig. 3   Interaction and Communication Networks

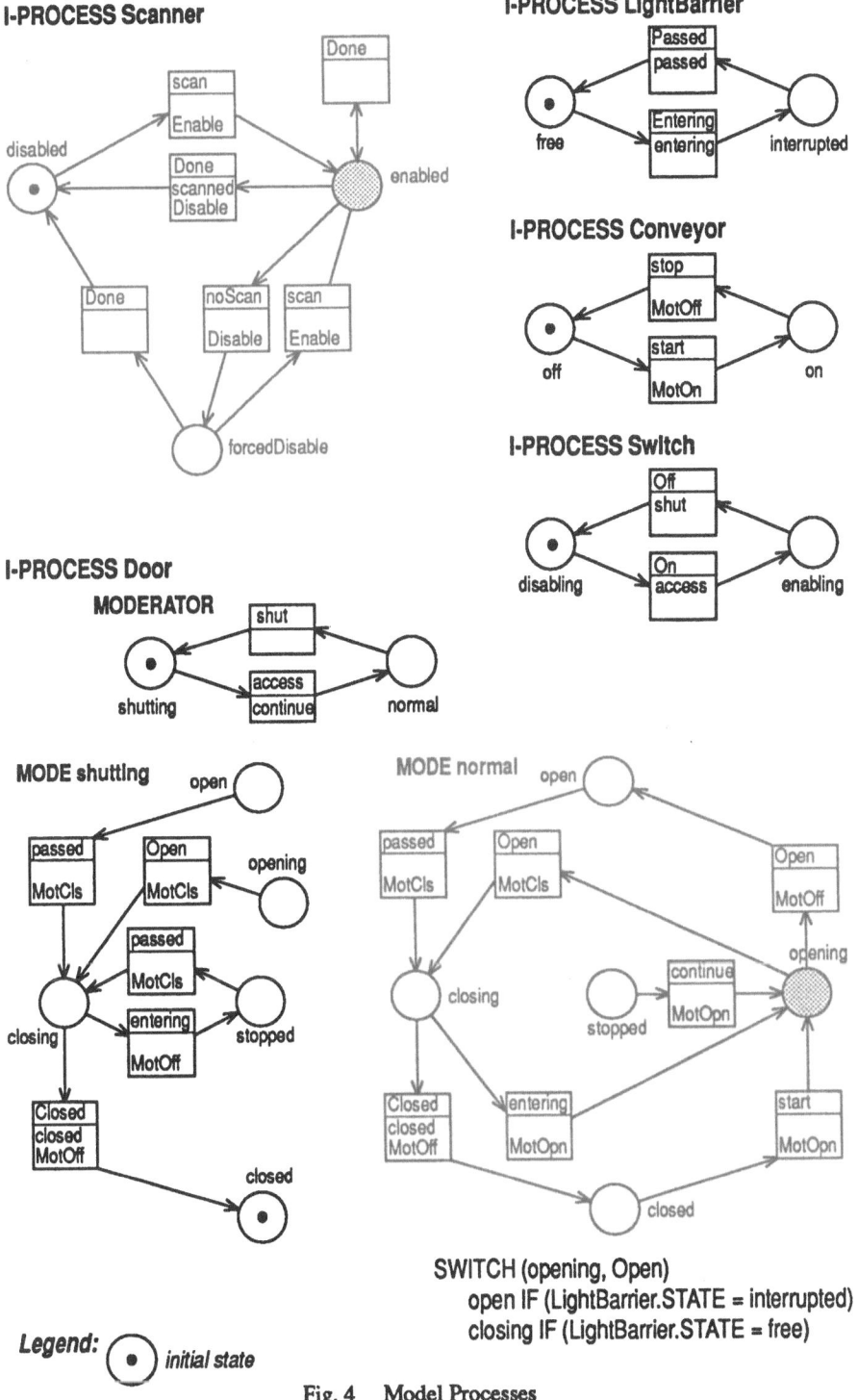

Fig. 4   Model Processes

I-PROCESS Controller

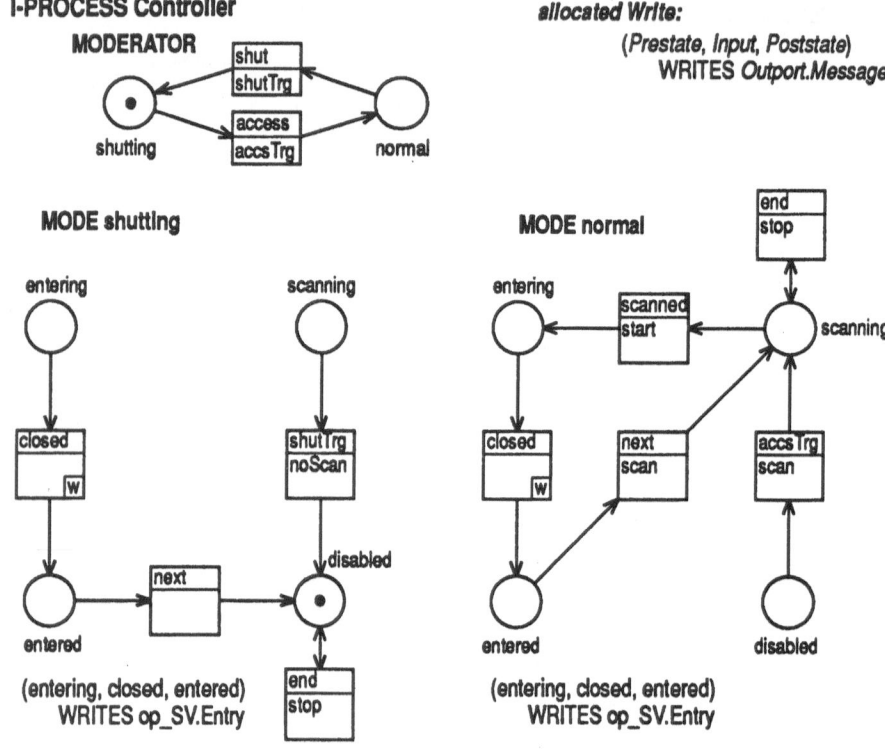

allocated Write:
(Prestate, Input, Poststate)
WRITES Outport.Message

(entering, closed, entered)
WRITES op_SV.Entry

(entering, closed, entered)
WRITES op_SV.Entry

C-PROCESS SuperVisor

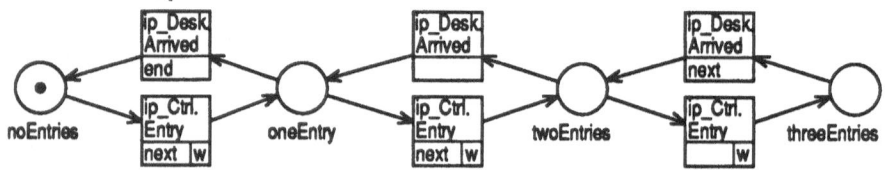

(noEntries, ip_Ctrl.Entry, oneEntry) WRITES op_Desk.Await
(oneEntry, ip_Ctrl.Entry, twoEntries) WRITES op_Desk.Await
(twoEntries, ip_Ctrl.Entry, threeEntries) WRITES op_Desk.Await

C-PROCESS Desk

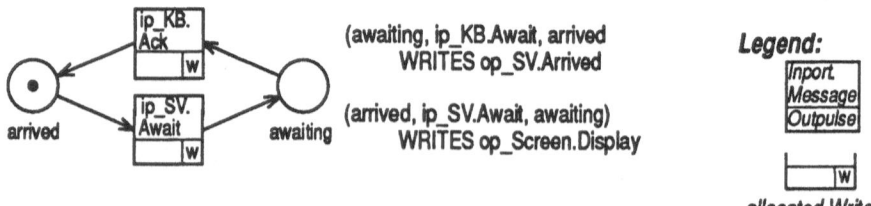

(awaiting, ip_KB.Await, arrived
WRITES op_SV.Arrived

(arrived, ip_SV.Await, awaiting)
WRITES op_Screen.Display

Legend:
Inport.
Message
Outpulse

allocated Writes

Fig. 5   Function Processes

4.4   Comments on the specified AccessControlSystem

COMMUNICATION NET

The system consists of the two concurrent clusters *Entrance* and *Reception*.

The cluster *Entrance* contains all model processes, the interaction driven process *Controller* and the communication driven process *SuperVisor*. The *Controller* informs the *SuperVisor* about entered visitors, while the *SuperVisor* is continously aware of the number of actual entries.

The cluster *Reception* consist of the communication process *Desk* only. *Desk* prompts for information from the reception desk (devices) in order to inform the *SuperVisor* about arrived visitors.

INTERACTION NET OF CLUSTER *Entrance*

The main function of the *Controller* consists of activating and deactivating the *Scanner*, the *Door* and the *Conveyor*. The control depends on the behaviour of the *Scanner* and the *Door* which in turn signalize occuring events to the *Controller* trough correspondingly transmitted pulses. The control of the *Conveyor* depends on the number of the actually entering visitors. The *Controller* obtains the necessary information trough pulses stemming from the *SuperVisor*.

The influence of the *LightBarrier* on the *Door* is realized trough direct interaction between these two model processes. For every occuring event of the *LightBarrier* a corresponding pulse is sent to the *Door*.

Enabling and disabling of the system through the *Switch* process influences the behaviour of the *Controller* and the *Door*.

CASCADES OF CLUSTER *Entrance*

The cascades represent partially ordered and cycle free subnets of the interaction net. Every casacade defines the possible pulse flow caused by an occurring event or a consumed message of its top process.

PROCESSES

The *Controller* and the *Door* are specified as interactively moderated processes. The state transition structures of their *shutting* and *normal* modes describe the corresponding alternative dynamical behaviour. The moderators of these processes are stimulated by the *shut* and *access* pulses of the *Switch* process. Trough emitted triggers (*shutTrg, accsTrg, continue*), a mode transition can activate a state transition of the new mode.

The non-deterministic branching (*opening, Open*) of the *normal* mode of the *Door* process is resolved trough an associated switch, which inspects the current state of the *LightBarrier*.

The non-deterministic branching (*enabled, Done*) of the *Scanner* process is not resolved within this specification. A corresponding specification based on extended state machines would associate a switch depending on the record data transmitted by the scanner event *Done*.

REMARK

For expository reasons we identified sent and received pulses and messages by their name. In order to support modularity, the CIP system description language associates outputs to inputs trough explicit translation functions. Furthermore, abstraction from the state inspection mecanism is obtained through the encapsulation of states and variables of inspected processes by specific state vector access procedures.

CIP Tool

A graphical specification tool has been developed which automatically tests recorded specifications for consistency. A code generator provides modules, which can be animated in the source code environment, or which can be ported on a target computer. In order to complete an implementation it suffices to write I/O-drivers for the transfer of physical events into logical ones and of logical actions into physical ones. In some instances communication links with the environment have to be created.

References

[1]   Backus J. Can Programming Be Liberated from the von Neumann Style? A Functional Style and Its Algebra of Programms. Comm. ACM 1978; 21, 8: 613-641

[2]   Berry G, Moisan S, Rigault J.P. ESTEREL: Towards a Synchronous and Semantically Sound High Level Language for Real Time Applications. Proc. IEEE Real-Time Systems Symposium 1983, pp 30-37

[3]   Bolognesi T, Brinksma E. Introduction to the ISO Specification Language LOTOS. In: van Eijk P.H.J, Vissers L.A, Diaz M (eds) The Formal Description Technique LOTOS. Elsevier Science Publishers B.V, North Holland, 1989, pp 23-73

[4]   Cameron J.R. The modelling phase of JSD. In: Cameron J.R (ed) JSP and JSD: The Jackson Approach to Software Development. IEEE Computer Society Press, 1989, pp 282-292

[5]   Caspi P, Pilaud D, Halbwachs N, Plaice J.A. LUSTRE: A declarative language for programming synchronous systems. In: Fourteenth Annual ACM SIGACT-SIGPLAN Symposium on Principles of Programming Languages, Munich 1987, pp 178-188

[6]   Dembinski P, Budkowski S. The Specification Language Estelle. In: Diaz M et al. (eds) The Formal Description Technique Estelle. Elsevier Science Publishers B.V, North Holland, 1989, pp 35-76

[7]   Fierz H. SCSM, a Synchronous Calculus on Sequential Machines. Submitted to Science of Computer Programming 1993.

[8]   Fierz H. The Synchronous System Description language IPL. To be published.

[9]   Fierz H. The Formal Semantics of the CIP Specification Language. To be published.

[10]   Harel D, Pnueli A. On the Development of Reactive Systems. In: Apt K.R (ed) Logics and Models of Concurrent Systems. Springer New York, 1985, pp 477-499

[11]   Harel D. Statecharts: A Visual Formalism for Complex Systems. Science of Computer Programming 1987; 8: 231-274.

[12]   Jones C.B. Systematic Software Development using VDM. Prentice-Hall International, Englewood Cliffs N. J, 1986

[13]   Peterson J.L. Petri net theory and the modelling of systems. Prentice-Hall Prentice-Hall International, Englewood Cliffs N. J, 1981

[14]   Zave P. The Operational Approach versus the Conventional Approach to Software Development. Comm. ACM 1984; 27, 2: 104-118

Exception Handling and Predictability in Hard Real-Time Systems

Matjaž Colnarič

Faculty of Technical Sciences, University of Maribor
Maribor, Slovenia

Wolfgang A. Halang

Department of Electrical Engineering, FernUniversität Hagen
Hagen, Germany

Abstract

The objective of this paper is to give some reflections about handling of exceptions in hard real-time environments, which is among the less elaborated topics in this domain.

A classification of possible exceptions in real-time systems is done, to identify the ones which can be prevented by certain design measures or avoided by specifying and servicing them within their contexts. A way to survive the remaining ones in a well-structured and predictable way, and as painlessly as possible, is proposed.

1   Introduction

In his reference paper [14], Stankovic is unmasking several misconceptions in the domain of hard real-time systems. It seems that the most characteristic one is that real-time computing is equal to fast computing. It is obvious that computer speed itself can not guarantee that the specified timing requirements will be met.

Instead, a different ultimate objective was set: predictability of temporal behaviour. Being able to assure that a process will be served within a predefined time frame is of utmost importance. In multiprogramming environments this condition can be expressed as schedulability: the ability to find a schedule such that each task will meet its deadline [16].

For schedulability analysis, execution times of tasks must be known in advance. These, however, can only be determined if the system functions predictably. To assure overall predictability, all levels of system design must be predictable in temporal sense, from the processor to the system architecture, language, operating system, and exception handling (layer-by-layer predictability, [15]).

In recent years, the domain of real-time systems substantially gained research interest. Certain sub-domains have been examined very thoroughly, such as scheduling and analysis of program execution times. It is typical that most of the research done was dedicated to the higher level topics and presumes that the underlying ones are fully predictable.

Exception handling is one of the most severe issues to be solved when a system is to behave predictably. By an exception, any intrusion in the normal program flow which can not be considered in schedulability analysis is meant and is usually related to residual specification and implementation errors, and failures. Anticipated timing events and events from the environment, which trigger associated processes do not belong to this category. They should be implemented in a way, which does not cause any non-deterministic delays in execution of the running task. That can be done by migrating event recognition and operating system services out of the main task processor, as is done in the Spring project [13], or proposed in [3].

When an exception occurs in a program, the latter is inevitably delayed causing a serious problem with respect to the a priori determined execution time. Therefore, exceptions should be prevented by all means, whenever and wherever it is possible [1]. If it is not possible to prevent them to happen, they should be handled in a consistent and safe way in conformity with the hard real-time systems design guidelines. The urge for consistent solution of exception problem is even increased by the fact that exceptions are often a result of some critical state of the system, which is when the computer control aid is needed most.

2 Classification of Exceptions

In this section we attempt to identify the exceptions appearing in the hard real-time environments. For that reason we classify them in two ways with regard (a) to their origin and (b) to whether they can be prevented or not.

2.1 Origins of Exceptions in Hard Real-Time Environments

Screening possible run-time errors in various programming environments and relying on our experience in real-time programming, we established the following classification of exceptions according to their origin:

a) **Exceptions caused by I/O operations**

- *I/O device errors*

- *Invalid device addressing (invalid unit identification, no such device)*

- *Exceptions caused by the file management (where provided)*

b) **Exceptions caused by invalid data**

- *Traps and errors concerning irregular results of operations (overflow, underflow, undefined)*

- *Arithmetic functions with illegal run-time parameter values (square root, logarithms etc.)*

- *Format declaration/run-time value mismatch or conversion errors in I/O operations*

- *Subscript (array or string index) out of range*

- *Invalid procedure parameter numbers or types*

c) **Errors preventable by imposing restrictions**

- *Errors connected with dynamic language features (insufficient memory due to recursion or pointers, dynamic formatting, dynamic function calls etc.)*

- *Problems concerning virtual addressing*

d) **Problems in tasking**

- *conflict situations like terminating, suspending or resuming a non-existent task etc.*

e) **System exceptions**

- *diagnostics, hardware and system alerts due to unit failures (e.g. bus error etc.)*

In the sequel we classify exceptions according to the criterion whether they can be prevented or not.

2.2 Preventable Exceptions

Some exceptions can be prevented by restricting the use of potentially dangerous features. Compliance with these restrictions must be checked by the compiler.

For example, only sequential file organisation and compile-time known file names and other parameters may be used. No dynamic features like recursion, references, virtual addressing etc. are allowed.

Another means to prevent exceptions is to implement strict type checking in the language, so that possible irregular operations can be reported at the compile time (as an example, see [6], supported by corresponding hardware [8]).

Since strict type checking seems impractical, we suggest to extend the input and output data types by two "irregular" values representing "signed infinity" to accommodate overflows and underflows, and "undefined" (a solution with "holes" in the domain and "bumps" in the range was already implemented in CLU [11]). The "undefined" value is used when a non-recoverable problem occurred in a calculation or during an I/O process rendering the results meaningless. A similar principle is followed in the IEEE 32-bit floating point standard [2] and implemented in the MC68881 co-processor [12], with a quiet or signalling "not-a-number (NaN)".

Thus, generated irregular values do not raise exceptions, but are propagated to the subsequent or higher-level blocks, which may be able to handle them (see the example of an implementation below). Any operation on irregular operands always yields a result of irregular type.

Intelligent I/O interfaces should react in a predefined way, if a final result, which is output to them, is irregular. Reactions on different irregular types can be different. The interfaces may tolerate them, they may be able of a local graceful degradation of their performance (if the action to be taken is not vital or can be pre-programmed for such cases) or, if inevitable, recognise a catastrophic situation. E.g., if a regulating system as a reaction to a disturbance requires "as intensive counter-response as possible" (like a D-regulator) an "infinite" value may be produced, resulting in a maximal possible control signal which may depend on a type of the implemented actuator.

2.3 Non-Preventable Exceptions

In the sequel we further classify the non-preventable exceptions into anticipated and non-anticipated ones. The former can be avoided, the latter, however, must be handled in a consistent and safe way. A reference study in the domain of non-preventable exceptions was done by Cristian [4, 5].

2.3.1 *Anticipated Exceptions*

If the potential danger of irregularity can be recognised during the design time, it has to be taken care of in the specifications.

For example, peripheral devices shall be intelligent, fault-tolerant and self-checking in order to be able to recognise their own malfunctions and to insulate the effects of the latter in "watertight" compartments. They shall react reasonably in conflict situations. When the error is recoverable, they should try to recover locally using fault- tolerance principles (self checking, redundancy etc.).

A number of exceptions resulting from irregular data can be avoided by prophylactic run-time checks before entering critical operations. Many tasking errors are also avoidable by previously using monadic operations to check the system state.

An obvious and frequently used way of avoiding critical failures in hard real-time systems design is redundancy. Redundant system components must be implemented according to thorough analysis of fault hypothesis. The latter should beside the physical faults in the operation also include errors in the design and implementation of hardware and software components. E.g., in avionics implementations of redundant systems can be found, based on different processors and done by different teams.

An example of consistent implementation of redundancy is the MARS system

[10]. Components possess self-checking properties and produce either correct or no results (fail silently); in the latter case the redundant component's results are taken. To determine where and to what extent redundancy should be applied, the Mars Reliability Predictor and Low-Cost Estimator (MARPLE) was implemented. Programs written in general-purpose design language for distributed systems are translated into reliability models, which are then analysed by the Symbolic Hierarchical Automated Reliability and Performance Evaluator (SHARPE) and several parameters are produced. Based on these parameters, dependability can be estimated. However, if a system is extremely safety critical, also the failure of redundant devices must be taken into account, in spite of the low probability of such an event.

2.3.2 *Non Anticipated Exceptions*

If there is no way to predict an error, the exception caused must be handled in order to survive it. These are situations when "the impossible happens" [1], in which programs do not follow their specifications due to hardware failures, residual software errors or wrong specifications. For example, failure of a part of memory can result in the change of constant values; an error in file management or on a disk is usually unexpected. In safety-critical control systems the non-anticipated exceptions may have catastrophic consequences. There it is especially important to implement a mechanism for their safe and consistent handling.

In his early paper, ·Goodenough [7] presented an idea of assigning default- or programmed exception handlers to every potentially dangerous operation. According to severity of the exception raised the running process was either terminated or suspended and resumed later. Similar mechanism although considerably more elaborated and adopted for use in hard real-time systems was implemented in Real-Time Euclid [9]. There, exception handlers were (optionally) located within block constructs and were executed in a case of an exception. If there were no exceptions they had no effect except for their impact on the block's execution time estimated by schedulability analyser thus making it more difficult to be scheduled. Exceptions may be raised by kill, terminate or except statements, to terminate a process entirely or only its frame, or to execute the handler without termination of the process, respectively.

3 Coping with Non-Preventable Exceptions

To handle catastrophes we propose a combination of preconditions, postconditions and modified recovery blocks implementing both backward and forward recovery. Its syntax is following:

block ::= block_begin block_tail

block_begin ::= BEGIN | PROCEDURE parameters & attributes; |
     TASK parameters & attributes; | parameters REPEAT

block_tail ::= [declaration sequence] [alternative_sequence] END;

alternative_sequence ::= {[ALTERNATIVE [PRE bool–exp;] [POST bool–exp;]]
[RESTORE] statement_sequence}

A block (task, procedure or other block structure) consists of alternative sequences of statements. Every alternative can have its own pre- and/or post-conditions, presented by Boolean expressions. When program flow enters the surrounding block, the initial state of the system is stacked if there is at least one alternative implementing backward recovery what is denoted by the keyword RESTORE. Then, the first alternative statement sequence, whose pre-condition (if it exists) is fulfilled, is executed. At the end, its post-condition is checked, and if this is also fulfilled, execution of the block is successfully terminated. If the post-condition is not fulfilled the next alternative is checked for its pre-condition and eventually executed. If backward recovery is requested, the initial state is restored.

The alternatives may contain independently designed and coded programs to comply with specifications and to eliminate possible implementation problems or residual software errors. They may also contain alternative design solutions or redundant resources, when problems are expected. A further possibility is to assert less restrictive pre- and/or post-conditions and to degrade performance gracefully. By the means presented in [17] it is also possible to bound the execution times of alternatives. If one of them fails to complete inside the predefined period, a less demanding alternative is taken.

If there is no alternative, whose pre- and post-conditions are fulfilled, the block execution was unsuccessful. If the block was nested inside an alternative on the next higher level, this alternative fails as well and the control is given to the next one, thus providing a chance to resolve the problem in a different way. On the highest level, the last alternative must not have any pre- or post-conditions. It must solve the problem by applying some conventional actions like employing fault tolerant measures or performing smooth power-down. Since the system is in extreme and unrecoverable catastrophic conditions, different control and timing policies may be in action, requesting safe termination of the process and possibly post-mortem diagnostics.

For embedded systems it is important to consider whether backward recovery of certain block is possible or not. If in an alternative block an action is triggered like commencing a peripheral process which causes an irreversible change of initial state, it cannot be restored for another try. In this case only forward recovery is possible, bringing the system to certain predefined, safe, and stable state.

The method inevitably yields pessimistic execution time estimation which is a sum of execution times of all alternatives together with pre- and post-condition evaluation times and administration overhead. However, this is not due to this specific method. In safety-critical hard real-time systems it is necessary to consider the worst case execution time, which must also imply exceptional conditions. Depending on the performance reserve of the system there may be implemented more or less alternatives, performing more or less degraded functions. In extremely time-critical systems there may be implemented a single alternative on the highest level block only providing a safe and smooth power-down.

To cope with that problem some further solutions are possible. Each subsequent alternative may be bounded to a half of the execution time of the previous one; thus, the block will terminate in at most double execution time of the primary alternative. Also, from a failure of an alternative it is possible to deduce which subsequent alternatives in next blocks are reasonable and which not, and accordingly set their pre-conditions. However, this requires a sophisticated run-time analyser.

4 Conclusion

In order to assure a predictable behaviour of real-time systems, it is necessary to determine a priori bounds for the task execution times. Exception handling represents the most severe obstacle to this end. Therefore, exceptions were investigated and classified with the objective of obtaining a remedy for this problem. It turned out that a large group of them can be either prevented by appropriate measures or avoided at run-time. The others are coped with in a well-structured environment by providing sequences of gradually more and more evasive software reactions. In either case, the run-time behaviour of real-time tasks becomes fully predictable.

References

[1] Andrew P. Black. Exception handling: The case against. Technical Report TR 82-01-02, Department Of Computer Science, University of Washington, May 1983. (originally submitted as a PhD thesis, University of Oxford, January 1982).

[2] W.J. Cody, J.T. Coonen, D.M. Gay, K. Hanson, D. Hough, W. Kahan, R. Karpinski, J. Palmer, F.N. Bis, and D. Stevenson. A proposed radix- and word-length-independent standard for floating-point arithmetic. *IEEE Micro*, 4(4):86–100, August 1984.

[3] Matjaž Colnarič and Wolfgang A. Halang. Architectural support for predictability in hard real-time systems. *Control Engineering Practice*, 1(1):51–59, February 1993.

[4] Flaviu Cristian. Exception handling and software fault tolerance. *IEEE Transactions on Computers*, 31(6):531–540, June 1982.

[5] Flaviu Cristian. Correct and robust programs. *IEEE Transactions on Software Engineering*, 10(2):163–174, March 1984.

[6] Ian F. Currie. NewSpeak: a reliable programming language. In *High-integrity Software*, pages 122–158. Pitman Publishing, London, 1988.

[7] John. B. Goodenough. Exception handling: Issues and a proposed notation. *Communication of the ACM*, 18(12):683–696, 1975.

[8] John Kershaw. The VIPER microprocessor. Technical Report 87014, Royal Signals And Radar Establishment, Malvern,Worcs, London: Her Majesties' Stationery Office, November 1987.

[9] Eugene Kligerman and Alexander Stoyenko. Real-time Euclid: A language for reliable real-time systems. *IEEE Transactions on Software Engineering*, 12(9):941–949, September 1986.

[10] Hermann Kopetz, A. Damm, Ch. Koza, M. Mulazzani, W. Schwabl, Ch. Senft, and R. Zainlinger. Distributed fault-tolerant real-time systems: The MARS approach. *IEEE Micro*, 9(1):25–40, February 1989.

[11] Barbara H. Liskov and Alan Snyder. Exception handling in CLU. *IEEE Transactions on Software Engineering*, 5(6):546–558, November 1979.

[12] Motorola. *MC68881 Floating-Point Coprocessor User's Manual*, first edition, 1985.

[13] Krithi Ramamritham and John A. Stankovic. Overview of the SPRING project. *Real-Time Systems Newsletter*, 5(1):79–87, Winter 1989.

[14] John A. Stankovic. Misconceptions about real–time computing. *IEEE Computer*, 21(10):10–19, October 1988.

[15] John A. Stankovic and Krithi Ramamritham. Editorial: What is predictability for real–time systems. *Real-Time Systems*, 2(4):246–254, November 1990.

[16] Alexander Stoyenko. *A Real-Time Language With A Schedulability Analyzer*. PhD thesis, University of Toronto, December 1987.

[17] Domen Verber and Matjaž Colnarič. A tool for estimation of real-time process execution times. In *Proceedings of Software Engineering for Real-Time Applications Workshop*, Cirencester, September 1993. IEE.

Development of a Fail-Safe Data Transmission System for use in Life-Critical Applications

M.B. Schrönen[†] and M.R. Inggs[‡]
Department of Electrical Engineering, University of Cape Town
Private Bag Rondebosch, South Africa, 7700

Abstract

It is essential that safe data transmission between systems in a distributed life-critical application occurs in a safe manner. To date this transmission has occurred in a parallel form, however rising material and labour costs have made this method of safe data transmission expensive. In this paper a development of Fail-Safe Data Transmission System (**FSDTS**), which can be used in these distributed applications, is presented.

1 Introduction

There are distributed life-critical applications that require safe data to be transmitted over large distances e.g. railway interlockings. To date, this transmission has usually occurred in a parallel manner, which has become expensive. The advent of microprocessors has made it possible to develop cost-effective solutions to overcome this problem. There is however, an antipathy to using microprocessor based systems in safety-critical applications because they introduce unidentified and often diverse factors which could result in life-threatening failures.

The aim of this paper is to introduce a FSDTS which was developed incorporating techniques that ensure safety when these complex devices are used in life or safety-critical applications.

1.1 Structure of Paper

The Fail-Safe Data Transmission Project (**FSDTP**) which includes the development of the FSDTS, has been broken up into a number of phases. This paper is divided into five sections, which correspond to the phases of the project that have been completed and are currently in progress. A breakdown of the paper is given below:

Section 1 (**Introduction**) will introduce the FSDTP project, giving an overview of the FSDTS with emphasis on the operating requirements and constraints.

---

† email: mschronen@eleceng.uct.ac.za

‡ email: mikings@eleceng.uct.ac.za

Section 2 (**Safe Serial Data Transmission**) delineates the method for achieving safety in the transmission of safe-data over a serial channel in the presence of noise. Issues relating to error control mechanisms with emphasis on the code and communication scheme selection will be discussed.

Section 3 (**Development of a Fail-Safe Data Transceiver**) describes how a Fail-Safe Data Transceiver (**FSDT**) can be designed. Here hardware safety will be qualified and the architecture of the FSDT will be presented.

Section 4 (**Ensuring "Error-Free" Software**) will give preliminary results of this research. The aims of this phase of the project will be presented with comments on the future research to be undertaken.

Section 5 (**Conclusions**) gives an overview of the project to date, future plans and the results from the FSDTs being monitored in the field.

1.2 FSDTP Background

The FSDTP was initiated to investigate a method of overcoming the heavy financial burden in maintaining the existing cabling infrastructure used to transmit safe data between various interlockings within the train movement control system (**TMCS**). Contributing factors to this were: cost of new cables, degradation of the existing cables, cost of maintaining the existing cables and the theft of cables.

The aim of the project was to develop a microprocessor-based serial data transmission system which could be used to transmit this safe information. The system would initially be used to transmit small amounts of data between interlockings within the TMCS. After a successful evaluation phase the transceivers would be used to transmit control information within the interlocking and to the train itself.

1.3 System Overview

The FSDTS must have the ability to utilise a variety of serial channels and be transparent to the safety-system, in the sense that the safety systems must still communicate in a parallel form.

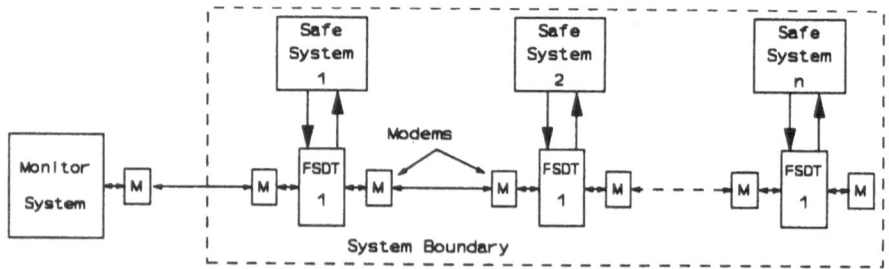

Figure 1: FSDTS in Routing Configuration

The FSDTS can either be used in a store and feed forward configuration shown in Figure 1 above or in a master slave configuration illustrated below in Figure 2.

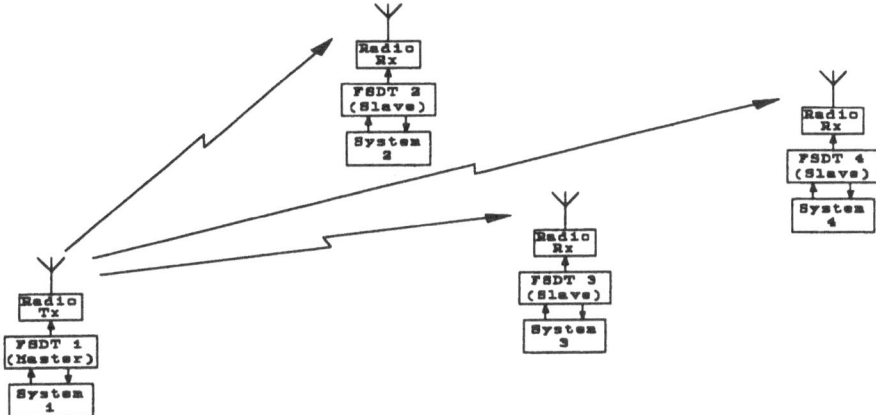

Figure 2: FSDTS in Master-Slave Configuration

1.4 FSDTS Requirements and Constraints

To offer the same safety as a parallel transmission system the FSDTS must adhere to the following requirements and constraints:

- Safety must be guaranteed from when the data leaves the transmitting safe system until it arrives at the receiving safe system.
- Due to the nature of the application, safe systems which are communicating with each other must ensure that the latest information is always transmitted to the remote safe systems.
- If there is a break in data received by a FSDT, its outputs must be set to a safe state.
- The integrity of the channel must be monitored to ensure that the acceptable noise level is not exceeded. If the level is exceeded the outputs of the receiving FSDT must be set to a safe state.
- Once a FSDT has shutdown, no safe data can be transmitted in the channel to which it is connected.
- All first faults which affect the safety of the system (including those of the fault detection circuitry) must be detected within approximately one second of occurrence. Faults which do not affect the safety or those which cause the system to revert to its safe state are also required to be detected within a reasonable time to fulfil the requirements of maintainability.
- Having detected the first fault (affecting safety), the system must automatically revert to safe (predetermined) state.
- The safe state to which the system reverts after the detection of the first fault must be irreversible i.e. the system must not be able to become unsafe even after the occurrence of further faults.

2 Safe Serial Data Transmission

Noise which is inherent in the transmission system and noise from the environment will corrupt data transmitted over the communication channel. To achieve safe data transmission it is necessary to combat the effects of this noise. In this section we will demonstrate that it is possible to achieve safe data transmission over a serial channel in the presence of noise, by incorporating an error control strategy. For this purpose we will consider only one link of the FSDTS.

2.1 Selecting an Error Control Strategy

Two strategies which are used to control errors in a communication channel are automatic repeat-request (**ARQ**) and forward-error correction (**FEC**). In this type of application an ARQ strategy is preferred if a reverse channel is available [1,2]. To implement effective coding techniques it is essential to obtain statistical data relating to the types of errors, and also their occurrence rate [3]. Once an analysis of the errors occurring in the communication channel is complete, it is possible to model the channel and select a coding scheme that will be the most effective against the predicted noise.

2.2 Expected Errors in Channel

Errors resulting from noise are classified into two main categories i.e. **random errors** and **burst errors**. Random errors result from random noise which is primarily gaussian noise and shot noise. These errors are normally inherent in the communication system. On the other hand, burst errors occur due to natural causes (e.g. lightening) and man-made sources (e.g. power systems and electrical machinery) [1,4]. As indicated by [5,6] both random and burst errors are expected to occur. There is strong evidence to suggest that long transmission periods will exist where few errors will occur (mainly random errors), followed by short periods where a great number of mainly burst errors will occur. This might be due to loss of synchronisation or from some form of burst noise.

2.3 Channel Modelling

There are two approaches to modelling the channel. Collected statistical data can be used to determine the parameters of the channel model (**"descriptive"** modelling) or a mathematical model can be adapted to real measurements of the channel (**"generative"** modelling) [3]. It is apparent that in both instances substantial statistical data is required to model the channel effectively.

Measured statistics of various channels are dependant on the type of equipment used and the environment in which the channels operate. With the lack of available statistical data and vast amounts of different channel equipment and environments found in the proximity of the trackside, it was impractical to gather sufficient statistical data to model all the channels with their associated equipment and environment. It was thus decided to make the following assumptions:

- The channel has no memory i.e. one bit inverted will not effect the next bit.
- The channel is a binary symmetrical channel (**BSC**).
- Errors occur randomly and each bit has the same probability of being incorrectly asserted.
- Messages are equiprobable i.e. each message has the same probability of containing as many "1's" as "0's".

By ignoring the effects of burst noise the results obtained for the probability of error would be optimistic as illustrated by [7]. For this reason a random error detecting/correcting code with good random and burst error capabilities will be used to implement the error coding in the error control system. Thereafter techniques to improves the code's burst error capabilities will be implemented.

2.4 Selecting an Error Control Code

Error control codes are broken up into two main types, being block codes and convolutional codes. In this application convolutional codes were not considered for reasons mentioned in [1,8]. A block code or cyclic code with good random and burst error detection capabilities will be selected and thereafter the codes's burst error capabilities will be improved. The code evaluation will be based on the probability of an undetected error $P_u(E)$.

2.4.1 Error Capabilities of Linear Block Codes

The error control capabilities of a linear block code used for error detection is determined by the code's minimum distance (d_{min}) known as the **Hamming distance** [9,10,11]. For a (n,k) linear block code, there are $(2^k - 1)$ possible undetected errors. This occurs when a code-vector is corrupted in such a way that it becomes another valid code-vector. If the weight (**Hamming Weight**) of the code is known, it is possible to calculate the probability of an undetected error for the code, if it is used on a BSC. This can be calculated using equation (1), as described in [11, 12, 13] as follows:

$$P_u(E) = \sum_{i=1}^{n} A_i \ p_i \ (1-p)^{n-i} \tag{1}$$

where A_i is the number of code vectors of weight i in the code and **p** is the channel bit error rate (**BER**). For large values of **n** and **k** however, it becomes almost impossible to calculate the weight distribution of a code. In some instances however, it is possible to calculate the weight distribution of the dual of the code. By using MacWilliams identity [14], it is possible to calculate the probability of an undetected error for the code. If this method is not possible, then equation (2) can be used.

$$P_u(E) = 2^{-(n-k)} B(1-2p) - (1-p)^n \quad \text{where} \quad B(1-2p) = \sum_{i=1}^{n} B_i(1-2p)^i \qquad (2)$$

where $(B_0, B_1,...B_n)$ is the weight distribution of the dual code. If neither of the above methods are possible the upper-bound as described in equation (3).

$$P_u(E) \le 2^{-(n-k)} \qquad (3)$$

where $P_u(E)$ is an upper bound for the average probability of an undetected error. It must be emphasized that this upper-bound is only valid for a few codes as illustrated in [15,16,17].

2.4.2 Code Evaluation

In order to satisfy the system safety requirement the code selected must improve the channel BER from 10^{-5} to 1.32×10^{-18} (with a transmission rate 2400 bps). By using the MacWilliams identity and tables in [18,19], it was found that certain Bose, Chaudhuri and Hocquenghem (BCH) codes, the most powerful known class of binary codes for correcting random errors [20], would satisfy the safety requirement.

2.4.3 Improving a Code's Burst noise Immunity

As a result of the assumptions made the theoretical values for $P_u(E)$ will be optimistic. By using techniques such as bit stuffing, interleaving, concatenated codes and a second code it is possible to improve the code's burst error capabilities. Of these methods interleaving offers the best improvement for overhead as described in [21,22].

2.5 Error Control Implementation

The code selected in this application is a interleaved BCH(15,7,2) with an interleaving degree λ of 5. In this application a modified ARQ communication scheme is used. Retransmissions are not required because the latest data is always transmitted. Received data that is corrupted is discarded. Encoding and decoding of the data is done by a lookup table. There are 2^7 valid code-vectors requiring 128 bytes of memory. The time taken to lookup one code-vector is $\pm 4\mu s$ excluding the software overhead.

3 Development of Fail-Safe Data Transceiver

When microprocessors are used in the design of equipment that is to be used in life-critical applications, it is not feasible to incorporate interlocking to detect all the possible microprocessor failure modes. To overcome this problem, the system is designed to have either a voting or comparative architecture. In both approaches designed to have either a voting or comparative architecture. In both approaches similar or diverse hardware can be used. In this section we present the formulas to quantify the system safety, and, discuss the FSDT.

3.1 Determining the Hardware Safety

Hardware safety is expressed as the mean time between wrong side failures (MTBWSF). The MTBWSF is derived from the mean time between failures (MTBF) of the components of the system. We will consider the two-out-of-two comparative architecture. The following assumptions are made:
- components are used in their useful life period (ie random failures occur).
- all first faults will result in an unsafe failure if undetected.

The instantaneous failure rate or hazard rate of a component is expressed as

$$\lambda(t) = \frac{f(t)}{R(t)} \qquad (4)$$

where $f(t)$ is the failure probability density function and $R(t)$ is the reliability of the component. When using components during their useful life, $\lambda(t)$ is constant and is expressed as λ. The mean time between system failure (MTBSF) as calculated in [23] expressed in equation (5) where $\lambda \ll \mu$ and $1/\mu$ is the mean repair time.

$$MTBSF = \frac{\mu}{2\lambda^2} \qquad (5)$$

Consider a two-processor system where the system output is controlled by a redundant management system which has the ability to shut the system down to a safe state should faults occur in either of the processors. The following assumptions are made:
- There are no design errors in the system which will render the redundant management system unable to detect an unsafe failure.
- Failures in each of the two processors are independent.
- Any failure occurring after the first undetected failure will result in the management system unable to revert the system to a safe state.

The system safety can be quantified in terms of the MTBWSF by redefining $f_R(t)$ as the probability density function of the fault detection process which has a fault detection time of τ. The safety can now be expressed as

$$MTBWSF = \frac{1}{\lambda} + \frac{1}{2\lambda^2\tau} \approx \frac{MTBF^2}{2\tau} \qquad (6)$$

When the two elements are configured as described above and used in a safe environment the two redundant elements together operate as a single element and the

It is thus possible to determine the safety of the system by calculating the MTBF of each system in a two-out-of-two comparative architecture. The MTBF of each system is derived from the MTBF of the individual components that make up each system.

3.2 Selecting the FSDT Architecture

After fault-tree and FMECA analyses were performed it was found that a two-out-of-two comparative architecture would meet the FSDT requirements. Similar as opposed to diverse hardware was selected. An illustration of the architecture is given in Figure 3.

3.3 Description of the FSDT Hardware Implementation

The FSDT comprises of two identical sub-systems, which are electrically isolated from each other to obtain statistical independent errors. The modules within the sub-system are microprocessor controlled and are configured in a master/slave configuration. Each sub-system comprises an input, output, comms and display module as illustrated in Figure 4. Each module can be configured as a master or a slave.

The slave performs only its primary function whereas a master performs its primary function, data collection and distribution and rendezvous with the other sub-system, for data. The two sub-system are loosely synchronised and re-adjustment occurs at each rendezvous.

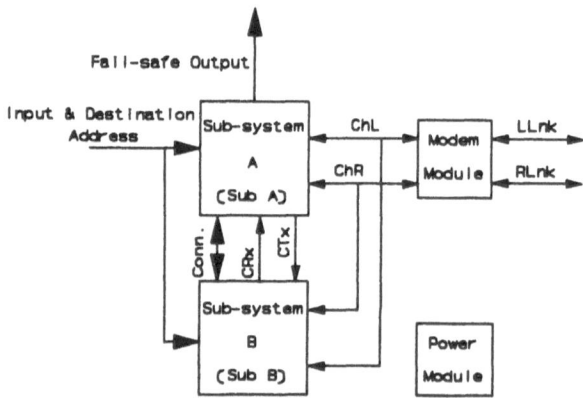

Figure 3: Fail-Safe Data Transceiver Architecture

All of the modules in both sub-systems have the ability to shut the system down on the detection of an error. This is done by blowing a fuse which isolates the FSDT from the safe system. Once the FSDT is shutdown, no data can pass between itself and the safe system. The serial channels can still be used to transmit maintenance information depending on the type of failure that caused the system to be shut down.

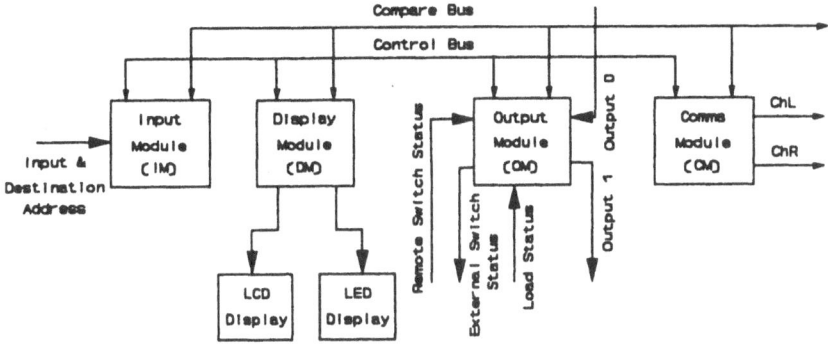

Figure 4: FSDT Sub-System Components

4 Ensuring "Error-Free" Software

The FSDTS that is under evaluation at present has software which is functionally correct. By this is meant that the functions performed by the software are correct, but the software has not been verified and has not been proven to be correct. The software was written in assembler and no interrupts have been used. The validation and verification of the software is part of the research that is currently being undertaken at the University of Cape Town. The following issues are being addressed:
- Use of fault-tree analysis and other methods to identify high risk areas and to provide input into the validation process.
- Generation of a meta language to facilitate the formulation of the system requirements into a formal specification.
- Using Statecharts to model and represent the system for input into a synchronous language such as Esterel.

5 Conclusions

To date 5 systems have been manufactured and are currently being tested in Cape Town. Three of the links used are leased telephone type circuits and the fourth is a microwave link. The telephone links are 4-15km in length and the microwave link is ± 100km in length. At present the FSDTs are monitored by Test Generation Modules (TGMs) and safety system simulators (SSS). To date the FSDTS has operated successfully and once the software research is completed the FSDTs will be used in live applications. At this stage it is anticipated that the systems will be ready for installation early in 1994.

6 References

1. Shanmugan KS. Digital and Analog Communication Systems. John Wiley & Sons, New York, 1979, pp 443-504

2. Chien RT. "Block-Coding Techniques for Reliable Data Transmission". IEEE Trans Commun Technol, Oct 1971; vol COM-19, no 5, pp 743-751

3. Kanal LN, Sastry ARK. "Models for Channels with Memory and Their Applications to Error Control". Proc IEEE July 1978; vol 66, no 7, pp 724-744

4. Lawrence, Apperley, Auclair et al. "Errors in Digital Transmission Systems", ORE Report RP5, 1985, pp 1-52

5. Fontaine AB, R.G. Gallager RG. "Error Statistics and Coding for Binary Transmission Over Telephone Circuits". Proc IRE, Jun 1961; vol 49, no 6, pp 1059-1065

6. Reiffen B, Schmidt WG, Yudkin HL. "The Design of an Error-Free Data Transmission System for Telephone Circuits", AIEE Trans Jul 1961; vol 80, Part 1, pp 224-231

7. Lawrence, Apperley, Auclair et al. "On proving the Safety of Transmissions Systems". ORE Report RP8, Apr 1986, pp 1-62

8. Berlekamp ER, Peile RE, Pope SP. "The Application of Error Control to Communications". IEEE Commun Mag, Apr 1987; vol 25, no 4

9. Hamming RW. Coding and Information Theory. New Jersey, Prentice-Hall Inc 1980, pp 21-49

10. Roden MS. Digital and Data Communication Systems, New Jersey, Prentice-Hall Inc 1982, pp 121-159

11. Peterson WW, Weldon EJ, Error-Correcting Codes 2nd Edition, Massachussets, MIT Press, 1972, pp 64-70

12. Berlekamp ER. Algebraic Coding Theory, New York, McGraw-Hill, 1968, pp 397-431

13. Peterson WW, Weldon EJ. Error-Correcting Codes 2nd Edition, Massachussets, MIT Press, 1972, pp 64-70

14. MacWilliams FJ, Sloane NJA, The Theory of Error-Correcting Codes. Amsterdam, North-Holland Mathematical Library, 1977, pp 125-132

15. Leung-Yan-Cheong SK, Hellman ME. "Concerning a Bound on Un-detected Error Probability", IEEE Trans Inform Theory, Jan 1979; vol IT-25, no 1, pp 235-237

16. Leung-Yan-Cheong SK, Barnes ER, Friedman DU. "On Some Properties of the Undetected Error Probability of Linear Codes". IEEE Trans Inform Theory, Jan 1979; vol IT-25, no 1, pp 110-112

17. Wolf JK, Michelson AM, Levesque AH. "On the Probability of Undetected Error for Linear Block Codes". IEEE Trans Commun Feb 1982; vol COM-30, no 2, pp 317-324

18. Costello Jr. DJ, Lin S. Error Control Coding: Fundamentals and Applications, New York, Prentice-Hall Inc, 1983, pp 141-183

19. MacWilliams FJ, Sloane NJA. The Theory of Error-Correcting Codes. Amsterdam, North-Holland Mathematical Library, 1977, pp 445-453

20. Wiggert D. Error Control Coding and Applications, Massachusetts, Artech House, 1975, pp 77-85

21. Costello Jr. DJ, Lin S. Error Control Coding: Fundamentals and Applications, New York, Prentice-Hall Inc, 1983, pp 271-272

22. Shanmugan KS. Digital and Analog Communication Systems, New York, John Wiley & Sons, 1979, pp 476-478

23. Liebowitz BH. Reliability Considerations for a Two Element Redundant System with Generalised Repair Times. Oper Res, 1966, pp 233-241

Author Index